10th International Conference on Numerical Modeling of Space Plasma Flows (ASTRONUM 2015)

Journal of Physics: Conference Series
Volume 719

Avignon, France
8 – 12 June 2015

Editors:

Edouard Audit
Nikolai V. Pogorelov
Gary P. Zank

ISBN: 978-1-5108-2456-0
ISSN: 1742-6588

Printed from e-media with permission by:

Curran Associates, Inc.
57 Morehouse Lane
Red Hook, NY 12571

Some format issues inherent in the e-media version may also appear in this print version.

Printed by Curran Associates, Inc. (2016)

For permission requests, please contact the Institute of Physics
at the address below.

Institute of Physics
Dirac House, Temple Back
Bristol BS1 6BE UK

Phone: 44 1 17 929 7481
Fax: 44 1 17 920 0979

techtracking@iop.org

Additional copies of this publication are available from:

Curran Associates, Inc.
57 Morehouse Lane
Red Hook, NY 12571 USA
Phone: 845-758-0400
Fax: 845-758-2633
Email: curran@proceedings.com
Web: www.proceedings.com

Table of contents

Volume 719

10th International Conference on Numerical Modeling of Space Plasma Flows: ASTRONUM-2015
8–12 June 2015, Avignon, France

Accepted papers received: 4 May 2016
Published online: 31 May 2016

Preface

Papers

Turbulence and Cosmic ray Transport

Astrophysical Flows

Space Plasma Flows

Numerical Methods

The Center for Space Plasma and Aeronomic Research (CSPAR) at the University of Alabama in Huntsville (UAH) and Maison de la Simulation at the French Alternative Energies and Atomic Energy Commission (Commissariat a l'energie atomique et aux energies alternatives, CEA) organized the 10th annual International Conference on Numerical Modeling of Space Plasma Flows (ASTRONUM-2015) on June 8–12, 2015 in Avignon, France.

The Program Committee consisted of Tahar Amari (CNRS Ecole Polytechnique, France), Edouard Audit (CEA/CNRS Maison de la Simulation, Gif-sur-Yvette, France, co-chair), Phillip Colella (Lawrence Berkeley National Laboratory, USA), José María Ibáñez, University of Valencia, Spain), Ewald Müller (Max-Planck-Institute for Astrophysics, Garching, Germany), Nikolai Pogorelov (University of Alabama in Huntsville/CSPAR, USA, chair), Kazunari Shibata (Kyoto University, Japan), James Stone (Princeton University, USA), Brian Van Straalen (Lawrence Berkeley National Laboratory, USA), and Gary P. Zank (University of Alabama in Huntsville/CSPAR).

The conference attracted 67 scientists representing different branches of the plasma simulation community. The distinctive feature of this conference is a combination of diverse research topics, all of which are essential for performing high-resolution, continuum mechanics and particle simulations of physical phenomena in space physics and astrophysics. Among such topics were software packages for modeling and analyzing plasma flows; advanced numerical methods for space, astrophysical and geophysical flows; large-scale fluid-based, kinetic, and hybrid simulations; turbulence and cosmic ray transport; and magnetohydrodynamics. The applications discussed included cosmology and galaxy formation, supernova explosions, physics of the Sun-heliosphere-magnetosphere environments, the interstellar medium and star formation, stellar physics, experimental plasma physics, astrophysical accretion, numerical methods for ideal and non-ideal, relativistic and nonrelativistic MHD, etc. The proceedings volume is structured so that it covers all of these topics.

A distinctive feature of the ASTRONUM conference series is based on the idea that modelers working in seemingly distant fields should have an opportunity to share their scientific achievements with the broad community of computational scientists performing numerical experiments. As in previous ASTRONUM meetings, we were interested in physical systems that are coupled across a multiplicity of spatial and temporal scales via collision-like integral terms that incorporate diverse physical processes.

The contributors to this volume are both young researchers and renowned experts in space physics and astrophysics, applied mathematics, and computer science, particularly data handling and visualization. This book describes the application of numerical methods and the algorithms themselves, allowing us to discuss the challenges that theory imposes on numerical schemes for solving partial differential equations describing collisional and collisionless processes in space and astrophysical plasmas.

We would like to thank the participants who submitted their papers to Proceedings of ASTRONUM-2015 and especially to those who reviewed manuscripts thus ensuring the high quality of this publication. We also are grateful to Valérie Belle and Adeline Holton for the excellent management of the conference.

The book will be useful to graduate and postgraduate students majoring in space physics, astrophysics, numerical, engineering, and applied mathematics. It is also aimed at specialists in applied mathematics, and various fields of physics that involve flows of partially ionized plasmas at both the collisional and collisionless levels.

Edouard Audit, Nikolai V. Pogorelov, and Gary P. Zank

Peer review statement

All papers published in this volume of *Journal of Physics: Conference Series* have been peer reviewed through processes administered by the proceedings Editors. Reviews were conducted by expert referees to the professional and scientific standards expected of a proceedings journal published by IOP Publishing.

Forward and Inverse Cascades in EMHD Turbulence

Jungyeon Cho

Department of Astronomy and Space Science, Chungnam National University, Daejeon, Korea (ROK)

E-mail: jcho@cnu.ac.kr

Abstract. Electron magnetohydrodynamics (EMHD) provides a simple fluid-like description of physics below the proton gyro-scale in collisionless plasmas, such as the solar wind. In this paper, we discuss forward and inverse cascades in EMHD turbulence in the presence of a strong mean magnetic field. Similar to Alfvén waves, EMHD waves, or EMHD perturbations, propagate along magnetic field lines. Therefore, two types of EMHD waves can exist: waves moving parallel to and waves moving anti-parallel to the the magnetic field lines. For energy cascade in EMHD turbulence, the relative amplitudes of opposite-traveling waves are important. When the amplitudes are balanced, we will see fully-developed forward cascade with a $k^{-7/3}$ energy spectrum and a scale-dependent anisotropy. On the other hand, when the amplitudes are imbalanced, we will see inverse cascade, as well as (presumably not fully developed) forward cascade. The underlying physics for the inverse cascade is magnetic helicity conservation.

1. Introduction

In the solar wind, it is well known that the observed magnetic energy spectrum shows a break near the ion gyro-scale [1]. Above the break, the spectrum is compatible with a $k^{-5/3}$ spectrum and, below the break, the spectrum is steeper than the $k^{-5/3}$ spectrum. In this paper, we focus on the nature of turbulence below the break.

Magnetohydrodynamics (MHD) is a powerful and relatively simple tool for physics on scales larger than the ion gyro-scale. Indeed, the observed $k^{-5/3}$ spectrum above the break is consistent with a theoretical prediction for MHD turbulence in a strongly magnetized medium [2]. However, it is obvious that MHD is invalid below the break because the fluid approximation, in which MHD is based on, is invalid there.

Although fluid approximation is violated below the break, it is still possible to construct a fluid-like model for the small scales. Electron MHD (EMHD) mode [3] is the simplest model among such fluid-like models for the small scales. The most fundamental assumption of the EMHD model is that, below the ion gyro-scale, ions provide only smooth charge background and electron 'fluid' moves and carries current.

As in MHD, waves propagate along magnetic field lines in the presence of a strong mean magnetic field and their interactions are important for turbulence in EMHD. Since perturbations propagate along magnetic field lines in EMHD, two types of waves can exist: waves moving parallel to and waves moving anti-parallel to the magnetic field lines. Then two types of interactions can exist: interactions between counter-propagating waves and self-interactions between waves moving in one direction. Note that interactions of the latter type are absent in MHD, but they are present in EMHD.

Journal of Physics: Conference Series **719** (2016) 012001 doi:10.1088/1742-6596/719/1/012001

Throughout the paper, let us call EMHD waves moving parallel to the magnetic field 'positive' waves and those moving anti-parallel to it 'negative' waves. Then, let us consider a collection of 'positive' and 'negative' EMHD wave packets with similar sizes. In this case, the relative amplitudes of the positive and the negative wave packets are important. In case two types of amplitudes are balanced, numerical studies have revealed that interactions between opposite-traveling wave packets are more important than interactions between wave packets moving in the same direction [4, 5]. On the other hand, if there is substantial imbalance in their amplitudes, then interactions between dominant wave packets, i.e. either positive wave packets or negative wave packets that have larger amplitudes, become important and result in inverse cascade [6, 7, 8].

In this paper, we review the scaling relations of forward cascade in balanced EMHD turbulence and the nature of inverse cascade in imbalanced EMHD turbulence. We describe numerical method in Section 2. We discuss properties of EMHD waves and their implications for turbulence cascade in Section 3. We present scaling relations and numerical results for forward cascade in Section 4 and for inverse cascade in Section 5. We give summary in Section 6.

2. Numerical Methods

2.1. Incompressible EMHD equation

In this subsection, we briefly describe derivation of the incompressible EMHD equation [3]. The EMHD model deals with physics below the ion gyro-scale and the most important assumption of the model is that the ions create only smooth motionless charge background. Therefore, in the EMHD model, we consider motions of electrons only (see Figure 1). Since motions of electrons create current density, we can write

$$\mathbf{v}_e = -\frac{\mathbf{J}}{n_e e} = -\frac{c}{4\pi n_e e}\nabla \times \mathbf{B}, \tag{1}$$

where $\mathbf{v}_e$ is the electron velocity, $\mathbf{J}$ is the electric current density, $\mathbf{B}$ is the magnetic field, n_e is the electron number density, and e is the absolute value of the electric charge.

The starting point for deriving the EMHD equation is the magnetic induction equation:

$$\frac{\partial \mathbf{B}}{\partial t} = \nabla \times (\mathbf{v}_e \times \mathbf{B}) + \eta \nabla^2 \mathbf{B}, \tag{2}$$

where we consider motions of electrons only and η is magnetic diffusivity. Substituting equation (1) into equation (2), we obtain the incompressible EMHD equation

$$\frac{\partial \mathbf{B}}{\partial t} = -\frac{c}{4\pi n_e e}\nabla \times [(\nabla \times \mathbf{B}) \times \mathbf{B}] + \eta \nabla^2 \mathbf{B}. \tag{3}$$

2.2. Numerical method

We use a pseudospectral code to solve the normalized 3-dimensional incompressible EMHD equation in a periodic box of size 2π:

$$\frac{\partial \mathbf{B}}{\partial t} = -\nabla \times [(\nabla \times \mathbf{B}) \times \mathbf{B}] + \eta' \nabla^2 \mathbf{B}, \tag{4}$$

where the magnetic field, time, and length are normalized by a mean field B_0, the whistler time $t_w = L^2(\omega_{pe}/c)^2/\Omega_e$ (ω_{pe} = electron plasma frequency; Ω_e= electron gyrofrequency), and a characteristic length scale L [9]. The resistivity η' in equation (4) is dimensionless. The magnetic field consists of a uniform background field and a fluctuating field: $\mathbf{B} = \mathbf{B}_0 + \mathbf{b}$. The strength of the uniform background field, B_0, is set to 1. We use either 256^3 or 512^3 grid points.

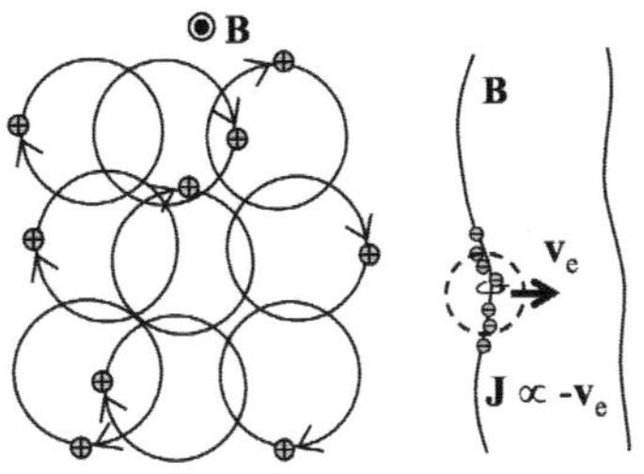

Figure 1. In EMHD, ions are assumed to provide uniform charge background below the ion gyro-scale. Therefore, only electrons move and carry current, which results in $\mathbf{J} \propto -\mathbf{v}_e$.

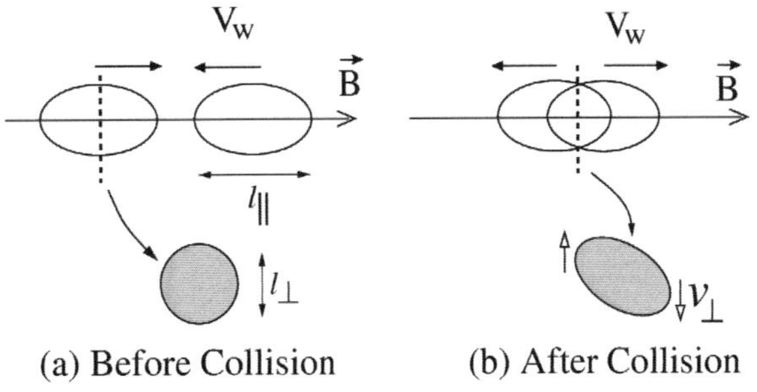

Figure 2. Turbulence is generated when two oppositely traveling EMHD wave packets (or eddies) collide. The shaded shapes represent cross-sectional views. After collision, an eddy is distorted due to the shear provides by the other eddy. Figure is modified from [10].

Hyperdiffusivity is used for the diffusion term. The power of hyperdiffusivity is set to 3, so that the dissipation term in the above equation is replaced with $\eta_3(\nabla^2)^3\mathbf{B}$, where the value of η_3 is approximately 3×10^{-10} for 256^3 and 1×10^{-11} for 512^3.

Throughout the paper, we use the following convention: the field **B** actually means the magnetic field divided by $(4\pi\rho)^{1/2}$, where ρ is density. Thus, the field **B** is, in fact, the Alfvén velocity, although we refer to it 'magnetic field' in some places.

3. Properties of EMHD waves and turbulence cascade

3.1. Properties of EMHD waves

In some respects, EMHD waves are similar to Alfvén waves. For example, they propagate along magnetic field lines. But, in many respects, EMHD waves are different from Alfvén waves. In this subsection, we briefly discuss properties of EMHD waves (see Ref. [8] for further discussions).

First, EMHD waves are dispersive. From equation (4), we can show that the dispersion relation for an EMHD wave is $\omega \propto kk_{||}B_0$, where k is the wavenumber, $k_{||}$ is the wavenumber parallel to the mean magnetic field, and B_0 is the Alfvén speed of the background magnetic field [11, 9, 7]. The dispersion relation indicates that an EMHD wave packet of size l ($\sim 1/k$) propagates along the mean magnetic field at the speed of

$$V_w \propto kB_0. \tag{5}$$

If we consider a group of different-size wave packets traveling in one direction along the mean magnetic field, the dispersion relation implies that smaller wave packets are faster than larger ones. Therefore, EMHD wave packets moving in the same direction can collide each other and produce small-scale structures through nonlinear interactions during collisions [12, 7].

Second, an EMHD wave is circularly polarized [13, 14] and it has a non-zero magnetic helicity. The sign of magnetic helicity depends on the direction of propagation. A 'positive' wave, which moves in the positive direction along the magnetic field, has a positive magnetic helicity. In fact, the magnetic helicity of a positive wave is b^2/k, where k is the wave number and b is the amplitude of the wave, which allows us to write

$$H(k) = E_b(k)/k, \tag{6}$$

where $H(k)$ is the magnetic helicity spectrum and $E_b(k)$ is the magnetic energy spectrum. On the other hand, a 'negative' wave, which moves in the negative direction along the magnetic field, has a negative magnetic helicity. Therefore, if there is imbalance in the amplitudes of positive and negative EMHD wave packets with similar sizes, then the net magnetic helicity of the system is not zero. For example, if positive wave packets have larger amplitudes than negative wave packets, then the net magnetic helicity is positive.

3.2. *Magnetic helicity and turbulence cascade*

Magnetic helicity is a more strictly conserved quantity than magnetic energy: it is very difficult to destroy magnetic helicity even in the presence of turbulence, while it is relatively easy to destroy magnetic energy through turbulence energy cascade. The reason for this is that magnetic helicity spectrum is steeper than magnetic energy spectrum[1].

Magnetic helicity conservation can play an important role in energy cascade in EMHD turbulence. Let us first consider two interacting wave packets moving in the same direction. We assume that their sizes are similar[2] and of order l ($\sim 1/k$). Let the magnetic energy of each packet be $\sim b^2$. Then, as we discussed in the previous subsection, each wave packet has magnetic helicity of order lb^2. The sign of magnetic helicity of the packets is same since they are moving in the same direction. Nonlinear interactions occur when two wave packets collide. Due to magnetic helicity conservation, the net magnetic helicity before and after the collision is unchanged and is of order

$$\sim 2lb^2. \tag{7}$$

When two wave packets interact, their energy should ultimately decrease. If this indeed happens, then the scale l should increase in order to conserve magnetic helicity. Therefore we expect inverse cascade, i.e. increase of the typical scale of wave packets. Note that the underlying physics for inverse cascade is magnetic helicity conservation.

In the previous paragraph, we considered interactions between wave packets moving in the same direction. Such interactions are important in imbalanced EMHD turbulence. In imbalanced turbulence, waves packets moving in one direction are stronger than those moving in the other direction. Therefore, imbalanced EMHD turbulence has two important properties. First, in imbalanced EMHD turbulence, it is possible that interactions between dominant wave packets are more efficient than interactions between dominant and sub-dominant wave packets. Second, in imbalanced EMHD turbulence, the net magnetic helicity is not zero. From the discussion in the previous paragraph, we expect inverse cascade in imbalanced EMHD turbulence due to conservation of magnetic helicity. We will discuss scaling relations of inverse cascade in imbalanced EMHD turbulence in Section 5.

Now let us consider balanced EMHD turbulence, in which opposite-traveling wave packets with similar sizes have similar amplitudes. In this case, interactions between two opposite-traveling wave packets with similar sizes should be more efficient than interactions between wave packets moving in the same direction[3]. If we have two balanced opposite-traveling wave

[1] If magnetic energy spectrum $E_b(k)$ is proportional to k^{-m} ($k_p \leq k \leq k_d$), then we have $b^2/(\dot{b^2}) \propto k_p E_b(k_p)/[k_d E_b(k_d) k_d^2] \propto (k_d/k_p)^{m-1}$, where we use the fact $b_l^2 \sim kE(k)$ (with $l \sim 1/k$). Note that $m \geq 1$ for most cases (e.g. m=5/3 for a Kolmorogov spectrum). Similarly, if magnetic helicity spectrum $H(k)$ is proportional to k^{-m_h} ($k_p \leq k \leq k_d$), we can show that $h^2/(\dot{h^2}) \propto (k_d/k_p)^{m_h-1}$. Since magnetic helicity spectrum is steeper than magnetic energy spectrum (see, for example, equation (6)), we have $m_h > m$ and $h^2/(\dot{h^2}) > b^2/(\dot{b^2})$. Therefore, dissipation timescale for magnetic helicity is relatively larger than that for magnetic energy.

[2] Note that nonlinear interactions between wave packets occur when they collide and, hence, same-size wave packets moving in the same direction do not interact efficiently because their propagation speed is same. Therefore, interactions between wave packets moving in the same direction happen between wave packets with slightly different sizes.

[3] Note that, in the latter, wave packets with slightly different sizes interacts. Therefore, if locality of interaction

packets with size $\sim l$ and energy $\sim b^2$, the net magnetic helicity is close to zero: $lb^2 - lb^2 = 0$. When they collide, either forward cascade or inverse cascade is allowed because it does not change the net magnetic helicity. Note that the net magnetic helicity after collision will still be close to zero as long as wave packets remain balanced. We know that collision of two balanced opposite-traveling wave packets results in forward cascade in MHD turbulence. Therefore, we expect that interactions between opposite-traveling wave packets in EMHD turbulence are likely to produce forward cascade. We will discuss scaling relations of forward cascade in balanced EMHD turbulence in Section 4.

4. Forward cascade in EMHD

4.1. Scaling Relations

In this subsection, we derive scaling relations of forward cascade in EMHD turbulence. As we mentioned in the previous section, the following scaling relations are based on interactions between opposite-traveling wave packets and will be valid for balanced EMHD turbulence.

The magnetic energy spectrum for forward cascade in EMHD turbulence has been known since 1970's [15, 11, 16] and scaling relations based on 'critical balance' [2] were first derived and numerically tested by Cho & Lazarian [4]. We can re-derive the scaling relations in a more intuitive way as follows: Let us consider two EMHD wave packets in a medium threaded by a uniform magnetic field $\mathbf{B}_0$. We assume that the packets are moving in opposite directions (see Figure 2). Note that, a strong (local) mean field naturally makes EMHD turbulence anisotropic: eddies are elongated along the (local) mean-field direction [17]. Therefore, in Figure 2, we draw that $l_\parallel > l_\perp$, where $l_\parallel$ and $l_\perp$ are the parallel and the perpendicular sizes of an eddy, respectively. Here 'parallel' and 'perpendicular' refer to the directions with respect to the (local) mean magnetic field.

As in most turbulence theories, let us assume locality of interactions: only collisions between similar-size eddies are important for energy cascade (see, however, Refs. [18] and [19]). When two oppositely traveling wave packets (of size $l_\perp$ and $l_\parallel$) collide, one wave packet provides shear and, thus, distortion to the other (Figure 2). The time scale for complete distortion is of the order $t_{eddy} \sim l_\perp/v_\perp$, where $v_\perp$ is the fluctuating velocity on the scale $l = l_\perp$. If the strengths of the fluctuating magnetic field is b_l, the fluctuating velocity $v_\perp$ is roughly

$$\mathbf{v} \propto \nabla \times \mathbf{b} \rightarrow v_\perp \propto b_l/l_\perp. \quad (8)$$

Therefore, we have

$$t_{eddy} \sim l_\perp/v_\perp \propto l_\perp^2/b_l \quad (9)$$

On the other hand, the duration of the collision is

$$t_{coll} \sim l_\parallel/V_w \propto l_\parallel/(kB_0). \quad (10)$$

As in the MHD turbulence case (see Ref. [2]), we can argue that there is a regime of turbulence where

$$t_{eddy} \sim t_{coll}, \quad \text{or} \quad l_\perp B_0/(l_\parallel b_l) \sim 1, \quad (11)$$

which is known as 'critical balance'. This type of turbulence is called 'strong' turbulence. Note that, in strong EMHD turbulence, the time scale for energy cascade, t_{cas}, can be either the collision time t_{coll} or the eddy turnover time t_{eddy}. When we combine this condition and the constancy of energy cascade rate,

$$b_l^2/t_{cas} \sim \text{constant}, \quad (12)$$

is a valid assumption in EMHD, we expect interactions between wave packets moving in one direction are less efficient than interactions between opposite-traveling wave packets.

Journal of Physics: Conference Series **719** (2016) 012001 doi:10.1088/1742-6596/719/1/012001

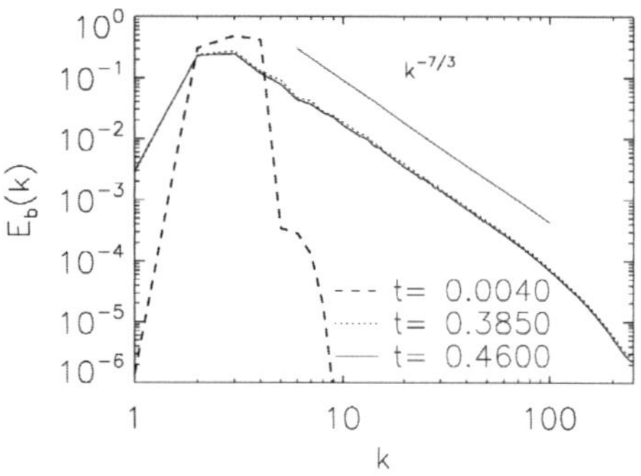

Figure 3. Spectrum of decaying balanced EMHD turbulence at different time points. At $t \approx 0$, only small-wavenumber Fourier modes are excited. At later times, energy gradually cascades down to large k regions and a $k^{-7/3}$ inertial range develops. From [5].

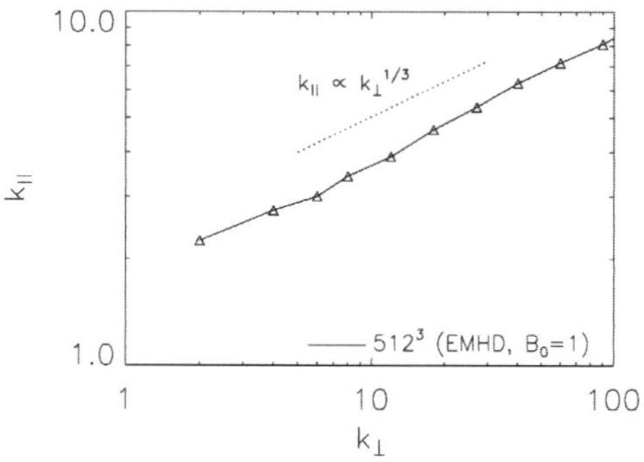

Figure 4. Anisotropy of decaying balanced EMHD turbulence. The relation $k_\parallel \propto k_\perp^{1/3}$ implies scale-dependent anisotropy: $l_\parallel \propto l_\perp^{1/3}$, where $l_\parallel$ and $l_\perp$ are the size of an eddy parallel and perpendicular to the (local) mean magnetic field, respectively. From [5].

we get the scaling relations

$$E_b(k) \propto k^{-7/3} \quad \text{and} \quad l_\parallel \propto l_\perp^{1/3}, \quad (\text{or, } k_\parallel \propto k_\perp^{1/3}) \tag{13}$$

where the latter implies scale-dependent anisotropy: smaller eddies are more elongated. Here, $k_\perp$ is the wavenumber perpendicular to the (local) mean magnetic field. Here, we used $b_l^2 \sim kE_b(k)$. Compared with the Goldreich-Sridhar scaling relations [2] for MHD turbulence, the energy spectrum is steeper and the anisotropy is stronger in EMHD turbulence.

4.2. Numerical results

Figure 3 shows magnetic energy spectrum in a decaying EMHD simulation with 512^3 grid points (see Ref. [5]). At t=0, Fourier modes with $2 \leq k < 5$ are excited. Here $k = \sqrt{k_\perp^2 + k_\parallel^2}$ and, thus, the initial perturbation is isotropic (i.e. $k_\parallel \sim k_\perp$). The strength of the mean magnetic field (B_0) is 1 and the rms value of the random magnetic field (b) is ~ 1.21. Therefore, the critical balance, $bk_\perp/B_0k_\parallel = 1$, is roughly satisfied at t=0. Figure 3 shows how the inertial range develops. The dashed curve in Figure 3 shows the initial spectrum. As the turbulence decays, the initial energy cascades down to smaller scales and, as a result, small-scale (i.e. large-k) modes are excited. When the energy reaches the dissipation scale at $k \sim 100$, the whole energy spectrum goes down without changing its slope (the dotted and the solid curves). The magnetic energy spectrum at this stage is very close to that of the predicted one:

$$E_b(k) \propto k^{-7/3}. \tag{14}$$

For the study of anisotropy, we use the data cube at $t \approx 0.46$. The solid curve in left panel of Figure 3 represents the spectrum at that time. Figure 4 shows the relation between $k_\parallel$ and $k_\perp$ (see [5] for the calculation method). The result is compatible with the expected anisotropy

$$k_\parallel \propto k_\perp^{1/3}. \tag{15}$$

5. Inverse cascade in EMHD

As we mentioned earlier, imbalanced EMHD turbulence shows inverse cascade. Let us consider two exemplary cases: freely decaying completely imbalanced EMHD turbulence and EMHD turbulence with completely imbalanced driving. Note that "completely imbalanced" in the former means that all EMHD waves are freely propagating in one direction at t=0 and "completely imbalanced driving" in the latter means ether positive waves or negative waves

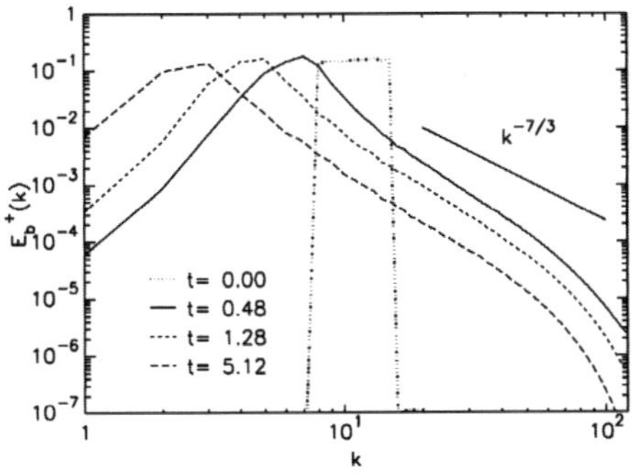

Figure 5. Magnetic energy spectra for a freely decaying imbalanced EMHD turbulence. At t=0, all EMHD waves move in one direction and their spectrum in shown in dotted line. As time goes on the peak of the spectrum moves to the right. We show the spectra of the dominant waves. From [8].

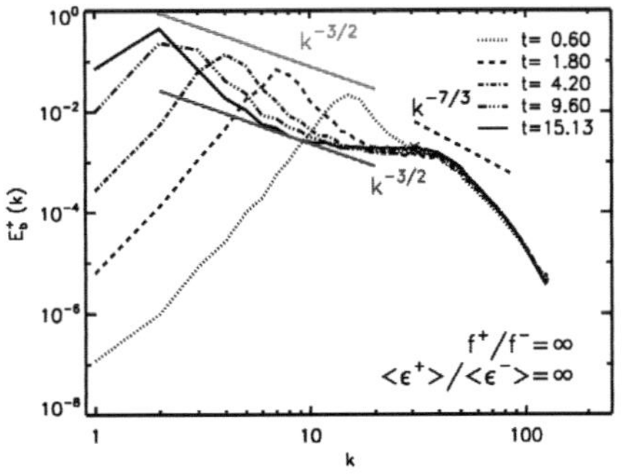

Figure 6. Magnetic energy spectra for driven imbalanced EMHD turbulence. We drive only the positive waves, i.e. waves moving parallel to the magnetic field. As time goes on the peak of the spectrum moves to the upper-right direction. We show the spectra of the dominant waves. From [8].

are driven. For the sake of simplicity, let us assume that only the positive waves are present at t=0 in the former case and only the positive waves are driven in the latter case.

In either case, the positive EMHD waves can interact with each other, generate opposite-traveling negative waves, develop small-scale structures, and dissipate magnetic energy. At the same time, they show inverse cascade due to magnetic helicity conservation. Note that magnetic helicity stays constant in the former case and it increases linearly in time in the latter case. In this section, we consider consequences of magnetic helicity conservation. Most discussions in the section are from Refs. [7, 8].

5.1. Freely decaying completely imbalanced EMHD turbulence

Let us first consider freely decaying completely imbalanced EMHD waves (i.e. EMHD waves moving in one direction at t=0). Suppose that they have a power-law energy spectrum between $k = k_p$ and k_d:

$$E_b(k) \propto k^{-m}, \quad k_p \le k \le k_d, \tag{16}$$

where k_d is the dissipation scale. Note that, even though they do not have a power-law spectrum at t=0, they develop a rough power-law-like spectrum at later times (see Figure 5). Then the magnetic energy and the magnetic helicity densities are approximately $\sim k_p E_b(k_p)$ and $\sim E_b(k_p)$, respectively. As the EMHD waves propagate, the helicity density stays the same, so that $E_b(k_p)$ is approximately constant. On the other hand, if there is no driving, the energy density, $\sim k_p E_b(k_p)$, decreases due to dissipation. Therefore k_p should decrease, which means that the peak of the energy spectrum moves to smaller wavenumbers.

Numerical simulations confirm the theoretical expectations in the previous paragraph. Figure 5 shows magnetic energy spectra of the dominant waves in a simulation with a completely imbalanced initial condition. At $t = 0$, Fourier modes between $k = 8$ and $k = 15$ are excited. The dotted curve in Figure 5 shows the initial spectrum. As time goes on, the initial energy cascades down to smaller scales and, as a result, a power-law-like spectrum forms for $k > 15$. But the magnetic energy spectrum for $k > 15$ is slightly steeper than $k^{-7/3}$. At the same time the peak of the energy spectrum moves to larger scales, so that the wavenumber at which the spectrum peaks, k_p, gets smaller. We clearly observe inverse cascade of magnetic energy.

5.2. EMHD turbulence with completely imbalanced driving

Now, let us consider EMHD turbulence with completely imbalanced driving. When driving injects magnetic energy into waves moving only in one direction, it also injects magnetic helicity into the system. Because the magnetic helicity is a well-conserved quantity, the net magnetic helicity density grows linearly in time in driven imbalanced EMHD turbulence:

$$H_{net} = \epsilon_h t \approx (\epsilon_b/k_f)t, \tag{17}$$

where k_f is the driving-scale wavenumber, ϵ_h is the magnetic helicity injection rate, and ϵ_b is the magnetic energy injection rate. On the other hand, the magnetic energy density follows

$$\dot{b^2} = \epsilon_b - D_M < \epsilon_b \quad \longrightarrow \quad b^2 < \epsilon_b t, \tag{18}$$

where D_M is the magnetic energy dissipation rate.

If the magnetic spectrum has a peak at k_p, then from equations (17) and (18) we have

$$k_p E_b(k_p) \ (\approx b^2) < \epsilon_b t \approx k_f H_{net} \approx k_f E_b(k_p), \tag{19}$$

which implies

$$k_p < k_f. \tag{20}$$

Here we used $H_{net} \approx E_b(k_p)$. We can show that k_p cannot be a fixed wavenumber: if we assume that k_p does not change in time, we will have a contradictory result. Recall that we are considering EMHD turbulence with completely imbalance driving. In this case, most of the energy near $k = k_p$ resides in the dominant waves, which naturally produce sub-dominant waves. Suppose that an amount of magnetic energy is transferred from the dominant to the sub-dominant waves near $k = k_p$. As a result of the energy transfer, the magnetic energy ($\sim k_p E_b(k_p)$) of the dominant waves near $k = k_p$ decreases, while the magnetic helicity ($\sim E_b(k_p)$) of the dominant waves near $k = k_p$ increases to compensate magnetic helicity transfer to the sub-dominant waves[4]. Therefore, k_p cannot be a constant and should decrease. In summary, when a continuous magnetic helicity injection is present, we expect that the peak of magnetic energy spectrum moves to smaller wavenumbers.

In numerical simulations of driven imbalanced EMHD turbulence, we can clearly see inverse cascade [8]. Figure 6 shows magnetic energy spectrum of the dominant waves in a simulation with completely imbalanced driving. The driving wavenumber k_f is approximately 40. The small-scale spectra of the dominant waves are very steep, partly due to our limited numerical resolution of 256^3. The large-scale magnetic energy spectrum clearly shows inverse cascade: the peak of the spectrum gradually moves to larger scales. Although we do not observe a well-defined power-law spectrum at a given time, the location of the peaks follows $E_b(k_p) \propto k_p^{-3/2}$. The fact that the magnetic energy spectrum for $k < k_f$ does not follow a single power-law implies that inverse cascade in EMHD turbulence is not a self-similar process. Therefore, the nature of the inverse cascade in EMHD turbulence can be very different from cascades in other types of turbulence (see [8] for details).

[4] When magnetic energy ΔE is transferred from the positive waves near $k = k_p$ to the negative waves near $k = k_p$, the magnetic helicity of the negative waves decreases: $H^- \rightarrow H^- - \Delta E/k_p$. This is because the sign of magnetic helicity of the negative waves is negative and, hence, increase of magnetic energy in negative waves means decrease of magnetic helicity. The net magnetic helicity should not be changed by the energy transfer. Therefore, the magnetic helicity of the positive waves should increase: $H^+ \rightarrow H^+ + \Delta E/k_p$.

6. Summary

Similar to Alfvén waves in MHD, EMHD waves propagate along magnetic field lines. Therefore, two types EMHD waves can exist: waves moving parallel to and waves moving anti-parallel to the magnetic field. The relative amplitudes of these two types of waves are important for turbulence cascade. In case their amplitudes are balanced, we observe efficient forward cascade. On the other hand, in case there is large imbalance in their amplitudes, inverse cascade becomes important.

Forward cascade in balanced EMHD turbulence has a steep magnetic energy spectrum, $E_b(k) \propto k^{-7/3}$, and a strong anisotropic eddy shapes, $l_\parallel \propto l_\perp^{1/3}$, where $l_\parallel$ and $l_\perp$ are the size of an eddy parallel and perpendicular to the (local) mean magnetic field, respectively. Inverse cascade in imbalanced EMHD turbulence does not show a well-defined power-law spectrum at a given time for the inverse cascade. However, the location of the peaks at different times follows $E_b(k_p) \propto k_p^{-3/2}$, where k_p is the wavenumber at which the magnetic energy spectrum peaks.

References

[1] Smith C W, Hamilton K, Vasquez B J and Leamon R J 2006 *Astrophysical J. Letters* **645** L85–L88
[2] Goldreich P and Sridhar S 1995 *Astrophysical J.* **438** 763–775
[3] Kingsep A S, Chukbar K V and Yan'kov V V 1990 *Reviews of Plasma Physics, Vol. 16* (Consultants Bureau, New York)
[4] Cho J and Lazarian A 2004 *Astrophys. J. Lett.* **615** L41–L44
[5] Cho J and Lazarian A 2009 *Astrophys. J.* **701** 236–252
[6] Schekochihin A A, Cowley S C, Dorland W, Hammett G W, Howes G G, Quataert E and Tatsuno T 2009 *Astrophysical J. Supp.* **182** 310–377 (*Preprint* 0704.0044)
[7] Cho J 2011 *Physical Review Letters* **106** 191104 (*Preprint* 1104.4148)
[8] Kim H and Cho J 2015 *Astrophysical J.* **801** 75
[9] Galtier S and Bhattacharjee A 2003 *Physics of Plasmas* **10** 3065–3076
[10] Cho J 2014 *Journal of Korean Physical Society* **65** 871–875
[11] Biskamp D, Schwarz E and Drake J F 1996 *Physical Review Letters* **76** 1264–1267
[12] Ng C S, Bhattacharjee A, Germaschewski K and Galtier S 2003 *Physics of Plasmas* **10** 1954–1962
[13] Mahajan S M and Krishan V 2005 *Monthly Not. Roy. Astron. Soc.* **359** L27–L29
[14] Galtier S 2006 *Journal of Plasma Physics* **72** 721–769 (*Preprint* arXiv:physics/0608227)
[15] Vaĭnshteĭn S I 1973 *Sov. Phys. JETP* **37** 73–+ [Zh. Eksp. Teor. Fiz. **64**, 139 (1973)]
[16] Biskamp D, Schwarz E, Zeiler A, Celani A and Drake J F 1999 *Physics of Plasmas* **6** 751–758
[17] Dastgeer S and Zank G P 2003 *Astrophys. J.* **599** 715–722
[18] Beresnyak A and Lazarian A 2010 *Astrophysical J. Lett.* **722** L110–L113 (*Preprint* 1002.2428)
[19] Cho J 2010 *Astrophysical J.* **725** 1786–1791 (*Preprint* 1005.0878)

Journal of Physics: Conference Series **719** (2016) 012002 doi:10.1088/1742-6596/719/1/012002

The role of turbulence, magnetic fields and feedback for star formation

Christoph Federrath

Research School of Astronomy and Astrophysics, Australian National University, Canberra, ACT 2611, Australia

E-mail: christoph.federrath@anu.edu.au

Abstract. Star formation is inefficient. Only a few percent of the available gas in molecular clouds forms stars, leading to the observed low star formation rate (SFR). The same holds when averaged over many molecular clouds, such that the SFR of whole galaxies is again surprisingly low. Indeed, considering the low temperatures, molecular clouds should be highly gravitationally unstable and collapse on their global mean freefall timescale. And yet, they are observed to live about 10–100 times longer, i.e., the SFR per freefall time ($\mathrm{SFR_{ff}}$) is only a few percent. Thus, other physical mechanisms must provide support against quick global collapse. Magnetic fields, turbulence and stellar feedback have been proposed as stabilising agents controlling star formation, but it is still unclear which of these processes is the most important and what their relative contributions are. Here I present high-resolution, adaptive-mesh-refinement simulations of star cluster formation that include turbulence, magnetic fields, and protostellar jet/outflow feedback. These simulations produce nearly realistic star formation rates consistent with observations, but only if turbulence, magnetic fields and feedback are included simultaneously.

Star formation is inefficient. For example, our entire home galaxy—the Milky Way—only produces about $2\,M_\odot\,\mathrm{yr}^{-1}$ [1], while there is plenty of gas available to form hundreds of stars per year [2, 3]. Based on the range of typical molecular cloud masses of $\sim 10^2$–$10^7\,M_\odot$ and sizes of ~ 1–$100\,\mathrm{pc}$, the average cloud mass density ρ is a few times 10^{-21}–$10^{-19}\,\mathrm{g\,cm^{-3}}$, corresponding to an average particle number density of about 10^3–$10^5\,\mathrm{cm^{-3}}$ for standard molecular composition. This would render the clouds highly unstable and immediately leads to a very short freefall time, $t_\mathrm{ff} = \sqrt{3\pi/(32G\rho)} = 0.1$–$1\,\mathrm{Myr}$, which is at least one order of magnitude shorter than the typical lifetime of molecular clouds. Thus, other mechanisms than thermal pressure must be at work to prevent molecular clouds from collapsing globally and from forming stars at a 10–100 times higher rate than observed.

For a long time, the standard picture of star formation was that magnetic fields provide the primary source of stabilizing pressure and tension, which acts against the quick global collapse. Only after the neutral species had slowly diffused through the charged particles over an 'ambipolar-diffusion' time-scale of about 10 Myr, would star formation proceed in the central regions of magnetized clouds [4, 5, 6]. This standard theory requires that clouds start their lives dominated by a strong magnetic field, rendering them initially subcritical to collapse. After about one ambipolar diffusion time-scale, magnetic flux was left behind in the cloud envelope, while the mass has increased in the cloud core such that some stars could form in centre. Thus, star formation regulated by ambipolar diffusion predicts a higher mass-to-flux ratio in the cores than in the envelopes of the clouds, which is—however—typically not observed [7, 8, 9, 10, 11, 12].

An alternative scenario is that clouds are in fact collapsing globally, thus initially forming stars at a very high rate, but that the ionization feedback from massive stars eventually terminates the global collapse and disperses the clouds [13].

A third alternative is that the observed supersonic random motions [2, 3, 14, 15, 16, 17, 18, 19, 20] regulate star formation. In this modern picture, turbulence plays a crucial dual role. On the one hand, the turbulent kinetic energy stabilizes the clouds on large scales and prevents global collapse, on the other hand, it induces local compressions in shocks, because the turbulence is supersonic [21, 22, 23]. This generates the initial conditions for star formation, because the local compressions typically produce filaments and dense cores at the intersections of filaments [24].

So which scenario is right or wrong? As it turns out, neither of the three is entirely wrong, nor do they explain everything by themselves. In the most recent years, we have come to a deeper understanding of the relative importance of magnetic fields and turbulence through numerical simulations. While the role of turbulence cannot be denied, because it naturally explains most of the observed velocity and density structure of molecular clouds, the role of magnetic fields remained less clear. Previous simulation work suggested that magnetic fields can suppress fragmentation on small scales and hence influence the shape of the initial mass function of stars [25, 26, 27] and that the SFR can be reduced by a factor of 2–3 compared to the non-magnetized case [28, 29, 30, 31].[1] Importantly however, magnetic fields also launch fast, powerful, mass-loaded jets and outflows from the protostellar accretion disc. These jets and outflows can drive turbulence and alter the cloud structure and dynamics so profoundly that the SFR and the initial mass function might be even more affected by this jet/outflow feedback mechanism [34, 35].

The aim of this study is to determine the physical processes that make star formation inefficient. We measure which physical mechanisms are relevant for bringing the SFR in agreement with observations and we determine their relative importance, i.e., by what amount they reduce the SFR individually and when acting all in concert.

Section 1 summarizes our numerical methods and simulations from which we measure the SFR. Our results are presented in Section 2, where we find that purely self-gravitating molecular clouds are indeed highly unstable, while the step-by-step inclusion of turbulence, magnetic fields, and jet/outflow feedback brings the SFR down by one to two orders of magnitude when all of these physical mechanisms act together. We summarize our findings and conclusions in Section 3. Some of the results of this work are published in Federrath (2015) [36], and a featured article about this work with comments and animations of the simulations presented here is posted on `http://astrobites.org/2015/04/28/why-is-star-formation-so-inefficient/`.

1. Numerical simulation techniques

We use the multi-physics, adaptive mesh refinement (AMR) [37] code FLASH [38, 39] in its latest version (v4), to solve the compressible magnetohydrodynamical (MHD) equations on three-dimensional (3D) periodic grids of fixed side length L, including turbulence, magnetic fields, self-gravity and outflow feedback. The positive-definite HLL5R Riemann solver [40] is used to guarantee stability and accuracy of the numerical solution of the MHD equations.

1.1. Turbulence driving

Turbulence is a key for star formation [21, 22, 23, 41, 42], so most of our simulations include a turbulence driving module that produces turbulence similar to what is observed in real molecular clouds, i.e., driving on the largest scales [43, 44] and with a power spectrum, $E(k) \sim k^{-2}$, consistent with supersonic, compressible turbulence [14, 18, 20] and confirmed by simulations

[1] See also earlier work on the effects of magnetic fields on the star formation efficiency [32, 33].

Table 1. Key simulation parameters and measured star formation rates (SFRs).

Model Name	Turbulence	σ_v(km/s)	$\mathcal{M}$	$B(\mu$G)	β	$\mathcal{M}_{\rm A}$	Jets/Outflows	$N_{\rm res}^3$	SFR($M_\odot$/yr)	SFR$_{\rm ff}$
(1)	(2)	(3)	(4)	(5)	(6)	(7)	(8)	(9)	(10)	(11)
G	None	0	0	0	∞	∞	No	1024^3	1.6×10^{-4}	0.47
GvsT	Mix	1.0	5.0	0	∞	∞	No	1024^3	8.3×10^{-5}	0.25
GvsTM	Mix	1.0	5.0	10	0.33	2.0	No	1024^3	2.8×10^{-5}	0.083
GvsTMJ	Mix	1.0	5.0	10	0.33	2.0	Yes	1024^3	1.4×10^{-5}	0.041
GvsTMJ512	Mix	1.0	5.0	10	0.33	2.0	Yes	512^3	1.3×10^{-5}	0.039
GvsTMJ256	Mix	1.0	5.0	10	0.33	2.0	Yes	256^3	8.9×10^{-6}	0.027

Notes. Column 1: simulation name. Columns 2–4: the type of turbulence driving [46], turbulent velocity dispersion, and turbulent rms sonic Mach number. Columns 5–7: magnetic field strength, the ratio of thermal to magnetic pressure (plasma β), and the Alfvén Mach number. Note that these magnetic quantities are based on the initial uniform magnetic field. However, the total (mean+turbulent) plasma β at the end of the simulations is $\beta_{\rm end} \sim 0.25$ and thus not significantly different from the initial value $\beta = 0.33$. This is because the field is relatively strong to begin with, such that turbulent tangling can only amplify the magnetic field by a factor of $(\beta/\beta_{\rm end})^{1/2} \sim 1.15$ over the course of the whole simulation. Column 8: whether jet and outflow feedback was included or not. Column 9: maximum effective grid resolution (note that refinement is based on the Jeans length with a minimum of 32 cells per Jeans length). Columns 10–11: the absolute SFR and the SFR per mean global freefall time. The standard four simulation models are in the first four rows. Two additional simulations with lower resolution, but otherwise identical to GvsTMJ, are listed in the last two rows, in order to check convergence of our results for the SFR.

[45, 46, 47]. We drive turbulence by applying a stochastic Ornstein-Uhlenbeck process [48, 49] to construct an acceleration field $\mathbf{F}_{\rm stir}$, which serves as a momentum and energy source term in the momentum equation of MHD. As suggested by observations, $\mathbf{F}_{\rm stir}$ only contains large-scale modes, $1 < |\mathbf{k}|\, L/2\pi < 3$, where most of the power is injected at the $k_{\rm inj} = 2$ mode in Fourier space, i.e., on half of the box size. The turbulence on smaller scales is not directly affected by the driving and develops self-consistently. The turbulence forcing module used here excites a natural mixture of solenoidal and compressible modes, corresponding to a turbulent driving parameter $b = 0.4$ [46], although some cloud-to-cloud variations in this parameter from $b \sim 1/3$ (purely solenoidal driving) to $b \sim 1$ (purely compressive driving) are expected for real clouds [50, 51].

1.2. Sink particles

In order to measure the SFR, we use the sink particle method [52]. Sink particles form dynamically in our simulations when a local region in the simulation domain undergoes gravitational collapse and forms stars. This is technically achieved by first flagging each computational cell that exceeds the Jeans resolution density,

$$\rho_{\rm sink} = \frac{\pi c_{\rm s}^2}{G\lambda_{\rm J}^2} = \frac{\pi c_{\rm s}^2}{4Gr_{\rm sink}^2}, \tag{1}$$

with the sound speed $c_{\rm s}$, the gravitational constant G and the local Jeans length $\lambda_{\rm J}$. Thus, the sink particle accretion radius is given by $r_{\rm sink} = \lambda_{\rm J}/2$ and set to 2.5 grid cell lengths in order to capture star formation and to avoid artificial fragmentation on the highest level of AMR [53]. If the gas density in a cell exceeds $\rho_{\rm sink}$, a spherical control volume with radius $r_{\rm sink}$ is constructed around that cell and it is checked that all the gas within the control volume is Jeans-unstable, gravitationally bound and collapsing towards the central cell. A sink particle is only formed in the central cell of the control volume, if all of these checks are passed. This avoids spurious formation of sink particles and guarantees that only bound and collapsing gas forms stars [52], which is important for accurately measuring the SFR.

On all the lower levels of AMR (except the highest level, where sink particles form), we use an adaptive grid refinement criterion based on the local Jeans length, such that λ_J is always resolved with at least 32 grid cell lengths in each of the three spatial directions of our 3D domain. This resolution criterion is very conservative and computationally costly, but guarantees that we resolve turbulence on the Jeans scale [54], potential dynamo amplification of the magnetic field in the dense cores [55], and capture the basic structure of accretion discs forming on the smallest scales [35]. If a cell within the accretion radius of an existing sink particle exceeds ρ_{sink} during the further evolution, is bound to the sink particle and is moving towards it, then we accrete the excess mass above ρ_{sink} on to the sink particle, conserving mass, momentum and angular momentum. We compute all contributions to the gravitational interactions between the gas on the grid [56] and the sink particles (by direct summation over all sink particles and grid cells). A second-order leapfrog integrator is used to advance the sink particles on a timestep that allows us to resolve close and highly eccentric orbits [52].

1.3. Outflow/Jet feedback

Powerful jets and outflows are launched from the protostellar accretion discs around newborn stars. These outflows carry enough mass, linear and angular momentum to transform the structure of their parent molecular cloud and to potentially control star formation itself. In order to take this most important mechanical feedback effect [34] into account, we recently extended the sink particle approach such that sink particles can launch fast collimated jets together with a wide-angle, lower-speed outflow component, to reproduce the global features of observed jets and outflows, as well as to be consistent with high-resolution simulations of the jet launching process and with theoretical predictions [35]. The most important feature of our jet/outflow feedback model is that it converges and produces the large-scale effects of jets and outflows already with relatively low resolution, such as with sink particle radii $r_{\mathrm{sink}} \sim 1000\,\mathrm{AU}$ used in our star cluster simulations here. Our module has been carefully tested and compared to previous implementations of jet/outflow feedback [57, 28, 58]. The most important difference to any previous implementation is that our feedback model includes angular momentum transfer, reproduces the fast collimated jet component and demonstrated convergence [35].

1.4. Simulation parameters

All our simulations share the same global properties: a cloud size $L = 2\,\mathrm{pc}$, a total cloud mass $M = 388\,M_{\odot}$ and a mean density $\rho_0 = 3.28 \times 10^{-21}\,\mathrm{g\,cm^{-3}}$, resulting in a global mean freefall time $t_{\mathrm{ff}} = 1.16\,\mathrm{Myr}$. Models including turbulence have a velocity dispersion $\sigma_v = 1\,\mathrm{km\,s^{-1}}$ and an rms Mach number of $\mathcal{M} = 5$, given the sound speed $c_{\mathrm{s}} = 0.2\,\mathrm{km\,s^{-1}}$, appropriate for molecular gas with temperature $T = 10\,\mathrm{K}$ over the wide range of densities that lead to dense core and eventually star formation [59]. Finally, models including a magnetic field start with a uniform initial field of $B = 10\,\mu\mathrm{G}$, which is subsequently compressed, tangled and twisted by the turbulence, similar to how it would be structured in real molecular clouds [60]. The magnetic field strength, the turbulent velocity dispersion and the mean density all follow typical values derived from observations of clouds with the given physical properties [16, 61]. This leads to the dimensionless virial ratio $\alpha_{\mathrm{vir}} = 1.0$ [62, 63] and to a plasma beta parameter (ratio of thermal to magnetic pressure) $\beta = 0.33$ or an Alfvén Mach number $\mathcal{M}_{\mathrm{A}} = 2.0$. An average Alfvén Mach number of about $\mathcal{M}_{\mathrm{A}} = 1.5$ has been measured in 14 different star-forming regions in the Milky

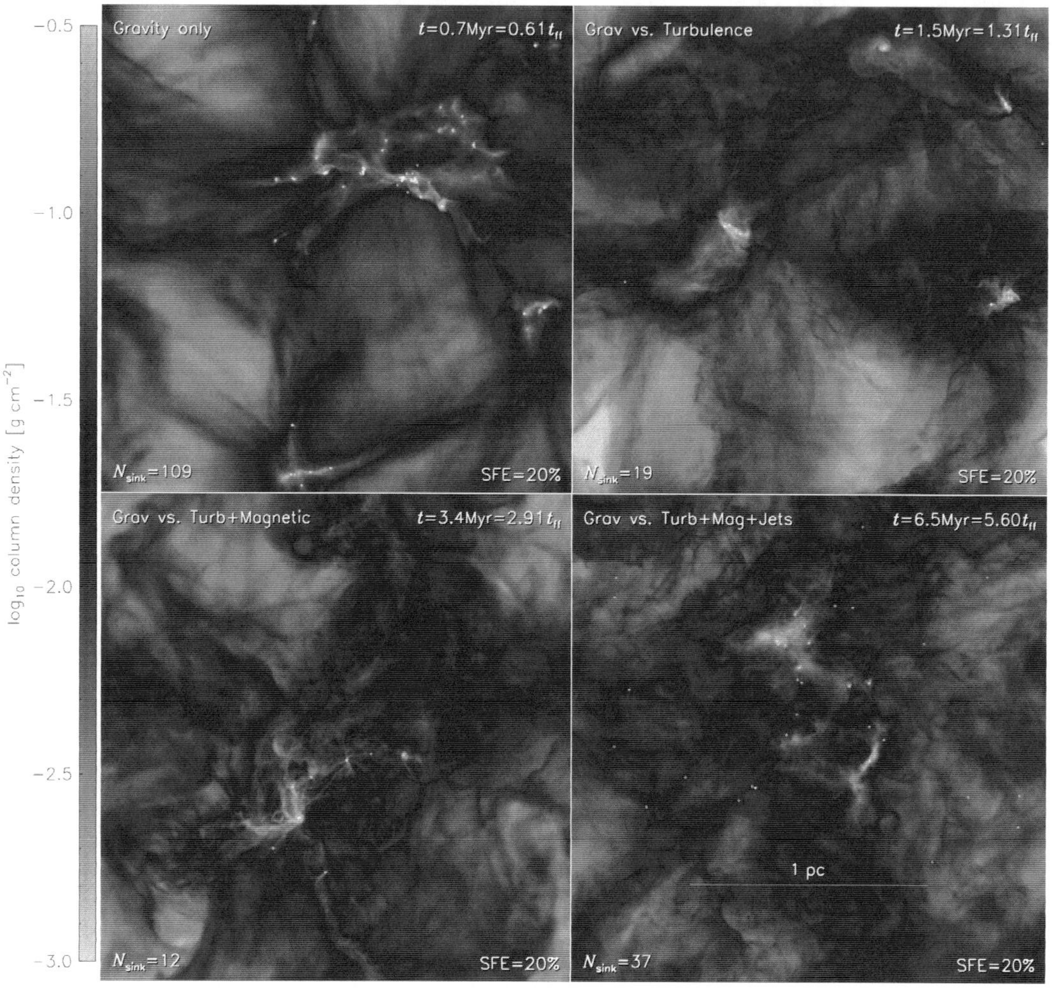

Figure 1. Column density projections at the end of each simulation model: Gravity only (top left), Gravity vs. Turbulence (top right), Gravity vs. Turbulence + Magnetic Fields (bottom left), and Gravity vs. Turbulence + Magnetic Fields + Jet/Outflow Feedback (bottom right). The time to reach a realistic SFE = 20% with small star clusters having formed, is shown in the top right corner of each panel. The time increases significantly for each model that adds physical processes opposing gravitational collapse. This demonstrates that only the combination of turbulence, magnetic fields and feedback can produces realistic SFRs, which is quantified below and summarized in Table 1. *An animation of these still frames is available at* `http://astrobites.org/2015/04/28/why-is-star-formation-so-inefficient/`

Way [64]. Thus, the assumed magnetic field in our simulation models is very close to the values typically observed in molecular clouds and cloud cores.

We run four basic models, which—step by step—include more physics. In the first simulation we only include self-gravity with a given initial density distribution resembling molecular cloud structure, but we do not include any turbulent velocities or magnetic field. In the second model, we include a typical level and mixture of molecular cloud turbulence (see §1.1). The third model is identical to the second model, but adds a standard magnetic field for the given cloud size and mass. Finally, the fourth model is identical to the third model, but additionally includes jet and outflow feedback (see §1.3). These four basic models were all run with a maximum effective grid resolution of 1024^3 cells. Their key parameters are listed in Table 1. We also run two additional models, which are identical to the fourth model (with jet/outflow feedback), but have a lower maximum effective resolution of 512^3 and 256^3 cells, respectively, in order to check numerical convergence of our results for the SFR. Those models are listed in the bottom two rows of Table 1.

2. Results

2.1. Cloud structure and stellar distribution

Figure 1 shows column density projections of our four basic models from Table 1. The simulation that only includes self-gravity (no turbulence, no magnetic fields and no feedback) is shown in the top left-hand panel. The gas structures we see in the figure resemble the typical distribution found in gravity-only simulations such as simulations of pure dark matter in the cosmological case. Gas from the voids continuously falls on to dense filaments. Stars form along filaments and in particular where filaments intersect. This seems to be true also for real molecular clouds to some degree [24], but the inter-filament gas is much more disturbed and turbulent in real molecular clouds compared to the gravity-only case. We stop the simulation when 20% of the gas is wound up in stars, i.e., the star formation efficiency $\mathrm{SFE} \equiv M_{\mathrm{stars}}/(M_{\mathrm{stars}}+M_{\mathrm{gas}}) = 20\%$, which is already reached within a fraction of a global mean freefall time, $t = 0.61\,t_{\mathrm{ff}}$, for the gravity-only simulation. Thus, the whole cloud forms stars in roughly a global freefall time, which is the expected—and unrealistic—outcome if no other physics but gravity is considered.

When we add a realistic level of turbulence to the cloud, shown in the top right-hand panel of Figure 1, we find that the time to reach SFE = 20% increases to $t = 1.31\,t_{\mathrm{ff}}$ or 1.5 Myr. The overall structure is now much closer to what we see in real molecular clouds, especially the turbulent velocity dispersion measured in observations is approximately reproduced in this model. However, the SFR is still an order of magnitude too large compared to typical observations and there is no magnetic field present, contrary to what is observed [61].

Thus, we add magnetic fields in the simulation shown in the bottom left-hand panel of Figure 1 and see that the end time is shifted by slightly more than a factor of 2, $t = 2.91\,t_{\mathrm{ff}} = 3.4\,\mathrm{Myr}$, compared to non-MHD case. Magnetic fields thus reduce the SFR by about a factor of 2–3, as found in previous simulations [28, 29, 65, 30]. Although we added a considerable set of physics (gravity, turbulence and magnetic fields) in this simulation model, the SFR is still much higher than inferred from observations, but many spatial and morphological features seen in real star-forming molecular clouds are well reproduced with this set of physics.

Finally, we add jet and outflow feedback in the simulation shown in the bottom right-hand panel of Figure 1. The overall cloud structure is similar to the turbulent MHD case without feedback, but the SFR is further reduced by about another factor of 2. Now it takes $6.5\,t_{\mathrm{ff}} = 5.6\,\mathrm{Myr}$ to reach a reasonable SFE = 20% in this small cluster-forming region of a molecular cloud. The most interesting morphological difference compared to all the no-feedback cases is that the oldest stars have wandered away from their formation sites and are now distributed over a much wider volume. Newborn stars are still always located on or very close to high-density peaks. A total of 37 stars have formed in the whole field, while some of those stars might actually represent close binaries or small multiple systems given the limited resolution available in these calculations. While the absolute fragmentation is not fully converged in these simulations, we emphasize that the SFR and the amount of mechanical feedback is well captured and is converged with our numerical resolution. This is shown in Table 1, where in the last two rows, we compare this feedback model (GvsTMJ) with two lower-resolution equivalents (GvsTMJ512 and GvsTMJ256), demonstrating convergence of the SFR to within 5%.

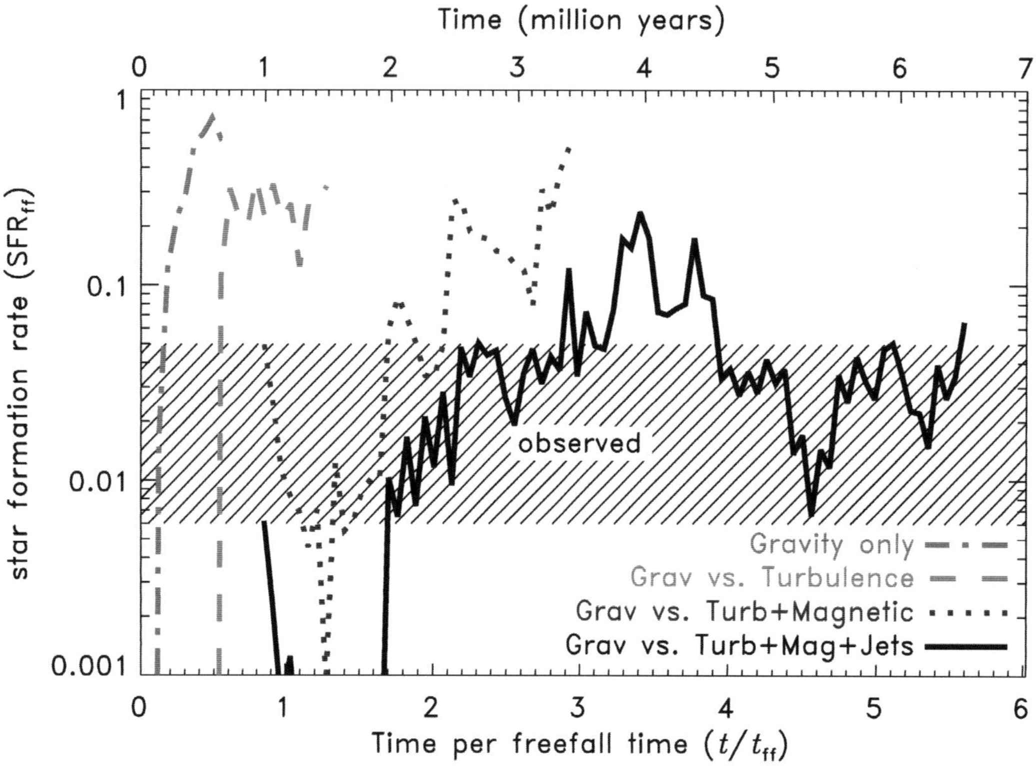

Figure 2. Time evolution of the SFR per freefall time, $\mathrm{SFR_{ff}} = d(\mathrm{SFE})/d(t/t_{\mathrm{ff}})$ for each of our four standard models from Table 1. We immediately see the drastic reduction in the SFR when turbulence, magnetic fields and jet/outflow feedback together counteract gravity, in which case the SFR is reduced to values consistent with the observed range of SFRs, shown as the hatched region.

2.2. Star formation rate

Now we concentrate on the quantitative determination of the SFR. Figure 2 shows the time evolution of the SFR in our four fiducial simulation models. While the gravity-only model (dot-dashed line) turns all the gas into stars in about a global mean freefall time, the SFR per freefall time defined as $\mathrm{SFR_{ff}} \equiv d(\mathrm{SFE})/d(t/t_{\mathrm{ff}})$ [66, 30] is reduced by about a factor of 2 when turbulence is included (dashed line). The $\mathrm{SFR_{ff}}$ is reduced by at least another factor of 2 when turbulence and magnetic fields are present (dotted line), and by another factor of 2 when jet and outflow feedback is included (solid line).

The hatched region in Figure 2 shows the observed range of $\mathrm{SFR_{ff}}$ based on the observational data compiled in Krumholz & Tan (2007) [67]. Our feedback model eventually settles into that observed range.

A very interesting and noteworthy feature of our feedback model is the evolution of the SFR with time, in particular the fact that the SFR slowly but steadily increases until about $t \sim 4.0\,\mathrm{Myr}$ and then drops significantly by almost an order of magnitude until $t \sim 5.3\,\mathrm{Myr}$, followed by a period of nearly constant SFR. This is a manifestation of self-regulation by feedback. As more material is accreted and the SFR increases, the amount of gas being re-injected into the interstellar medium in the form of fast mass-loaded jets and outflows increases proportionally. Eventually, the feedback enhances the turbulence and the amount of kinetic energy in the system, thereby increasing the virial parameter, $\alpha_{\text{vir}} = 2E_{\text{kin}}/E_{\text{grav}}$, such that star formation is rapidly quenched. This quenching of the SFR in turn reduces the amount of feedback such that the system eventually settles into a self-regulated state of star formation in which any intermittent increase in accretion triggers feedback that regulates the SFR down to a nearly constant level.

The last two columns in Table 1 list the time-averaged SFRs in each simulation model. For the gravity-only model, we find unrealistically high $\text{SFR}_{\text{ff}} = 0.47$, while the feedback model including turbulence and magnetic fields has a time-averaged $\text{SFR}_{\text{ff}} = 0.041$. Thus,cd the combination of turbulence, magnetic fields, and jet/outflow feedback reduces the SFR by more than an order of magnitude and brings the SFR into the observed range.

3. Summary and conclusions

We presented high-resolution hydrodynamical simulations of star cluster formation including turbulence, magnetic fields and jet/outflow feedback. Although our simulation models with turbulence, magnetic fields and jet/outflow feedback produce $\text{SFR}_{\text{ff}} \sim 0.04^{+0.04}_{-0.02}$, which is currently the closest available match to observations, they still overproduce stars by a factor of 2–4 compared to the observed average $\text{SFR}_{\text{ff}} \sim 0.01$. This discrepancy might be resolved by considering other types of feedback in addition to jet/outflow feedback, such as radiation pressure, which seems to be capable of reducing the SFR further by another factor of two [68]. We conclude that only the combination of strong turbulence with virial parameters above unity, strong magnetic fields and mechanical plus radiative feedback can produce realistic SFRs consistent with observations.

Acknowledgements

C.F. acknowledges funding provided by the Australian Research Council's Discovery Projects (grants DP130102078 and DP150104329). The author gratefully acknowledges the Jülich Supercomputing Centre (grant hhd20), the Leibniz Rechenzentrum and the Gauss Centre for Supercomputing (grants pr32lo, pr48pi and GCS Large-scale project 10391), the Partnership for Advanced Computing in Europe (PRACE grant pr89mu), and the National Computational Infrastructure (grant ek9), supported by the Australian Government. This work was further supported by resources provided by the Pawsey Supercomputing Centre with funding from the Australian Government and the Government of Western Australia. The software used in this work was in part developed by the DOE-supported Flash Center for Computational Science at the University of Chicago.

References

[1] Chomiuk L and Povich M S Dec 2011 *AJ* **142** 197
[2] Zuckerman B and Palmer P 1974 *ARAA* **12** 279
[3] Zuckerman B and Evans II N J Sep 1974 *ApJ* **192** L149
[4] Mestel L and Spitzer Jr. L 1956 *MNRAS* **116** 503
[5] Mouschovias T C Jul 1976 *ApJ* **207** 141
[6] Shu F H Oct 1983 *ApJ* **273** 202
[7] Crutcher R M, Hakobian N and Troland T H Feb 2009 *ApJ* **692** 844
[8] Mouschovias T C and Tassis K Nov 2009 *MNRAS* **400** L15
[9] Lunttila T, Padoan P, Juvela M and Nordlund Å Sep 2009 *ApJ* **702** L37
[10] Santos-Lima R, Lazarian A, de Gouveia Dal Pino E M and Cho J May 2010 *ApJ* **714** 442
[11] Lazarian A, Esquivel A and Crutcher R Oct 2012 *ApJ* **757** 154
[12] Bertram E, Federrath C, Banerjee R and Klessen R S Mar 2012 *MNRAS* **420** 3163
[13] Vázquez-Semadeni E 2015 in *Astrophysics and Space Science Library*, eds. A Lazarian, E M de Gouveia Dal Pino and C Melioli vol. 407 of *Astrophysics and Space Science Library* p. 401
[14] Larson R B Mar 1981 *MNRAS* **194** 809

[15] Solomon P M, Rivolo A R, Barrett J and Yahil A Aug 1987 *ApJ* **319** 730
[16] Falgarone E, Puget J L and Perault M Apr 1992 *A&A* **257** 715
[17] Ossenkopf V and Mac Low M M Jul 2002 *A&A* **390** 307
[18] Heyer M H and Brunt C M Nov 2004 *ApJ* **615** L45
[19] Schneider N et al. May 2011 *A&A* **529** A1
[20] Roman-Duval J, Federrath C, Brunt C, Heyer M, Jackson J and Klessen R S Oct 2011 *ApJ* **740** 120
[21] Mac Low M M and Klessen R S Jan 2004 *RvMP* **76** 125
[22] Elmegreen B G and Scalo J Sep 2004 *ARAA* **42** 211
[23] McKee C F and Ostriker E C Sep 2007 *ARAA* **45** 565
[24] Schneider N et al. Apr 2012 *A&A* **540** L11
[25] Price D J and Bate M R May 2007 *MNRAS* **377** 77
[26] Hennebelle P and Teyssier R Jan 2008 *A&A* **477** 25
[27] Peters T, Banerjee R, Klessen R S and Mac Low M Mar 2011 *ApJ* **729** 72
[28] Wang P, Li Z Y, Abel T and Nakamura F Jan 2010 *ApJ* **709** 27
[29] Padoan P and Nordlund Å Mar 2011 *ApJ* **730** 40
[30] Federrath C and Klessen R S Dec 2012 *ApJ* **761** 156
[31] Myers A T, Klein R I, Krumholz M R and McKee C F Feb 2014 *MNRAS* **439** 3420
[32] Vázquez-Semadeni E, Kim J and Ballesteros-Paredes J Sep 2005 *ApJ* **630** L49
[33] Nakamura F and Li Z Y Sep 2005 *ApJ* **631** 411
[34] Krumholz M R, Bate M R, Arce H G, Dale J E, Gutermuth R, Klein R I, Li Z Y, Nakamura F and Zhang Q 2014 *Protostars and Planets VI* 243
[35] Federrath C, Schrön M, Banerjee R and Klessen R S Aug 2014 *ApJ* **790** 128
[36] Federrath C Jul 2015 *MNRAS* **450** 4035
[37] Berger M J and Colella P May 1989 *Journal of Computational Physics* **82** 64
[38] Fryxell B et al. Nov 2000 *ApJ* **131** 273
[39] Dubey A et al. Apr 2008 in *Numerical Modeling of Space Plasma Flows*, eds. N V Pogorelov, E Audit and G P Zank vol. 385 of *Astronomical Society of the Pacific Conference Series* p. 145
[40] Waagan K, Federrath C and Klingenberg C May 2011 *Journal of Computational Physics* **230** 3331
[41] Krumholz M R Jun 2014 *Physics Reports* **539** 49
[42] Padoan P, Federrath C, Chabrier G, Evans II N J, Johnstone D, Jørgensen J K, McKee C F and Nordlund Å 2014 *Protostars and Planets VI* 77
[43] Heyer M H, Williams J P and Brunt C M Jun 2006 *ApJ* **643** 956
[44] Brunt C M, Heyer M H and Mac Low M Sep 2009 *A&A* **504** 883
[45] Kritsuk A G, Norman M L, Padoan P and Wagner R Aug 2007 *ApJ* **665** 416
[46] Federrath C, Roman-Duval J, Klessen R S, Schmidt W and Mac Low M Mar 2010 *A&A* **512** A81
[47] Federrath C Dec 2013 *MNRAS* **436** 1245
[48] Eswaran V and Pope S B 1988 *CF* **16** 257
[49] Schmidt W, Hillebrandt W and Niemeyer J C 2006 *CF* **35** 353
[50] Price D J, Federrath C and Brunt C M Jan 2011 *ApJ* **727** L21
[51] Kainulainen J, Federrath C and Henning T May 2013 *A&A* **553** L8
[52] Federrath C, Banerjee R, Clark P C and Klessen R S Apr 2010 *ApJ* **713** 269
[53] Truelove J K, Klein R I, McKee C F, Holliman II J H, Howell L H and Greenough J A Nov 1997 *ApJ* **489** L179
[54] Federrath C, Sur S, Schleicher D R G, Banerjee R and Klessen R S Apr 2011 *ApJ* **731** 62
[55] Sur S, Schleicher D R G, Banerjee R, Federrath C and Klessen R S Oct 2010 *ApJ* **721** L134
[56] Ricker P M May 2008 *ApJ* **176** 293
[57] Nakamura F and Li Z Y Jun 2007 *ApJ* **662** 395
[58] Cunningham A J, Klein R I, Krumholz M R and McKee C F Oct 2011 *ApJ* **740** 107
[59] Omukai K, Tsuribe T, Schneider R and Ferrara A Jun 2005 *ApJ* **626** 627
[60] Planck Collaboration et al. Feb 2015 *A&A* Accepted (arXiv:1502.04123)
[61] Crutcher R M, Wandelt B, Heiles C, Falgarone E and Troland T H Dec 2010 *ApJ* **725** 466
[62] Kauffmann J, Pillai T and Goldsmith P F Dec 2013 *ApJ* **779** 185
[63] Hernandez A K and Tan J C Aug 2015 *ApJ* **809** 154
[64] Falgarone E, Troland T H, Crutcher R M and Paubert G Aug 2008 *A&A* **487** 247
[65] Padoan P, Haugbølle T and Nordlund Å Nov 2012 *ApJ* **759** L27
[66] Krumholz M R and McKee C F Sep 2005 *ApJ* **630** 250
[67] Krumholz M R and Tan J C Jan 2007 *ApJ* **654** 304
[68] MacLachlan J M, Bonnell I A, Wood K and Dale J E Jan 2015 *A&A* **573** A112

The small-scale turbulent dynamo in smoothed particle magnetohydrodynamics

T S Tricco[1,2,3], **D J Price**[3] **and C Federrath**[4,3]

[1] Canadian Institute for Theoretical Astrophysics, University of Toronto, 60 St. George Street, Toronto, ON M5S 3H8, Canada
[2] School of Physics, University of Exeter, Stocker Road, Exeter, EX4 4QL, United Kingdom
[3] Monash Centre for Astrophysics, Monash University, Clayton, VIC 3800, Australia
[4] Research School of Astronomy and Astrophysics, The Australian National University, Canberra, ACT 2611, Australia

E-mail: `ttricco@cita.utoronto.ca`

Abstract. Supersonic turbulence is believed to be at the heart of star formation. We have performed smoothed particle magnetohydrodynamics (SPMHD) simulations of the small-scale dynamo amplification of magnetic fields in supersonic turbulence. The calculations use isothermal gas driven at rms velocity of Mach 10 so that conditions are representative of star-forming molecular clouds in the Milky Way. The growth of magnetic energy is followed for 10 orders in magnitude until it reaches saturation, a few percent of the kinetic energy. The results of our dynamo calculations are compared with results from grid-based methods, finding excellent agreement on their statistics and their qualitative behaviour. The simulations utilise the latest algorithmic developments we have developed, in particular, a new divergence cleaning approach to maintain the solenoidal constraint on the magnetic field and a method to reduce the numerical dissipation of the magnetic shock capturing scheme. We demonstrate that our divergence cleaning method may be used to achieve $\nabla \cdot \mathbf{B} = 0$ to machine precision, albeit at significant computational expense.

1. Introduction

There are three broad classes of methods used to simulate astrophysical hydrodynamics. Generally speaking, these are:

- Eulerian, grid-based methods which discretise the volume the fluid occupies into a collection of cells. The cells keep their location fixed in space, calculating the transfer of fluid between adjacent cells. Examples of such codes include PLUTO [1; 2], ATHENA [3], RAMSES [4], and FLASH [5; 6].
- Lagrangian, particle-based methods that discretise the fluid into a set of particles. The particles move according to the fluid flow, mimicing fluid behaviour. The most common approach is smoothed particle hydrodynamics (SPH), with codes such as GADGET2 [7; 8], GASOLINE [9] and PHANTOM [10].
- Methods that combine elements of Eulerian and Lagrangian methods, for example, the moving-mesh code AREPO [11], the mesh-free method GIZMO [12], and the Lagrangian-Eulerian re-map code LARE3D [13].

In this work, we use the second approach, employing smoothed particle magnetohydrodynamics (SPMHD). Right from the beginning when SPH was first formulated in 1977, there was optimism that magnetic fields could be included as part of the method, "*... magnetic fields may be included without difficulty ...*" [14]. However, this turned out not to be quite so easy in practice – the conservative SPMHD method is unstable when the magnetic pressure exceeds the thermodynamic pressure [15]. This was investigated further over the next 20 years by [16–18] and others, with the modern SPMHD method given in the series of papers by Price & Monaghan [19–21].

Nowadays, SPMHD has achieved a level of maturity that allows for practical simulations, for example, the study of jets & outflows during protostar formation [22–25]. A significant development has been the constrained hyperbolic divergence cleaning method to handle the divergence-free constraint of the magnetic field [26].

In this work, we apply SPMHD to the study of small-scale dynamo amplification of magnetic fields in supersonic turbulence, the first time such a regime has been explored with SPMHD. Purely hydrodynamic supersonic turbulence has been studied with SPH by [10], who compared the statistics of driven, supersonic turbulence between SPH and grid-based methods. They found excellent agreement on the qualitative behaviour and statistics of the turbulence. Our current work extends that comparison to include magnetic fields, thus we can be confident that the hydrodynamics are well-modelled and we can focus on differences introduced by magnetic fields.

2. Recent developments in smoothed particle magnetohydrodynamics

The set of simulations performed in this work solve the magnetohydrodynamics (MHD) equations, given by

$$\frac{\mathrm{d}\rho}{\mathrm{d}t} = -\rho\nabla\cdot\mathbf{v}, \tag{1}$$

$$\frac{\mathrm{d}\mathbf{v}}{\mathrm{d}t} = -\frac{1}{\rho}\nabla\left(P + \frac{B^2}{2\mu_0}\right) + \frac{1}{\mu_0\rho}\nabla\cdot(\mathbf{B}\mathbf{B}), \tag{2}$$

$$\frac{\mathrm{d}\mathbf{B}}{\mathrm{d}t} = (\mathbf{B}\cdot\nabla)\,\mathbf{v} - (\nabla\cdot\mathbf{v})\,\mathbf{B}, \tag{3}$$

$$\nabla\cdot\mathbf{B} = 0, \tag{4}$$

which, since we will solve them using a Lagrangian scheme, are written using the Lagrangian time derivative, $\mathrm{d}/\mathrm{d}t = \partial/\partial t + (\mathbf{v}\cdot\nabla)$. The variables have their usual meaning: ρ is the density, $\mathbf{v}$ is the velocity, $\mathbf{B}$ is the magnetic field, P is the thermodynamic pressure and μ_0 the permeability of free space. In SPMHD, the density is obtained by summation and solved iteratively with the smoothing length [27]. The magnetic tension force is implemented using the approach of [18]. A full review of SPMHD is given by [28; 29].

2.1. 'Constrained' hyperbolic/parabolic divergence cleaning

The problem with the induction equation (Eq. 3) is that it makes no guarantee to preserve the divergence-free nature of the magnetic field ($\nabla\cdot\mathbf{B} = 0$). Numerical errors will introduce monopoles into the magnetic field, which may then grow over the long-term and lead to unphysical results.

We use the 'constrained' hyperbolic/parabolic divergence cleaning method [26] to remove divergence errors in the magnetic field. The method is an SPMHD adaptation of the method by [30], and cleans the magnetic field by coupling a scalar field, ψ, to the magnetic field according to

$$\frac{\mathrm{d}\mathbf{B}}{\mathrm{d}t} = -\nabla\psi, \tag{5}$$

$$\frac{\mathrm{d}\psi}{\mathrm{d}t} = -c_{\mathrm{h}}^2\nabla\cdot\mathbf{B} - \frac{\psi}{\tau} - \frac{\psi}{2}(\nabla\cdot\mathbf{v}). \tag{6}$$

Together, they propagate the divergence of the magnetic field as a damped wave with characteristic speed c_{h} set equal to the local fast MHD wave speed. The damping timescale is $\tau = h/\sigma c_{\mathrm{h}}$, where h is the smoothing length and $\sigma = 1.0$ in 3D is a dimensionless constant. The third term in Eq. 6 was introduced to account for changes in ψ due to compression and rarefaction of the gas [26].

The method is 'constrained' in that it is built using the energy conserving formalism of SPH. Defining the specific energy content of the ψ field to be $e_\psi = \psi^2/2\mu_0\rho c_{\mathrm{h}}^2$, the discretised hyperbolic cleaning equations may be obtained such that the total energy (e_ψ plus magnetic) is constrained to stay constant. This remedies problems that would otherwise occur at sharp density contrasts and free boundaries when the same differential operator is used for $\nabla\psi$ and $\nabla\cdot\mathbf{B}$ [26], and when coupled with parabolic diffusion, means the divergence cleaning can only ever lead to reductions in divergence error.

2.2. Aritificial resistivity switch
Discontinuities in the magnetic field are captured by adding an artificial resistivity [19; 21]. This smooths the magnetic field on the resolution scale so that it remains single-valued. However, it is equivalent to adding diffusion in the form of $\eta_{\mathrm{AR}}\nabla^2\mathbf{B}$ with coefficient $\eta_{\mathrm{AR}} \sim \alpha_{\mathrm{B}} v_{\mathrm{sig}} h$, where α_{B} is a dimensionless constant of order unity and v_{sig} is the characteristic speed of the shock. Such diffusion is only meant to treat discontinuities, and the trick is to detect where those discontinuities are located so that artificial resistivity can be switched off when not needed in order to minimise numerical dissipation.

Price & Monaghan [21] used a switch to detect discontinuities, evolving α_{B} using a differential equation with a source term proportional to $\nabla\times\mathbf{B}$. However, this approach fails for strong shocks with very weak magnetic fields, not setting α_{B} to sufficient levels to keep the magnetic field single-valued [31].

Our new approach [31] is to directly set $\alpha_{\mathrm{B}} = h|\nabla\mathbf{B}|/|\mathbf{B}|$. This measures the relative size of the jump in the magnetic field compared to the absolute value of the field strength, giving a measure which works for any magnetic field strength. This is particularly relevant for the magnetised, supersonic turbulence comparison on which this paper is focused, allowing shocks to be detected during magnetic field amplification [31].

3. Enhancing the divergence cleaning
The divergence cleaning method can be enhanced in two ways. One is by explicitly increasing the cleaning wave speed. Increasing the wave speed by an over-cleaning factor of 10 would set $c_{\mathrm{h}} \to 10\times c_{\mathrm{h}}$. This approach is easy to implement and has found some use in practice [23], but is computationally expensive because it directly reduces the timestep due to the Courant condition. The second approach is to sub-cycle the cleaning equations between timesteps, pausing the rest of the simulation and holding the particles fixed while the cleaning equations are solved for a number of sub-steps. This is much more computationally efficient than over-cleaning, however is more complicated to implement in code, particularly when dealing with individual timesteps.

We have tested over-cleaning and sub-cycling of the divergence cleaning equations using the Orszag-Tang vortex [32]. In general, we find that over-cleaning and sub-cycling yield similar results for equivalent over-cleaning factors and numbers of sub-cycles (Fig. 1). That is, an over-cleaning factor of 10 produces results on par with performing 10 sub-cycles every timestep. Each

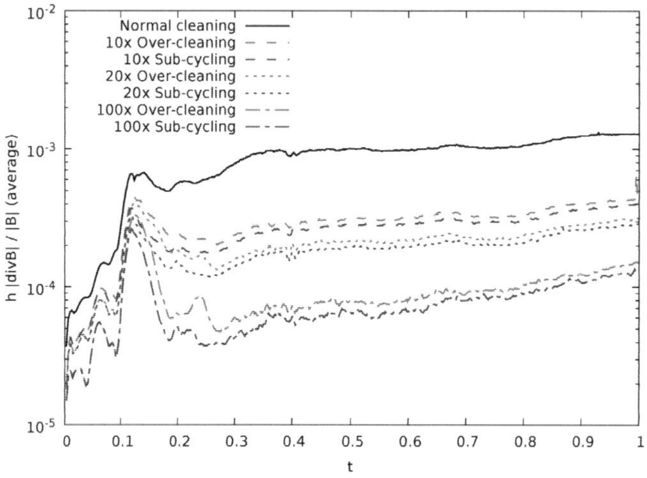

Figure 1. Average divergence error for the Orszag-Tang vortex using enhanced divergence cleaning techniques – 10×, 20×, and 100× over-cleaning and sub-cycling with 10, 20, and 100 iterations. Both approaches yield comparable levels of divergence error when the over-cleaning factor and sub-cycling iteration count are the same.

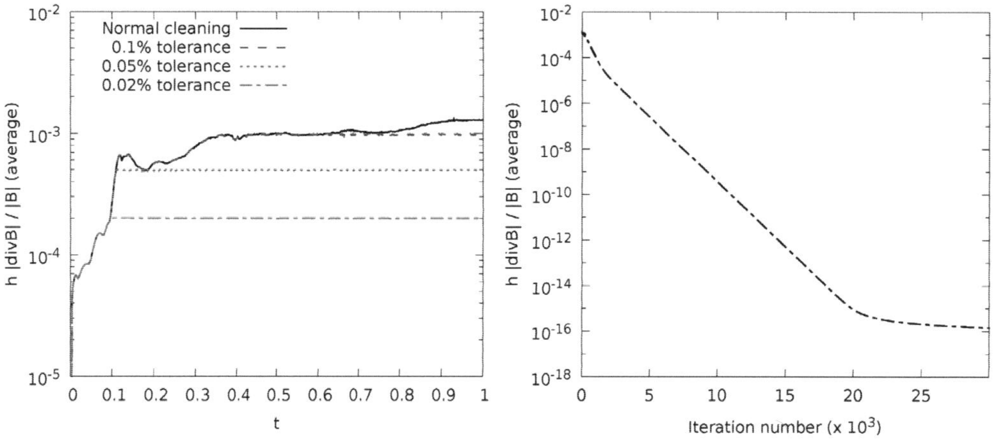

Figure 2. *Left:* Sub-cycling the divergence cleaning equations is performed each timestep on the Orszag-Tang vortex until the average error is below the specified tolerance, achieving tolerance based divergence cleaning. *Right:* The number of iterations required to reduce the average divergence error to specified tolerances. It is possible to clean the divergence of the magnetic field to machine precision (10^{-16}–10^{-18}), achieving $\nabla \cdot \mathbf{B} = 0$, but to do so requires on the order of a few tens of thousands of iterations.

factor of 10 increase in over-cleaning factor and number of sub-cycles produced half an order of magnitude reduction in average $h|\nabla \cdot \mathbf{B}|/|\mathbf{B}|$ divergence error.

An advantage that sub-cycling offers is tolerance based divergence cleaning. Instead of prescribing a fixed number of iterations, the cleaning equations may be sub-cycled until a particular criterion is met, such as a threshold on average level of divergence error. The left panel of Fig. 2 shows the average $h|\nabla \cdot \mathbf{B}|/|\mathbf{B}|$ divergence error with tolerances set to 0.1%, 0.05% and 0.02%, showing that tolerance based divergence cleaning with sub-cycling is possible.

It is also possible to set the tolerance level as strict as desired, and in the right panel of Fig. 2 we show that 'constrained' hyperbolic divergence cleaning may be used to reduce the divergence of the magnetic field all the way to zero, formally achieving $\nabla \cdot \mathbf{B} = 0$. That is not really feasible in practice though, as the number of sub-cycles required to achieve this is prohibitively expensive on the order of tens of thousands of sub-steps (right panel, Fig. 2).

4. Small-scale turbulent dynamo: SPH compared to Grid

Molecular clouds are supersonically turbulent, generating dense filaments that become the birth sites of protostars [33–36]. We have performed simulations of molecular cloud turbulence involving magnetic fields to test our SPMHD methods. The key piece of physics studied is the small-scale dynamo, an example of a *fast* dynamo that amplifies magnetic energy at an exponential rate [37]. By stretching, twisting and winding magnetic fields near the dissipation scale (hence 'small-scale'), kinetic energy is converted into magnetic energy.

The growth of the magnetic field due to small-scale dynamo undergoes three phases. First is the steady, exponential amplification of magnetic energy. Second, once the magnetic energy has saturated on the smallest scale, the dynamo enters a slow linear or quadratic growth phase while the medium and large scale magnetic fields continue to be amplified [38–40]. The third phase is the fully saturated phase once the magnetic field is saturated at all spatial scales.

The simulations are performed using the SPMHD code PHANTOM, and the grid-based code FLASH [5; 6], comparing results between the two distinct methods.

4.1. Magnetised turbulence-in-a-box

The simulations use a periodic box of length $L = 1$. The initial density is $\rho = 1$, velocity $\mathbf{v} = 0$ and an isothermal equation of state is used, $P = c_s^2 \rho$, with $c_s = 1$. The initial magnetic field is $B_z = 10^{-5}$ such that the plasma beta is $\beta = 10^{10}$. These simple initial conditions ensure that both codes start from identical states. The turbulence is driven by stirring the gas at large scales ($1 < k < 3$ peaked at $k = 2$) through an added solenoidal force generated through an Ornstein-Uhlenbeck procedure [41–43]. This keeps the rms velocity of the gas steady at Mach 10 ($\mathcal{M} = 10$), and maintains the reservoir of kinetic energy for the dynamo to draw upon. The simulation is run for 100 turbulent crossing times, defined as $t_c = L/2\mathcal{M}c_s$, capturing the full dynamo amplification phase and the statistical steady-state of the saturated phase. The simulations are performed for resolutions of 128^3, 256^3, and 512^3 particles and grid cells.

4.2. Results

The evolution of the kinetic and magnetic energy for all calculations at all resolutions is shown in Fig. 3. In all cases, the kinetic energy saturates within one crossing time ($E_k \sim 50$), and is maintained it at a near constant level by the driving routine. The three phases of dynamo amplification of magnetic energy are evident: the steady, exponential amplification of magnetic energy, transitioning to the slow growth of magnetic energy ($E_m \gtrsim 10^{-1}$), followed by the fully saturated magnetic field with a constant energy level ($E_m \sim 10^0$). The exponential growth rate has little variation for the FLASH calculations, yet exhibits an increasingly higher growth rate for higher resolution with PHANTOM. This can be traced to the differing shock capturing techniques used between the methods, as the numerical dissipation of the shock capturing is the primary factor in setting the growth rate of the dynamo for these simulations. All simulations saturate the magnetic energy at a similar level (2–4% of the mean kinetic energy).

Fig. 4 shows the column density and column integrated magnetic field strength for the 256^3 resolution calculations. At early times, both codes have similar density fields, but diverge at later times due to the non-linearity of the turbulence. The densest regions contain the strongest magnetic fields, yet the magnetic field throughout the whole domain is growing in strength.

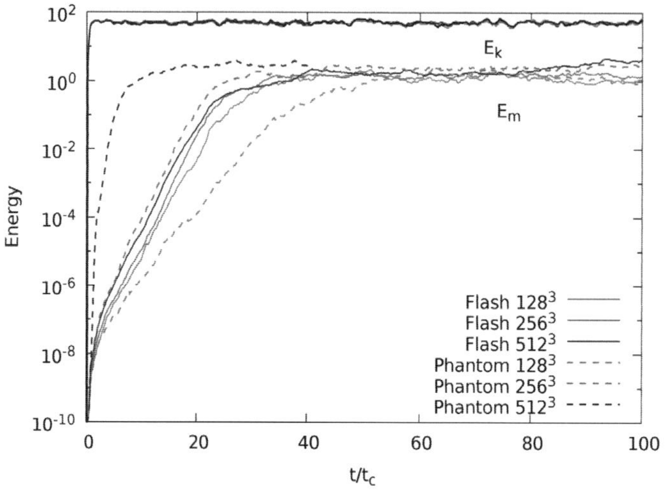

Figure 3. The kinetic and magnetic energy for grid (FLASH, solid lines) and SPMHD (PHANTOM, dashed lines) at resolutions of 128^3 (red), 256^3 (green), and 512^3 (blue) grid points and particles. The kinetic energy for all calculations is maintained at a near constant level by the driving routine. Both codes amplify the magnetic energy at an exponential rate at early times, with a slow growth of magnetic energy as it nears saturation. For all calculations, the magnetic energy saturates at a similar level. FLASH has similar magnetic energy growth rates across the resolutions simulated, while PHANTOM exhibits faster growth rates with increasing resolution. This resolution dependence is a consequence of the artificial dissipation terms used for shock capturing.

The growth of the magnetic field at different spatial scales can be quantified by looking at the power spectra of magnetic energy. As shown in Fig. 5, both codes exhibit the same behaviour for all resolutions, with the magnetic energy initially increasing uniformly at all spatial scales with little change in shape. This corresponds to the exponential growth phase, and is characteristic for the small-scale dynamo [37; 44]. The power spectra converge first on the smallest spatial scales, at which point the dynamo enters the slow growth phase [44]. This may be seen by the time delay for the larger scales of the power spectra to converge.

Fig. 6 shows the probability distribution function (PDF) of B^2, with each curve representing a single snapshot in time and the red line the time-averaged PDF in the saturation phase. At early times, during the exponential growth phase, both PHANTOM and FLASH have log-normal distributions of similar widths. The distributions maintain their width and shape as the magnetic field is amplified, with the mean increasing at a steady rate. As the dynamo nears saturation, the strongest magnetic fields no longer are amplified and the high-end tail of the distribution anchors in place. However, the low-end tail continues to be amplified (the slow growth phase), and the distribution takes on a lop-sided shape. Both codes exhibit this behaviour, consistent with previous numerical studies [45; 46], and have similar maximum magnetic field strengths.

5. Summary

We have performed simulations of the small-scale dynamo amplification of magnetic fields in supersonic turbulence, comparing results between SPMHD and grid-based methods. The latest developments in SPMHD are utilised, namely the 'constrained' hyperbolic divergence cleaning method to handle the divergence-free constraint of the magnetic field and a new switch to reduce numerical dissipation from artificial resistivity. The simulations are performed for 100

Journal of Physics: Conference Series **719** (2016) 012003 doi:10.1088/1742-6596/719/1/012003

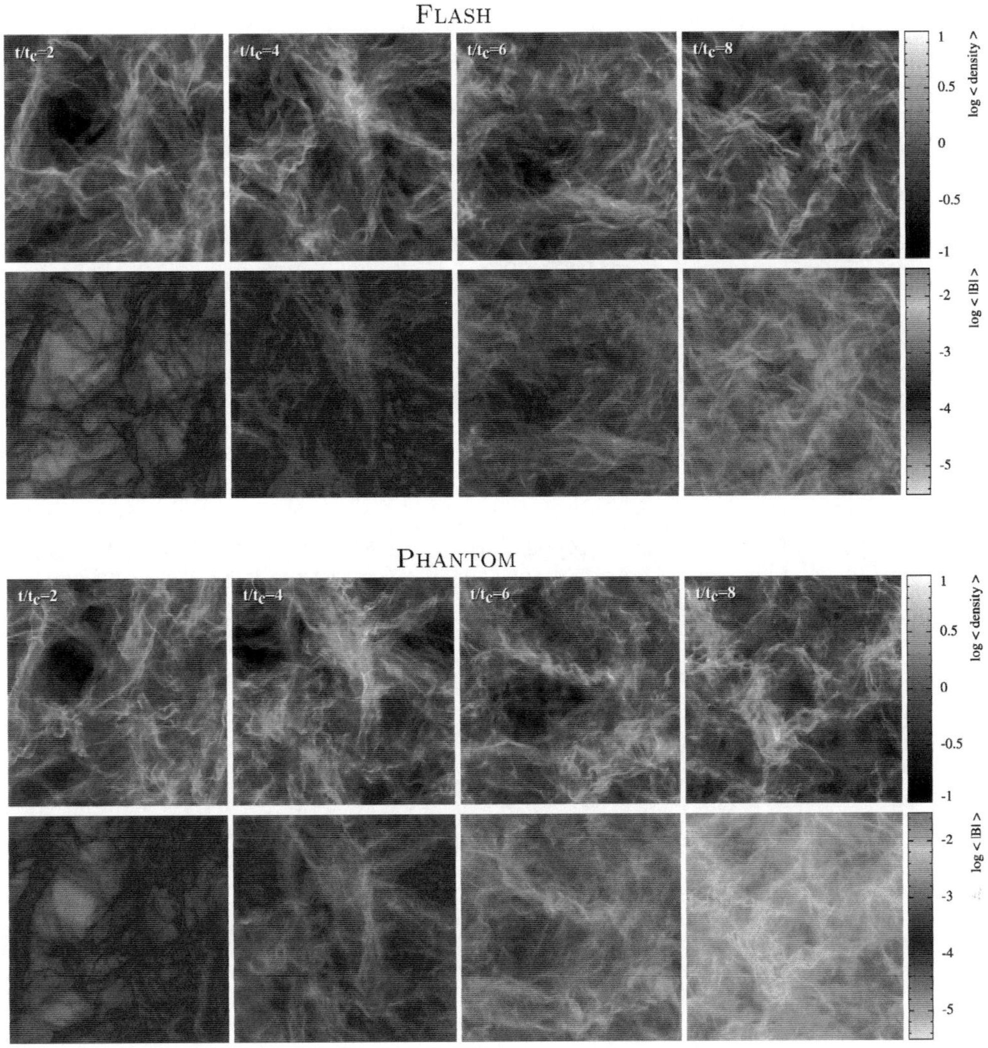

Figure 4. z-column integrated ρ and $|B|$, defined $< B >= \int |B| \mathrm{d}z / \int \mathrm{d}z$, for FLASH (top) and PHANTOM (bottom) at resolutions of 256^3 for $t/t_\mathrm{c} = 2, 4, 6, 8$. The density field has similar structure in both codes at early times, but diverge at late times due to the non-linear behaviour of the turbulence. The magnetic field is strongest in the densest regions, while the mean magnetic field strength throughout the domain increases with time.

turbulent crossing times, allowing study of the full growth and saturation of magnetic energy due to dynamo amplification. Excellent agreement is found on the qualitative behaviour and properties of the magnetic field between SPMHD and the grid code. The exponential growth rate of magnetic energy has different resolution dependence, with SPMHD yielding faster growth rates for higher resolution, caused by the different approaches used to treat shocks. Overall, we conclude that SPMHD can correctly model the small-scale dynamo amplification of magnetic fields in supersonic turbulence.

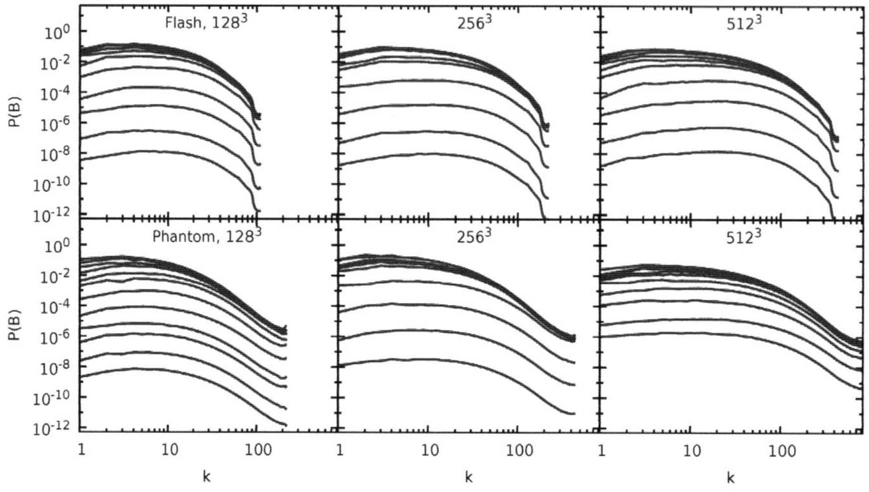

Figure 5. Spectra of the magnetic energy during the growth phase for FLASH (top row) and PHANTOM (bottom row) for resolutions of 128^3, 256^3, and 512^3 (left to right). Each spectral line is sampled at intervals of $5t/t_c$ up to $t/t_c = 50$, except for the 512^3 PHANTOM run which is sampled every t/t_c (from $t/t_c = 2$–12). Both codes exhibit the same qualitative behaviour: the magnetic energy spectra grow uniformly on all spatial scales with little change in shape, saturating first on the smallest scales with a time delay before the larger scales saturate.

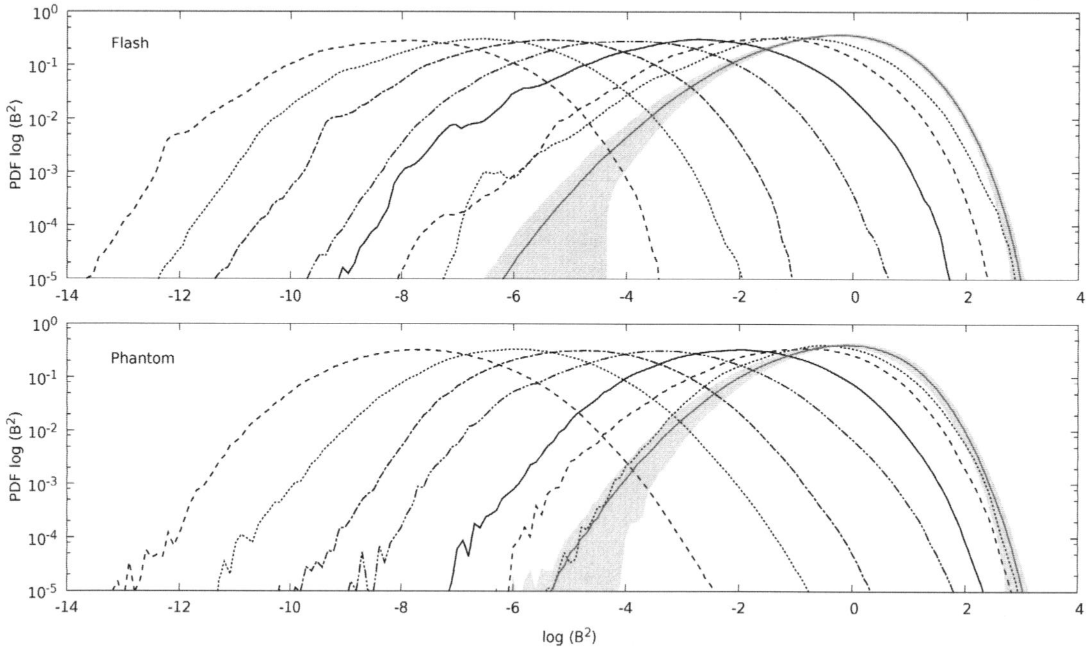

Figure 6. PDF of $\log(B^2)$ during the growth phase, with the red line time averaged during the saturation phase (top panel: FLASH, bottom panel: PHANTOM). The PDFs have log-normal distributions during the growth phase, maintaining their width while their peaks smoothly translate to higher magnetic field strengths. As the strongest magnetic fields saturate, the PDFs become lop-sided.

Acknowledgments
TST is supported by a CITA Post-doctoral Research Fellowship. The majority of this work was carried out while TST was still a PhD student, during which he was supported by an Australian Postgraduate Award and Endeavour International Postgraduate Research Scholarship. DJP is supported by ARC Future Fellowship FTI130100034, and acknowledges funding provided by the ARCs Discovery Projects (grant no. DP130102078). CF acknowledges funding provided by the Australian Research Councils Discovery Projects (grants DP130102078 and DP150104329). This research was undertaken with the assistance of resources provided at the Multi-modal Australian ScienceS Imaging and Visualisation Environment (MASSIVE) through the National Computational Merit Allocation Scheme supported by the Australian Government. We gratefully acknowledge the Jülich Supercomputing Centre (grant hhd20), the Leibniz Rechenzentrum and the Gauss Centre for Supercomputing (grant pr32lo), the Partnership for Advanced Computing in Europe (PRACE grant pr89mu), and the Australian National Computing Infrastructure (grant ek9). Some figures were created using the interactive SPH visualisation tool SPLASH [47].

References

[1] Mignone A, Bodo G, Massaglia S, Matsakos T, Tesileanu O, Zanni C and Ferrari A 2007 *ApJS* **170** 228–242 (*Preprint* astro-ph/0701854)

[2] Mignone A, Zanni C, Tzeferacos P, van Straalen B, Colella P and Bodo G 2012 *ApJS* **198** 7 (*Preprint* 1110.0740)

[3] Stone J M, Gardiner T A, Teuben P, Hawley J F and Simon J B 2008 *ApJS* **178** 137–177 (*Preprint* 0804.0402)

[4] Fromang S, Hennebelle P and Teyssier R 2006 *A&A* **457** 371–384 (*Preprint* astro-ph/0607230)

[5] Fryxell B, Olson K, Ricker P, Timmes F X, Zingale M, Lamb D Q, MacNeice P, Rosner R, Truran J W and Tufo H 2000 *ApJS* **131** 273–334

[6] Dubey A, Fisher R, Graziani C, Jordan IV G C, Lamb D Q, Reid L B, Rich P, Sheeler D, Townsley D and Weide K 2008 *Numerical Modeling of Space Plasma Flows* (*Astronomical Society of the Pacific Conference Series* vol 385) ed Pogorelov N V, Audit E and Zank G P p 145

[7] Springel V 2005 *MNRAS* **364** 1105–1134 (*Preprint* astro-ph/0505010)

[8] Dolag K and Stasyszyn F 2009 *MNRAS* **398** 1678–1697 (*Preprint* 0807.3553)

[9] Wadsley J W, Stadel J and Quinn T 2004 *New A* **9** 137–158 (*Preprint* astro-ph/0303521)

[10] Price D J and Federrath C 2010 *MNRAS* **406** 1659–1674 (*Preprint* 1004.1446)

[11] Springel V 2010 *MNRAS* **401** 791–851 (*Preprint* 0901.4107)

[12] Hopkins P F 2015 *MNRAS* **450** 53–110 (*Preprint* 1409.7395)

[13] Arber T D, Longbottom A W, Gerrard C L and Milne A M 2001 *J. Comput. Phys.* **171** 151–181

[14] Gingold R A and Monaghan J J 1977 *MNRAS* **181** 375–389

[15] Phillips G J and Monaghan J J 1985 *MNRAS* **216** 883–895

[16] Morris J P 1996 *Analysis of Smoothed Particle Hydrodynamics with Applications* Ph.D. thesis Monash University

[17] Meglicki Z, Wickramasinghe D and Dewar R L 1995 *MNRAS* **272** 717–729

[18] Børve S, Omang M and Trulsen J 2001 *ApJ* **561** 82–93

[19] Price D J and Monaghan J J 2004 *MNRAS* **348** 123–138 (*Preprint* astro-ph/0310789)

[20] Price D J and Monaghan J J 2004 *MNRAS* **348** 139–152 (*Preprint* astro-ph/0310790)

[21] Price D J and Monaghan J J 2005 *MNRAS* **364** 384–406 (*Preprint* astro-ph/0509083)

[22] Price D J, Tricco T S and Bate M R 2012 *MNRAS* **423** L45–L49 (*Preprint* 1203.2933)

[23] Bate M R, Tricco T S and Price D J 2014 *MNRAS* **437** 77–95 (*Preprint* 1310.1092)

[24] Lewis B T, Bate M R and Price D J 2015 *MNRAS* **451** 288–299 (*Preprint* 1504.08322)

[25] Wurster J, Price D J and Bate M R 2016 *MNRAS* **457** 1037–1061 (*Preprint* 1512.01597)

[26] Tricco T S and Price D J 2012 *J. Comput. Phys.* **231** 7214–7236 (*Preprint* 1206.6159)

[27] Price D J and Monaghan J J 2007 *MNRAS* **374** 1347–1358 (*Preprint* astro-ph/0610872)

[28] Price D J 2012 *J. Comput. Phys.* **231** 759–794 (*Preprint* 1012.1885)

[29] Tricco T S 2015 *Simulating astrophysical magnetic fields with smoothed particle magnetohydrodynamics* Ph.D. thesis School of Mathematical Sciences, Faculty of Science, Monash University

[30] Dedner A, Kemm F, Kröner D, Munz C D, Schnitzer T and Wesenberg M 2002 *J. Comput. Phys.* **175** 645–673

[31] Tricco T S and Price D J 2013 *MNRAS* **436** 2810–2817 (*Preprint* 1309.5437)

[32] Orszag S A and Tang C M 1979 *J. Fluid Mech.* **90** 129–143

[33] Mac Low M M and Klessen R S 2004 *Rev. Mod. Phys.* **76** 125–194 (*Preprint* astro-ph/0301093)

[34] McKee C F and Ostriker E C 2007 *ARA&A* **45** 565–687 (*Preprint* 0707.3514)

[35] Federrath C and Klessen R S 2012 *ApJ* **761** 156 (*Preprint* 1209.2856)

[36] Padoan P, Federrath C, Chabrier G, Evans II N J, Johnstone D, Jørgensen J K, McKee C F and Nordlund Å 2014 *Protostars and Planets VI* 77–100 (*Preprint* 1312.5365)

[37] Brandenburg A and Subramanian K 2005 *Phys. Rep.* **417** 1–209 (*Preprint* astro-ph/0405052)

[38] Cho J, Vishniac E T, Beresnyak A, Lazarian A and Ryu D 2009 *ApJ* **693** 1449–1461 (*Preprint* 0812.0817)

[39] Federrath C, Chabrier G, Schober J, Banerjee R, Klessen R S and Schleicher D R G 2011 *Phys. Rev. Lett.* **107** 114504 (*Preprint* 1109.1760)

[40] Schleicher D R G, Schober J, Federrath C, Bovino S and Schmidt W 2013 *New J. Phys.* **15** 023017 (*Preprint* 1301.4371)

[41] Eswaran V and Pope S B 1988 *Comput. Fluids* **16** 257–278

[42] Schmidt W, Federrath C, Hupp M, Kern S and Niemeyer J C 2009 *A&A* **494** 127–145 (*Preprint* 0809.1321)

[43] Federrath C, Roman-Duval J, Klessen R S, Schmidt W and Mac Low M M 2010 *A&A* **512** A81 (*Preprint* 0905.1060)

[44] Federrath C, Schober J, Bovino S and Schleicher D R G 2014 *ApJ* **797** L19 (*Preprint* 1411.4707)

[45] Schekochihin A A, Cowley S C, Maron J L and McWilliams J C 2004 *Phys. Rev. Lett.* **92** 064501 (*Preprint* nlin/0306059)

[46] Schekochihin A A, Cowley S C, Taylor S F, Maron J L and McWilliams J C 2004 *ApJ* **612** 276–307 (*Preprint* astro-ph/0312046)

[47] Price D J 2007 *PASA* **24** 159–173 (*Preprint* 0709.0832)

Journal of Physics: Conference Series **719** (2016) 012004 doi:10.1088/1742-6596/719/1/012004

Modeling the structure of magnetic fields in Neutron Stars: from the interior to the magnetosphere

Niccolò Bucciantini[1,2,3], Antonio G. Pili[2,1,3] and Luca Del Zanna[2,1,3]

[1] INAF - Osservatorio Astrofisico di Arcetri, L.go Fermi 5, 50125, Firenze, Italy
[2] Univ. Firenze - Dip. di Fisica e Astronomia , via G.Sansone 1, 50019, Sesto F.no, Italy
[3] INFN - Sezione di Firenze, via G.Sansone 1, 50019, Sesto F.no, Italy

E-mail: niccolo@arcetri.astro.it

Abstract. The phenomenology of the emission of pulsars and magnetars depends dramatically on the structure and properties of their magnetic field. In particular it is believed that the outbursting and flaring activity observed in AXPs and SRGs is strongly related to their internal magnetic field. Recent observations have moreover shown that charges are present in their magnetospheres supporting the idea that their magnetic field is tightly twisted in the vicinity of the star. In principle these objects offer a unique opportunity to investigate physics in a regime beyond what can be obtained in the laboratory. We will discuss the properties of equilibrium models of magnetized neutron stars, and we will show how internal and external currents can be related. These magnetic field configurations will be discussed considering also their stability, relevant for their origin and possibly connected to events like SNe and GRBs. We will also show what kind of deformations they induce in the star, that could lead to emission of gravitational waves. In the case of a twisted magnetosphere we will show how the amount of twist regulates their general topology. A general formalism based on the simultaneous numerical solution of the general relativistic Grad-Shafranov equation and Einstein equations will be presented.

1. Introduction

Neutrons Stars (NSs) are known to show a diverse phenomenology, that allows us to separate them into several classes. Among these classes, Anomalous X-Ray Pulsars (AXPs) and Soft Gamma-ray Repeaters (SGRs) have attracted attention because of their extraordinary energetic properties: a persistent X-ray emission with luminosities $L_X \sim 10^{33} - 10^{36}$ erg s^{-1}; flaring activity with X-ray bursts lasting $\sim 0.1 - 1$ s and with peak luminosities $\sim 10^{40} - 10^{41}$ erg s^{-1}, and in a few cases violent events, known as giant flares, during which an amount of energy $\sim 10^{44} - 10^{46}$ erg is released [1, 2, 3].

Today SGRs and AXPs are grouped in the same class of NSs called *magnetars* [4, 5]. These are young (with a typical age of 10^4 yr), isolated NSs with rotational period in the range $\sim 2 - 12$ s, and with a typical magnetic field in the range $10^{14} - 10^{15}$ G [6]. Since they are slow rotators, spin-down energy losses cannot power their emission, which is instead believed to come from the dissipation and rearrangement of their magnetic energy.

The magnetic field has important implication on the way NSs manifest themselves in the electromagnetic spectrum. In particular for Pulsars, models of the outer magnetosphere have been developed since the '60 [7], up to the present day [8]. On the other hand, the interest on the interior structure has been mostly driven by questions of nuclear and theoretical physics, especially their Equation of State (EoS) [9, 10] and their cooling properties [11, 12].

The simultaneous presence of high density, strong gravity, and strong magnetic fields makes magnetars a unique environment. The study of the properties and geometry of the internal magnetic field in NSs is thus an important step towards a complete understanding of magnetars. The analysis of equilibrium configurations has mainly focused on understanding the effects of the magnetic field on the structure of the star. Strong magnetic fields deform the star, and such deformations, in conjunction with fast rotation, could lead to emission of Gravitational Waves (GW) [13, 14]. Particular efforts have been recently aimed at investigating the stability of various magnetic configurations. However due to the complexity of the problem, models have been worked either in Newtonian regime [15, 16, 17], or in GR with a perturbative approach [18, 19, 20], and with currents purely confined to the interior. Only recently [21, 22] they have been worked out in the fully non-linear GR regime, including the possibility of currents extending outside into a magnetosphere [23].

We present and review here a study conducted by our group, aiming at modeling magnetized NSs in general relativity. Our approach is based on the simultaneous solution of the Einstein equations for the GR metric, of the general-relativistic Euler equation for the hydromagnetic equilibrium, and the general-relativistic Grad-Shafranov equation for the magnetic field structure, using a formalism that allows electric currents to flow outside the star. We have investigated several new functional forms for the current distribution, and studied the properties of the resulting magnetic field.

2. Solving Einstein equations

Given a generic spacetime, the line element can be written as [24, 25]:

$$ds^2 = -\alpha^2 dt^2 + \gamma_{ij}(dx^i + \beta^i dt)(dx^j + \beta^j dt), \tag{1}$$

where α is called the *lapse* function, β^i is the *shift vector*, γ_{ij} is the *three-metric*, and $i, j = r, \theta, \phi$, if a spherical coordinate system $x^\mu = (t, r, \theta, \phi)$ is chosen. The assumptions of *stationarity* and *axisymmetry* imply that all metric terms are only a function of r and θ.

For neutron stars endowed with an axisymmetric magnetic field the metric can be approximated to a high degree of accuracy as *conformally flat* [26], such that in spherical coordinates:

$$ds^2 = -\alpha^2 dt^2 + \psi^4(dr^2 + r^2 d\theta^2 + r^2 \sin\theta^2 d\phi^2), \tag{2}$$

where ψ is the conformal factor, and for simplicity we have neglected frame dragging given that magnetar are typically slow rotators.

The conformally flat condition (CFC) allows us to cast Einstein's equations in a simpler and numerically stable form and, as a consequence, it can handle stronger fields and deformations without compromising the accuracy of the results as we will discuss. With this approximation Einstein's equations reduce to the following Poisson-like equations:

$$\Delta\psi = -[2\pi\psi^6(e + B^2/2)]\psi^{-1}, \tag{3}$$

$$\Delta(\alpha\psi) = [2\pi\psi^6(e + B^2/2) + 2\pi\psi^6(6p + B^2)\psi^{-2}](\alpha\psi), \tag{4}$$

where Δ is the standard Laplacian operator in spherical coordinate while e, p and B^2 are respectively the energy density, the pressure and the magnetic field energy density as measured in the lab frame.

For the non-linear Poisson-like equations Eq. 3-4 we employ the XNS algorithm [27, 28, 29]. The code is designed to compute hydromagnetic equilibria even in the presence of rotation.

Axisymmetric solutions are searched in terms of a series of spherical harmonics $Y_l(\theta)$:

$$q(r,\theta) := \sum_{l=0}^{\infty}[A_l(r)Y_l(\theta)]. \tag{5}$$

The Laplacian can then be reduced to a series of radial 2nd order boundary value ODEs for the coefficients $A_l(r)$ of each harmonic, which are then solved using tridiagonal matrix inversion. This procedure is repeated until convergence, using in the source term the value of the solution computed at the previous iteration. We found that 20 spherical harmonics for the elliptic solvers and a grid in spherical coordinates in the domain $r = [0, 30]$km, $\theta = [0, \pi]$, with 250 points in the radial direction and 100 points in the angular one, are sufficient to achieve a numerical accuracy of the order of 10^{-3}. Comparison with results already present in literature for fast unmagnetized rotators, shows that the CFC approximation is correct to an accuracy $\sim 10^{-3}$ even for highly deformed stars.

3. Magnetic equilibria

The only non-vanishing equation of the static GRMHD system is the Euler equation in the presence of an external electromagnetic field:

$$\partial_i p + (e+p)\,\partial_i \ln\alpha = L_i := \epsilon_{ijk}J^j B^k, \tag{6}$$

with $i = r, \theta$, and where L_i is the Lorentz force and $J^i = \alpha^{-1}\epsilon^{ijk}\partial_j(\alpha B_k)$ is the conduction current. Assuming a *barotropic* EoS $p = p(\rho)$, $e = e(\rho)$ (ρ is the rest mass density), we find

$$\partial_i \ln h + \partial_i \ln\alpha = \frac{L_i}{\rho h}, \tag{7}$$

where the specific enthalpy is $h := (e+p)/\rho$. Integrability requires the existence of a scalar function $\mathcal{M}$ such that $L_i = \rho h \partial_i \mathcal{M}$.

In the case of a purely toroidal field, the Lorentz force is conveniently written in terms of αB_ϕ, and the Euler equation reads

$$\partial_i \ln h + \partial_i \ln\alpha + \frac{\alpha B_\phi \partial_i(\alpha B_\phi)}{\rho h\, \alpha^2\psi^4 r^2 \sin^2\theta} = 0, \tag{8}$$

such that

$$B_\phi = \alpha^{-1}\mathcal{I}(\mathcal{G}), \quad \mathcal{M}(\mathcal{G}) = -\int \frac{\mathcal{I}}{\mathcal{G}}\frac{d\mathcal{I}}{d\mathcal{G}}d\mathcal{G}, \quad \text{with} \quad \mathcal{G} = \rho h\, \alpha^2\psi^4 r^2 \sin^2\theta. \tag{9}$$

Among all possible functional choices we have selected a *magnetic barotropic law*:

$$\mathcal{I}(\mathcal{G}) = K_m \mathcal{G}^m, \quad \mathcal{M}(\mathcal{G}) = -\frac{mK_m^2}{2m-1}\mathcal{G}^{2m-1}. \tag{10}$$

where the parameter m plays the role of the polytropic, index for magnetic pressure.

In the case where a poloidal magnetic field is present a formulation based on the so-called *Grad-Shafranov equation* for the toroidal component of the vector potential A_ϕ is more convenient. In this case:

$$L_i = \rho h\, \partial_i \mathcal{M} = \rho h \frac{d\mathcal{M}}{dA_\phi}\partial_i A_\phi \to \ln\left(\frac{h}{h_c}\right) + \ln\left(\frac{\alpha}{\alpha_c}\right) - \mathcal{M}(A_\phi) = 0, \tag{11}$$

where constants are calculated at the stellar center. Moreover, the ϕ component of the Lorentz force, which must also vanish, implies

$$B_\phi = \alpha^{-1}\mathcal{I}(A_\phi). \tag{12}$$

Introducing $\sigma := \alpha^2\psi^4 r^2 \sin^2\theta$ and $\tilde{A}_\phi := A_\phi/(r\sin\theta)$ and the new operator

$$\tilde{\Delta}_3 := \Delta - \frac{1}{r^2\sin^2\theta} = \partial_r^2 + \frac{2}{r}\partial_r + \frac{1}{r^2}\partial_\theta^2 + \frac{1}{r^2\tan\theta}\partial_\theta - \frac{1}{r^2\sin^2\theta}, \tag{13}$$

for which $\tilde{\Delta}_3\tilde{A}_\phi = \Delta_* A_\phi/(r\sin\theta)$ (it coincides with the ϕ component of the *vector laplacian* in spherical coordinates), after some calculations we retrieve the *Grad-Shafranov* equation for the magnetic flux function A_ϕ:

$$\tilde{\Delta}_3\tilde{A}_\phi + \frac{\partial A_\phi \partial \ln(\alpha\psi^{-2})}{r\sin\theta} + \psi^8 r\sin\theta\left(\rho h\frac{d\mathcal{M}}{dA_\phi} + \frac{\mathcal{I}}{\sigma}\frac{d\mathcal{I}}{dA_\phi}\right) = 0. \tag{14}$$

Again we have investigated several functional forms for the scalar functions $\mathcal{M}$ and $\mathcal{I}$, including non-linear terms that allow us to model different current distributions inside the star. In principle one can set the function $\mathcal{I}$ to vanish outside the star, leading to the so called *twisted torus configurations*, where a twisted magnetic flux rope is present inside the star. On the other hand if one allows $\mathcal{I}$ to be non-zero also outside the star, one finds models with a so called *twisted magnetosphere*.

Interestingly the Grad-Shafranov equation Eq. 14 can be reduced to the solution of a non-linear vector Poisson equation, which is formally equivalent to the elliptic equations that one needs to solve in the CFC approximation. It is thus possible to use the same algorithm, with a combination of vector spherical harmonics decomposition for the angular part, and matrix inversion for the radial part, that was used in the metric solver [27].

4. Results

We begin by describing the results for purely toroidal magnetic fields. By modifying the currents profile it is possible to obtain different configurations where the magnetic field can be concentrated either toward the center or toward the edge of the star. In Fig. 3 we show the outcome for different magnetic field configurations. In all cases the maximum strength of the magnetic field is the same as well as the mass of the Neutron Star. However it is immediately evident that the deformation of the star is quite different, being much stronger for fields concentrated at the center. Given that a toroidally dominated configuration is expected to result from the core-collapse of a rapidly rotating core, due to differential rotation during infall, our results show that the important parameter is not just the global strength of the differential rotation, but its profile, which can be related to the pre-collapse evolution of the progenitor.

In general for purely toroidal models we find that, as the magnetic field increases the star inflates and its radius grows. Interestingly this leads to a saturation of the maximum value that the magnetic field can reach inside the star. Increasing the magnetization leads to a bigger star and not a stronger field. We also find that the effect of the magnetic field is more pronounced in low mass stars, while more compact configurations, of higher central density, tend to show smaller deformations, but can support strongher fields.

Models with purely poloidal field instead tend to show an oblate deformation. Again it is possible to modify the current distribution to achieve configurations where the field is distributed differently. Interestingly now, for fields concentrated toward the edge of the star, the deformation tends to be larger than in the case of magnetic field concentrated toward the center. It is also possible to obtain configurations where the magnetic field at the surface can differ substantially

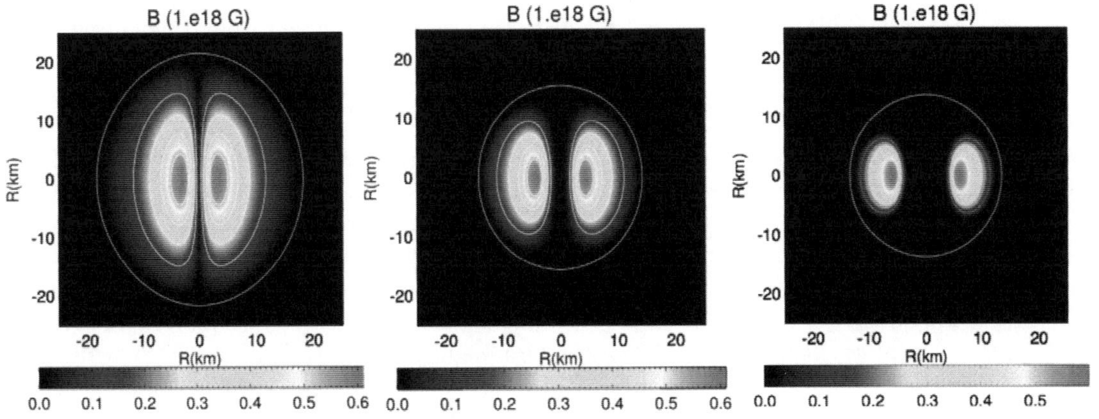

Figure 1. Neutron stars with purely toroidal field. Paneles show the strength of the magnetic field for a star of mass $M = 1.68M_{\odot}$. From left to right $m = 1, 2, 4$. The blue line is the stellar surface.

from a simple dipole shape. By adopting diferent prescriptions, we can make it either higher at the equator, or concentrated toward the axis, in a configuration where the bulk of the star is demagnetized. Such configurations are reminiscent of recent full multidimensional results in core-collapse simulations, where demagnetized neutron stars are obtained due to turbulent flux expulsion [30]. For very strong magnetic fields it is possible to obtain even configurations where the density maximum is displaced from the center due to magnetic field support.

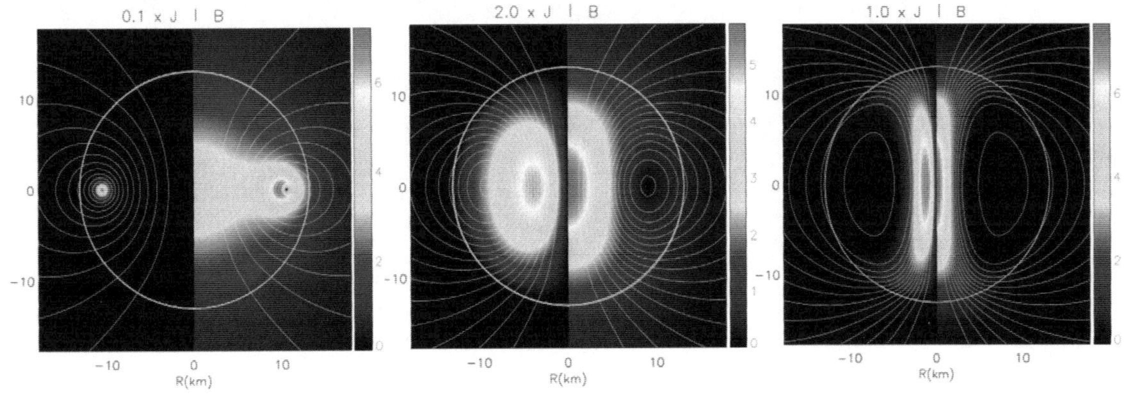

Figure 2. Neutron stars with purely poloidal field. Panels show on the right the strength of the magnetic field (units 10^{14}G), and on the left the strength of the currents, for a star of mass $M = 1.68M_{\odot}$. The left panel shows the effect of additive non-linear currents that concentrate the field toward the surface, the right one the effect of subtractive currents that concentrate the field toward the axis. The blue line is the stellar surface.

We have computed Twisted-Torus configurations in the non-perturbative regime. These mixed configurations are favored based on stability arguments. The toroidal component can reach a strength comparable with the poloidal one but it is always energetically subdominant. The deformations are almost completely due to the poloidal field, acting on the interior. Several functional forms for the current distribution were used, but the system was always found to saturate to configurations where the energy of the toroidal component is at most 10% of the total magnetic energy. The main reason for this is that to increase the toroidal magnetic field, one needs to rise also the toroidal currents (the two being related due to equilibrium requirement in the GS equation), which act to change the structure of the poloidal field, reducing the volume occupied by the toroidal field. It is possible to overcome this problem only by sacrificing global dynamical equilibrium in the outer layers of the star, or in the case of non-barotropic EoS. Interestingly it looks like such energy ratio depends more on the stratification of the NS than on the current distribution.

More recenlty [23] models were built with currents extending also outside the neutron star into a twisted magnetosphere. In the case of magnetars there is an increasing set of observational evidence pointing to the fact that their magnetosphere is endowed with a highly twisted magnetic field. This is strongly suggested by the features of their persistent X-ray spectra, which are well fitted by a blackbody-like component at $kT \sim 0.5$ keV, likely thermal emission from the neutron star surface, joined with a power-law tail that becomes dominant above 10 keV [31]. The latter can be explained in terms of resonant cyclotron scattering of the thermal photons by magnetospheric particles [32, 33]. Typically, in the standard reference model [32] the magnetosphere is described in terms of a self-similar, globally twisted, dipolar magnetic field. This model has been refined to account for higher order multipoles [34], in response to observational indications of a local, rather than global, twist in the magnetosphere [35, 36]. Recently this scenario has been strengthened also by the detection of a proton cyclotron feature in the X-ray spectrum of the "low-field" magnetar SGR 0418+5729 which is compatible with a strong, but localized, toroidal field of the order of 10^{15} G [37].

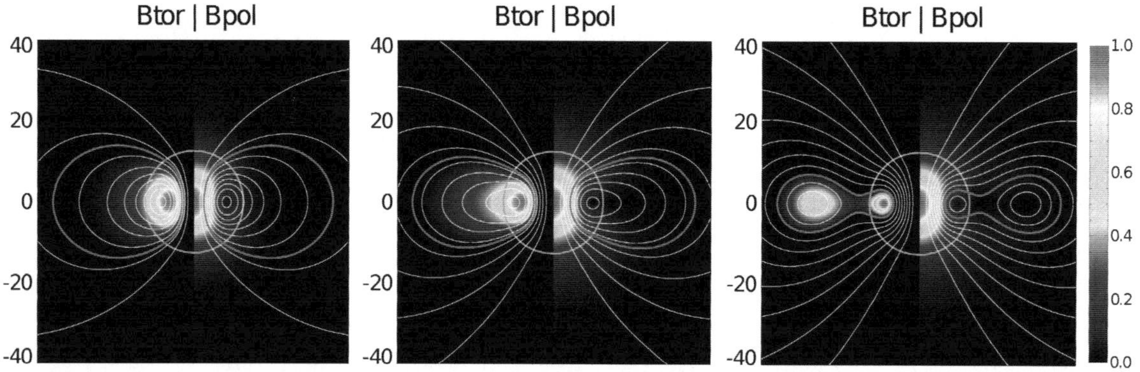

Figure 3. Neutron stars with twisted magnetospheres. Panels show on the left the strength of the toroidal component of the magnetic field, and on the right the strength of the poloidal component (normalized to the maxima). From left to right models have progressively higher level of twist, showing the transition from a configuration of topologically connected magnetic regions to a configuration where a detached magnetospheric rope is realized. The blue line is the stellar surface.

All the equilibrium models we obtained are energetically dominated by the poloidal component of the magnetic field, the energy of the external toroidal magnetic field is, at most, 25% of the total magnetic energy of the magnetosphere. This result is similar to what is found when the twisted field is fully confined within the star. The amount of twist in the magnetosphere can be regulated essentially by the amount of poloidal non-linear currents that are imposed to the system. When the non-linear current terms are weak, the magnetic field lines are inflated outward by the toroidal magnetic field pressure and the twist of the field lines extends also to higher latitudes. The result is a single magnetically connected region, where all field lines have footpoints attached to the stellar surface. As the currents increase the effects of the non-linearity get stronger. This not only increases the twist of the near-surface magnetic field but also leads to the formation of a disconnected magnetic island, reminiscent of the so-called plasmoids often found in simulations of the solar corona. This regime and these topologies are very likely to be unstable. Only magnetic ropes confined close to the stellar surface satisfy the Kruskal-Shafranov condition for stability against the development of kink. For all the configurations computed, the internal linear current is always greater than the external one reaching similar values only for configurations where the energy ratio reaches a maximum. Apparently, as one tries to rise the external currents, the system self-regulates inducing a change in the topology of the distribution of the magnetic field and the associated external current. As a consequence there is a maximum twist that can be imposed to the magnetosphere, before reconnection and plasmoid formation sets in. We found moreover that external currents contribute to the net dipole without affecting too much the strength of the magnetic field at the surface.

Magnetized rapidly rotating neutron stars have been invoked as a promising engine for both long duration GRBs [38, 39], and for short duration GRBs [40]. The energy losses due to the emission of a relativistic magnetically driven wind can explain the bulk of known events, but requires in general a high efficiency to convert the rotational energy of the proto-neutron star into wind kinetic energy. The presence of a strong magnetic field $\geq 10^{15}$G, is however expected to induce deformations in the neutron star, that can lead to copious emission of gravitational waves for rapid rotators [41, 13]. The energy loss via gravitational waves will then compete with the electromagnetic emission of the wind [14]. It is thus crucial to evaluate the level of quadrupolar deformation induced by the magnetic field. We have estimated the magnitude of this quadrupolar deformation for various cases, both for poloidal and toroidal magnetic fields, and various current and magnetic field distributions.

We find that, for a NS with a typical mass of $\sim 1.5M_{\odot}$, in the case of a purely toroidal magnetic field, the quadrupolar deformation scales as $\|\epsilon_B\| \simeq 5 \times 10^{-5} B_{16}^2/m$, in term of the maximum value of the magnetic field in units of 10^{16}G. Clearly models with a large m where the magnetic field is concentrated toward the NS surface, can have deformations that are even an order of magnitude smaller than the case $m = 1$ where the field is more concentrated toward the center. For neutron stars with a purely poloidal magnetic field, the quadrupolar deformation is found to scale as $\|\epsilon_B\| \simeq (1-5) \times 10^{-5} B_{16}^2$, where the upper limit is for the case of additive currents, with magnetic field concentrated in the outer stellar layers, and the lower bound for subtractive currents where the field is concentrated toward the axis, and the bulk of the star is weakly magnetized. This is the opposite trend with respect to the one found for the toroidal case. In our mixed twisted torus configurations, the deformation is always oblate, and dominated by the poloidal field, which is the energetically dominant component. However, such limitation is probably due to the barotropicity assumption in the equation of state. If this assumption is relaxed, in principle systems can be build with a large fraction of magnetic energy into the toroidal component. This cannot be tested with our algorithm, where barotropicity is a key assumption. However, if we compare the deformation derived from purely toroidal field, with the one in the purely poloidal case, it is evident, that typical twisted torus configurations, where the toroidal field is confined to small regions close to the surface (reminiscent of cases with large $m > 10$), while the poloidal component threads the entire star, will likely have a quadrupolar deformation dominated by the poloidal magnetic field unless the toroidal component is at least a few times stronger than the poloidal one. Substantial quadrupolar deformation in this case will likely require a toroidal field at least an order of magnitude stronger that the poloidal one.

5. Conclusions

We have presented here some recent results in the development of numerical models for equilibrium configurations of magnetized neutron stars in the fully non-linear general relativistic regime. This allows us to go beyond the simplified perturbative regime (where often only the fluid variables are perturbed while the metric terms are kept fixed), and to derive the correct absolute scaling that we can then extrapolate to those regimes more representative of physical situations. The algorithm we have developed has proved to be robust and efficient, also in handling extremely deformed objects, giving results that are in agreement with more sophisticated tecniques. Based on this approach a publicly available software written in `FORTRAN90` has been released, together with visualization and data reduction tools written in `IDL`, and can be downloaded at `www.arcetri.astro.it/science/ahead/XNS/`. Here we have shown how, by choosing different functional forms for the distribution of currents, it is possible to realize different magnetic field configurations. We have investigated how the magnetic field distribution affects the deformation of the star, in the context of possible gravitational wave emission from fast rotators. We have shown that, at least within a barotropic formalism for the equation of state, it is not possible to build configurations of mixed poloidal-toroidal field, where the toroidal component is energetically dominant, and that the equation of state itself (in particular the stratification in the outer part of the neutron star) is important. Given that toroidal dominated configurations are expected following the collapse of rapidly rotating cores, this suggests that either NSs are out of magnetic equilibrium, or the EoS cannot be described in terms of zero temperature matter at nuclear equilibrium. This might have important consequences for the magneto-thermal evolution of NS, and magnetars in particular. Finally we have presented also models where the currents extend outside the neutron star into a twisted magnetosphere. These configurations are thought to be more representative of the physical regime characterizing magnetars. The code we have developed can also model rotating systems, again both for poloidal and toroidal configurations (in the latter case it is possible to model also a differentially rotating profile) taking into account the fact that in GR rotating systems have an induced electric field that cannot be neglected, as it is usually done for non-relativistic MHD. We hope in the future to be able to evaluate the observable signatures of the magnetic field distributions that we have computed, either in the form of cyclotron absorption features, or in the polarized pattern that the emitted radiation acquires as it travels in the magnetosphere. This will allow us to check if it is e possible to constrain the magnetic geometry or maybe even GR effects with future missions for X-ray polarimetry [42].

References

[1] Mereghetti S 2008 *A&A Rev.* **15** 225–287 (*Preprint* 0804.0250)
[2] Rea N and Esposito P 2011 *High-Energy Emission from Pulsars and their Systems* ed Torres D F and Rea N p 247 (*Preprint* 1101.4472)
[3] Turolla R and Esposito P 2013 *International Journal of Modern Physics D* **22** 1330024 (*Preprint* 1303.6052)
[4] Duncan R C and Thompson C 1992 *ApJLett* **392** L9–L13
[5] Thompson C and Duncan R C 1993 *ApJ* **408** 194–217
[6] Kouveliotou C, Strohmayer T, Hurley K, van Paradijs J, Finger M H, Dieters S, Woods P, Thompson C and Duncan R C 1999 *ApJLett* **510** L115–L118 (*Preprint* astro-ph/9809140)
[7] Goldreich P and Julian W H 1969 *ApJ* **157** 869
[8] Tchekhovskoy A, Spitkovsky A and Li J G 2013 *MNRAS* **435** L1–L5 (*Preprint* 1211.2803)
[9] Chamel N and Haensel P 2008 *Living Reviews in Relativity* **11** 10 (*Preprint* 0812.3955)
[10] Lattimer J M 2012 *Annual Review of Nuclear and Particle Science* **62** 485–515 (*Preprint* 1305.3510)
[11] Yakovlev D G and Pethick C J 2004 *ARA&A* **42** 169–210 (*Preprint* astro-ph/0402143)
[12] Yakovlev D G, Gnedin O Y, Gusakov M E, Kaminker A D, Levenfish K P and Potekhin A Y 2005 *Nuclear Physics A* **752** 590–599 (*Preprint* astro-ph/0409751)
[13] Mastrano A, Lasky P D and Melatos A 2013 *MNRAS* **434** 1658–1667 (*Preprint* 1306.4503)
[14] Dall'Osso S and Stella L 2007 *Ap&SS* **308** 119–124 (*Preprint* arXiv:astro-ph/0702075)
[15] Lander S K and Jones D I 2009 *MNRAS* **395** 2162–2176 (*Preprint* 0903.0827)
[16] Glampedakis K, Andersson N and Lander S K 2012 *MNRAS* **420** 1263–1272 (*Preprint* 1106.6330)
[17] Fujisawa K, Yoshida S and Eriguchi Y 2012 *MNRAS* **422** 434–448 (*Preprint* 1204.5830)
[18] Ciolfi R, Ferrari V, Gualtieri L and Pons J A 2009 *MNRAS* **397** 913–924 (*Preprint* 0903.0556)

[19] Ciolfi R, Ferrari V and Gualtieri L 2010 *MNRAS* **406** 2540–2548 (*Preprint* 1003.2148)
[20] Ciolfi R and Rezzolla L 2013 *MNRAS* (*Preprint* 1306.2803)
[21] Pili A G, Bucciantini N and Del Zanna L 2014 *MNRAS* **439** 3541–3563 (*Preprint* 1401.4308)
[22] Bucciantini N, Pili A G and Del Zanna L 2015 *MNRAS* **447** 3278–3290 (*Preprint* 1412.5347)
[23] Pili A G, Bucciantini N and Del Zanna L 2015 *MNRAS* **447** 2821–2835 (*Preprint* 1412.4036)
[24] Alcubierre M 2008 *Introduction to 3+1 Numerical Relativity* (Oxford University Press)
[25] Gourgoulhon E (ed) 2012 *3+1 Formalism in General Relativity* (*Lecture Notes in Physics, Berlin Springer Verlag* vol 846)
[26] Wilson J R and Mathews G J 2003 *Relativistic Numerical Hydrodynamics*
[27] Bucciantini N and Del Zanna L 2011 *A&A* **528** A101 (*Preprint* 1010.3532)
[28] Del Zanna L, Zanotti O, Bucciantini N and Londrillo P 2007 *A&A* **473** 11–30 (*Preprint* 0704.3206)
[29] Bucciantini N and Del Zanna L 2013 *MNRAS* **428** 71–85 (*Preprint* 1205.2951)
[30] Obergaulinger M, Janka H T and Aloy M A 2014 *MNRAS* **445** 3169–3199 (*Preprint* 1405.7466)
[31] Kuiper L, Hermsen W, den Hartog P R and Collmar W 2006 *ApJ* **645** 556–575 (*Preprint* astro-ph/0603467)
[32] Thompson C, Lyutikov M and Kulkarni S R 2002 *ApJ* **574** 332–355 (*Preprint* astro-ph/0110677)
[33] Beloborodov A M and Thompson C 2007 *Ap&SS* **308** 631–639 (*Preprint* astro-ph/0608372)
[34] Pavan L, Turolla R, Zane S and Nobili L 2009 *MNRAS* **395** 753–763 (*Preprint* 0902.0720)
[35] Woods P M, Kouveliotou C, Finger M H, Göğüş E, Wilson C A, Patel S K, Hurley K and Swank J H 2007 *ApJ* **654** 470–486 (*Preprint* astro-ph/0602402)
[36] Perna R and Gotthelf E V 2008 *ApJ* **681** 522–529 (*Preprint* 0803.2042)
[37] Tiengo A, Esposito P, Mereghetti S, Turolla R, Nobili L, Gastaldello F, Götz D, Israel G L, Rea N, Stella L, Zane S and Bignami G F 2013 *Nature* **500** 312–314 (*Preprint* 1308.4987)
[38] Bucciantini N, Quataert E, Metzger B D, Thompson T A, Arons J and Del Zanna L 2009 *MNRAS* **396** 2038–2050 (*Preprint* 0901.3801)
[39] Metzger B D, Giannios D, Thompson T A, Bucciantini N and Quataert E 2011 *MNRAS* **413** 2031–2056 (*Preprint* 1012.0001)
[40] Bucciantini N, Metzger B D, Thompson T A and Quataert E 2012 *MNRAS* **419** 1537–1545 (*Preprint* 1106.4668)
[41] Mastrano A, Melatos A, Reisenegger A and Akgün T 2011 *MNRAS* **417** 2288–2299 (*Preprint* 1108.0219)
[42] Soffitta P, Barcons X, Bellazzini R, Braga J, Costa E, Fraser G W, Gburek S, Huovelin J, Matt G, Pearce M, Poutanen J, Reglero V, Santangelo A, Sunyaev R A, Tagliaferri G, Weisskopf M, Aloisio R, Amato E, Attiná P, Axelsson M, Baldini L, Basso S, Bianchi S, Blasi P, Bregeon J, Brez A, Bucciantini N, Burderi L, Burwitz V, Casella P, Churazov E, Civitani M, Covino S, Curado da Silva R M, Cusumano G, Dadina M, D'Amico F, De Rosa A, Di Cosimo S, Di Persio G, Di Salvo T, Dovciak M, Elsner R, Eyles C J, Fabian A C, Fabiani S, Feng H, Giarrusso S, Goosmann R W, Grandi P, Grosso N, Israel G, Jackson M, Kaaret P, Karas V, Kuss M, Lai D, Rosa G L, Larsson J, Larsson S, Latronico L, Maggio A, Maia J, Marin F, Massai M M, Mineo T, Minuti M, Moretti E, Muleri F, O'Dell S L, Pareschi G, Peres G, Pesce M, Petrucci P O, Pinchera M, Porquet D, Ramsey B, Rea N, Reale F, Rodrigo J M, Różańska A, Rubini A, Rudawy P, Ryde F, Salvati M, de Santiago V A, Sazonov S, Sgró C, Silver E, Spandre G, Spiga D, Stella L, Tamagawa T, Tamborra F, Tavecchio F, Teixeira Dias T, van Adelsberg M, Wu K and Zane S 2013 *Experimental Astronomy* **36** 523–567 (*Preprint* 1309.6995)

Journal of Physics: Conference Series **719** (2016) 012005 doi:10.1088/1742-6596/719/1/012005

Multi-Color Model for the Protoplanetary Disks HL Tau and HD142527

Tomoyuki Hanawa[1] **and Kazuki Mochida**[2]

[1]Center for Frontier Science, Chiba University, Yayoi-cho 1-33, Inage-ku, Chiba, 263-8522. Japan

[2]Graduate School of Science, Chiba University, Yayoi-cho 1-33, Inage-ku, Chiba, 263-8522. Japan

E-mail: hanawa@faculty.chiba-u.jp,mochidakz@astro.s.chiba-u.ac.jp

Abstract. Protoplanetary disks are circumstellar disks of gas and dust, from which planets may eventually form or be in the process of forming. Recent direct imaging of them has enabled us to derive the density and temperature distributions. Interestingly they often show quite different features depending on the wavelength observed. The near-infrared emission is dominated by scattering of stellar light while the mm- and submm- wave emissions are dominated by thermal dust emission. Thus, the near-infrared emission traces a low density surface layer where stellar light is scattered toward us. The mm- and submm- wave emission trace the high density part of the disk near the mid plane. In order to explain the wavelength-dependent images, we have constructed a passive disk model for HL Tau and HD142527. The former shows concentric rings in the ALMA image while the latter shows a highly asymmetric arc. Our models are based on the multi-color radiation transfer calculation. It takes account of radiation ranging from 100 nm to 3.16 mm. We used the M1 model to solve the radiative equilibrium. Our model gives some constraints on the optical properties of the dust.

1. Introduction

Newly formed stars are often associated with rotating disks consisting of gas and dust. They are named protoplanetary disks since planets are thought to be formed therefrom. They are extended more than a hundred AU from the host stars and their structures are directly observed with large ground based telescopes in the infrared and with ALMA in the radio. The brightness distributions enable us to derive the density and temperature of the dust therein if the dust opacity is given. The dust distribution is valuable for assessing planet formation in the disks. If they are measured at multiple wavelengths, we can get some constraints on the dust opacity. Thus the multi-color images are important probes into the disks.

Main heating source is radiation from the host stars in the protoplanetary disks. The disk surface layers absorb irradiation from the host star while scattering a part of it. They reemit mid-infrared radiation to heat up inner layers of the disk. The inner dense layers are responsible for brightness in the mm- and sub-mm images taken with ALMA. In short, each layer absorbs short wavelength radiation and reemits long wavelength radiation to achieve thermal equilibrium. The absorption, emission, and scattering are dominated by dust particles. Thus we need to take account of the radiative processes ranging from optical to radio in modeling the protoplanetary disks.

With these in mind, we have constructed a multi-color model for protoplanetary disks. Our model takes account of the radiative processes by solving the M1 model[1] in which only the zeroth and first moments of radiation, i.e., the energy density and flux, are the unknowns. It covers wide range of radiation from 0.1 μm to 3.16 mm with 226 colors. We summarize the methods of modeling and application to two interesting objects, HL Tau and HD142527. The former disk consists of multiple rings while the latter has a banana shape. The methods are given in §2. Application to HD142527 based on Muto et al.[2] is shown in §3 while that to HL Tau is given in §4. Short summary is given in §5.

2. Numerical Methods

We use the cylindrical coordinates, (r, z, φ), in which the host star is located at the origin and the disk mid plane coincides with $z = 0$. For simplicity we assume the disk is symmetric with respect to the rotation axis $(\partial/\partial\varphi = 0)$ and the mid plane $(z = 0)$. This assumption is valid for HL Tau since it consists of concentric rings as seen in the ALMA image [3]. It is also reasonable for HD142527 since the disk has a small radial extent and the azimuthal variation is mild.

In our model the radiative energy density (E_ν) and flux ($\boldsymbol{F}_\nu$) at frequency ν are expressed as

$$E_\nu(r, z) = E_{\nu,\mathrm{star}}(r, z) + E_{\nu,\mathrm{disk}}(r, z), \tag{1}$$

$$\boldsymbol{F}_\nu(r, z) = \boldsymbol{F}_{\nu,\mathrm{star}}(r, z) + \boldsymbol{F}_{\nu,\mathrm{disk}}(r, z), \tag{2}$$

respectively, where the subscripts star and disk denote those emitted from the star and disk, respectively. They are evaluated in the frquency range of 3.00×10^{15} Hz $\geq \nu \geq 9.48\times10^{10}$ Hz, i.e., in the wavelength range 0.1 μm $\leq \lambda \leq$ 3.16 mm with the frequency resolution of $\Delta \log \nu = 0.02$. The former is evaluated to be

$$E_{\nu,\mathrm{star}}(r, z) = \frac{R_{\mathrm{star}}^2}{c(r^2 + z^2)} B_\nu(T_{\mathrm{star}}) \exp\left[-\tau_\nu(r, z)\right], \tag{3}$$

$$\boldsymbol{F}_{\nu,\mathrm{star}}(r, z) = \frac{c}{\sqrt{r^2 + z^2}} \begin{pmatrix} r \\ z \end{pmatrix} E_{\nu,\mathrm{star}}(r, z), \tag{4}$$

$$\tau_\nu(r, z) = \int_0^r (\kappa_{\nu,\mathrm{abs}} + \kappa_{\nu,\mathrm{sca}}) \rho\left(r', \frac{zr'}{r}\right) \sqrt{1 + \left(\frac{z}{r}\right)^2} dr', \tag{5}$$

where $\kappa_{\nu,\mathrm{abs}}$, and $\kappa_{\nu,\mathrm{sca}}$ denote absorption and scattering opacity at frequency ν, respectively. The symbols, B_ν and c, denote the Planck function and the speed of light, respectively. The symbols, R_{star} and T_{star}, denote the radius and effective temperature of the host star, respectively. The scattering opacity is defined as

$$\kappa_{\nu,\mathrm{sca}} \equiv \int \frac{\partial \sigma_{\mathrm{sca}}}{d\Omega} (1 - \cos\theta)\, d\Omega, \tag{6}$$

where $d\sigma_{\mathrm{sca}}/d\Omega$ and θ denote the differential scattering cross section and the angle between the incident and scattered photons, respectively. We use this definition to take account of anisotropy scattering in the frame work of moment equations.

The radiation from the disk is computed by the M1 model equations,

$$\frac{\partial E_{\nu,\mathrm{disk}}}{\partial t} + \boldsymbol{\nabla} \cdot \boldsymbol{F}_{\nu,\mathrm{disk}} = -\kappa_{\nu,\mathrm{abs}}\rho \left[E_{\nu,\mathrm{disk}} - \frac{4\pi B_\nu(T)}{c}\right] + \kappa_{\nu,\mathrm{sca}} E_{\nu,\mathrm{star}}, \tag{7}$$

$$\frac{\partial \boldsymbol{F}_{\nu,\mathrm{disk}}}{\partial t} + c^2 \boldsymbol{\nabla} \boldsymbol{P}_{\nu,\mathrm{disk}} = -(\kappa_{\nu,\mathrm{abs}} + \kappa_{\nu,\mathrm{sca}}) \rho \boldsymbol{F}_{\nu,\mathrm{disk}}, \tag{8}$$

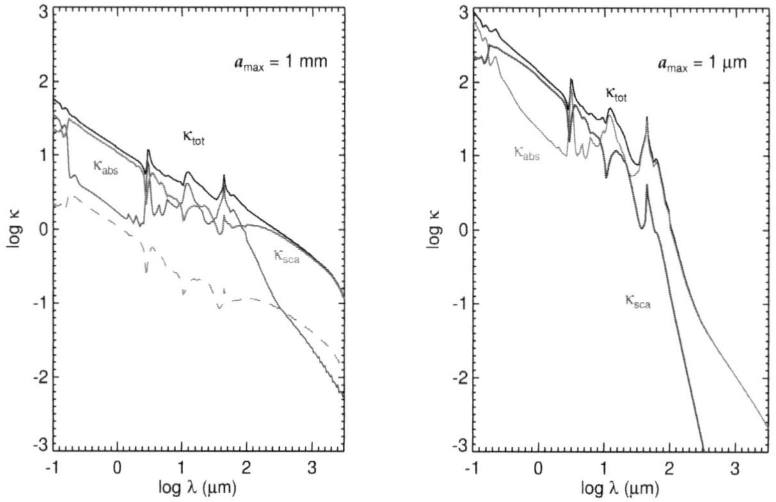

Figure 1. Left panel shows model opacity in which the dust is assumed to have the power law size distribution with $a_{\rm max} = 1$ mm. The blue, red, and black curves denote $\kappa_{\rm abs}$, $\kappa_{\rm sca}$, and $\kappa_{\rm tot} = \kappa_{\rm abs} + \kappa_{\rm sca}$, respectively, as a function of the wavelength, λ. Right panel is the same but for model opacity with $a_{\rm max} = 1\ \mu$m.

where

$$\boldsymbol{P}_{\nu,\rm disk} = \left(\frac{1-\chi_{\nu,\rm disk}}{2}\boldsymbol{I} + \frac{3\chi_{\nu,\rm disk}-1}{2}\boldsymbol{n}_{\nu,\rm disk}\boldsymbol{n}_{\nu,\rm disk}\right), \quad (9)$$

$$\chi_{\nu,\rm disk} = \frac{3+4\left(f_{\nu,\rm disk}\right)^2}{5+\sqrt{4-3\left(f_{\nu,\rm disk}\right)^2}}, \quad (10)$$

$$f_{\nu,\rm disk} = \frac{|\boldsymbol{F}_{\nu,\rm disk}|}{cE_{\nu,\rm disk}}, \quad (11)$$

$$\boldsymbol{n}_{\nu,\rm disk} = \frac{\boldsymbol{F}_{\nu,\rm disk}}{|\boldsymbol{F}_{\nu,\rm disk}|}. \quad (12)$$

The dust temperature, T, is chosen to satisfy the condition for thermal equilibrium,

$$\int \kappa_{\nu,\rm abs}\left\{E_\nu(r,z) - \frac{4\pi B_\nu\left[T(r,z)\right]}{c}\right\} d\nu = 0, \quad (13)$$

at each location.

In this paper, we assume that the gas to dust ratio is constant at $\rho/\rho_{\rm dust} = 100$ for simplicity. In other words, we ignore dust sedimentation. The density distribution is given so that the gas is in hydrostatic equilibrium in the vertical direction,

$$\frac{\partial P}{\partial z} = -\frac{GM_{\rm star}z}{(r^2+z^2)^{3/2}}\rho(r,z), \quad (14)$$

where G and $M_{\rm star}$ denote the gravitational constant and the mass of the host star, respectively. The gas pressure is expressed as

$$P(r,z) = \frac{k}{\mu m_{\rm H}}\rho(r,z)T(r,z), \quad (15)$$

where k and $m_{\rm H}$ denote the Boltzmann constant and the mass of H atom, respectively. The mean molecular weight is assume to be $\mu = 2.339$.

We use the model opacity shown in Figure 1. Left panel shows the model opacity in which the dust is assumed to have the power law size distribution, $a^{-3.5}$ with $a_{\rm max} = 1$ mm, where a denotes the radius of a dust particle. The dust is assumed to consist of silicate, carbonaceous grains, and water ice having mass fractional abundances of $\zeta_{\rm sil} = 0.0043$, $\zeta_{\rm carbon} = 0.0030$, and $\zeta_{\rm ice} = 0.0094$, [4] respectively. We reduce the scattering opacity artificially by a factor of 10 (the red dashed curve) in some models in order to evaluate the effects of scattering. We also use the model opacity shown in the right panel of Figure 1 in some other models. The latter is obtained by assuming $a_{\rm max} = 1\ \mu$m. The methods for computation of the dust opacity are given in Numura & Millar[5].

We have obtained radiative equilibria by following the time evolution of the M1 model equations in an explicit scheme given in Kanno et al. [6]. We have evaluated the expected brightness by solving the equation of radiative transfer,

$$\frac{dI_\nu}{ds} = -\kappa_{\nu,{\rm abs}} \left[I_\nu - B_\nu(T) \right] - \kappa_{\nu,{\rm sca}} \left(I_\nu - \frac{cE_\nu}{4\pi} \right), \tag{16}$$

along the line of sight, s, where the values of T and E_ν are taken from an equilibrium model.

3. HD 142527

The protoplanetry disk of HD142527 is thought to be in the transitional phase from gas-rich one to gas-poor debris one. It has an inner hole of dust emission and shows strong azimuthal asymmetry in the radio continuum (see, e.g., Fukagawa et al.[7] and references therin). The northern part of the disk is more than 20 times brighter at 336 GHz than the southern part. This means that the dust is concentrated strongly in the northern part.

We have made models to evaluate the dust surface density in the northern and southern parts of the disk. Although the brightness has a large contrast, it varies smoothly and more slowly in the azimuthal direction than in the radial direction. Hence we apply our 2D symmetric model to each of the northern and southern narrow sectors of the disk. We define the northern sector as the region of $11^\circ <$PA$< 31^\circ$ and the southern one as that of $211^\circ <$PA$< 231^\circ$. 336 GHz continuum emission is strongest in the northern sector and weakest in the southern sector. The peak brightness is 1.2 Jy arcsec^{-2} and 0.050 Jy arcsec^{-2} in the northern and souther sectors, respectively[2].

We use the opacity model of $a_{\rm max} = 1$ mm, since the model opacity of $a_{\rm max} = 1\ \mu$m cannot reproduce the brightness observed in the northern sector. This is because the opacity is very small at 336 GHz when $a_{\rm max} = 1\ \mu$m. The central star of HD 142527 is assumed to have the model parameters shown in Table 1.

Table 1. Model parameters for the host stars

Star	$T_{\rm star}$	$R_{\rm star}$	$M_{\rm star}$
HD 142527	6250 K	3.8 $R_\odot$	2.2 $M_\odot$
HL Tau	4000 K	6.9 $R_\odot$	0.5 $M_\odot$

Figure 2 shows the density and temperature distributions in the best fit models for the northern and southern sectors. The models covers the rectangular region of 30 AU $\leq r \leq$ 410 AU and $|z| \leq$ 120 AU with the resolution of $\Delta r = \Delta z = 2$ AU. The dust surface density distribution

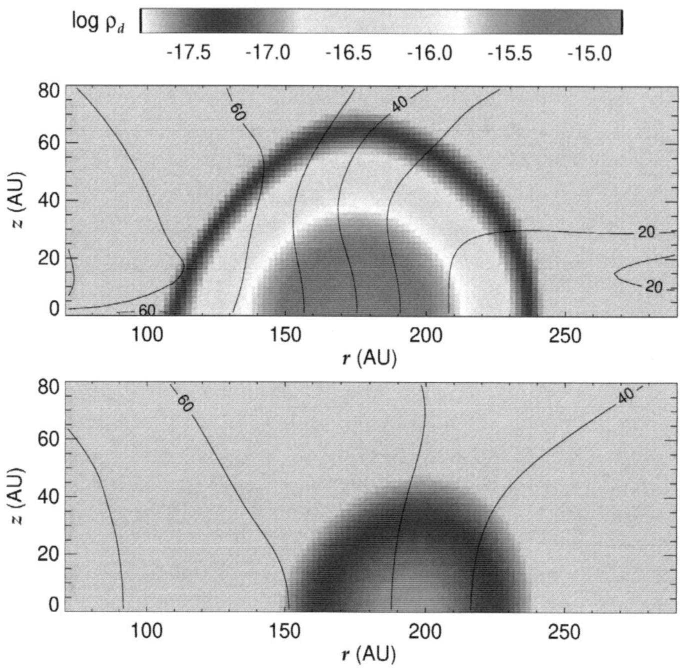

Figure 2. Dust density and temperature distributions in the best fit models for HD142527. The upper and lower panels are for PA = $11° - 31°$ and $221° - 23°$, respectively. Reproduction of Fig. 8 of Muto et al.[2].

is assumed to be

$$\Sigma_{\rm d}(r) \quad = \quad \Sigma_{\rm d,0} \exp\left\{ - \max\left[\left(\frac{r - r_0}{w} \right)^2, 10 \right] \right\}, \tag{17}$$

where the values of Σ_0, r_0, and w are given in Table 2. The temperature decreases with increase in the radius. It is lower in the northern sector than in the southern, since the surface density is higher and hence the optical depth is larger.

Table 2. Best fit models for HD 142527

Sector	Σ_0	r_0	w
North	0.6 g cm^{-2}	173 AU	26 AU
South	8.45×10^{-3} g cm^{-2}	196 AU	34 AU

Note that the contrast in Σ_0 is much larger than that in the brightness. This larger contrast is explained as follows. The brightness is roughly proportional to $T(1 - e^{-\tau})$, where τ denotes the optical depth. The brightness is not proportional to the surface density when the optical depth is appreciable. The northern sector has a lower temperature and is semi-transparent even at 336 GHz. Thus the contrast in the surface density (~ 70) is higher than that in the brightness (~ 24).

It should be also noted that the location of the peak surface density is about 20 AU further from the host star than that of peak brightness. This is due to the radial temperature gradient. The inner disk is radiating more energy per unit mass because it has a higher temperature, thus the brightness peak is shifted radially inwards from the peak surface density as the increase in brightness per unit mass dominates the radial decrease in surface density from the peak.

More detailed comparison with observation is given in Muto et al[2]. They have computed expected line emission from the above mentioned model. Gas is distributed over a wider region from several tens AU to $\sim$ 400 AU in the radial direction, although the line emission has a peak inside the dust disk. In other words, the dust is more concentrated in the radial direction.

4. HL Tau

HL Tau is a very young star located in a molecular ridge and its central star has not been detected in the optical wavelength. The optical and near infrared images depict conical reflection nebula extending up to several hundreds AU and bipolar jets in the center of the nebula. HL Tau is also associated with disk and infalling envelope observed in the radio. Recently Atacama Large Millimeter/submillimeter Array (ALMA) has clarified that the disk is geometrically thin and consists of concentric rings when observed at wavelengths of 2.9, 1.3, and 0.87 mm[3]. The disk inclination and position angle are derived to be $i = 46.°72 \pm 0.°05$ and PA $= 138.°0.2 \pm 0.°0.7$, respectively.

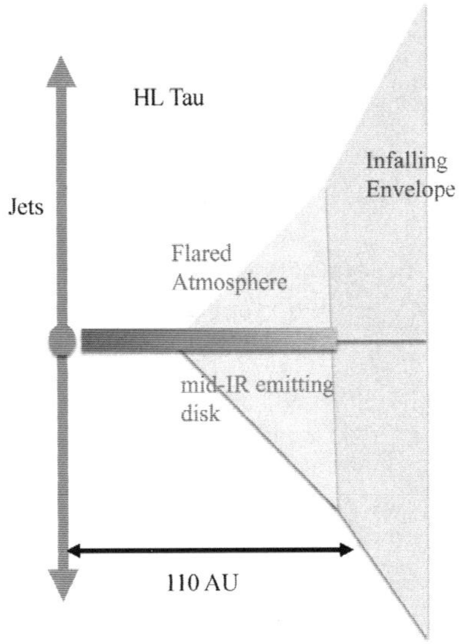

Figure 3. An illustration for jets, disk and envelope around HL Tau.

Figure 3 illustrates the geomety of jets, disk and envelope around HL Tau. Dust particles should be concentrated near the mid plane, since the ALMA images show clear gaps in the disk.[1] On the other hand, the optical and near-infrared images indicate that an appreciable amount of dust particles are distributed well above the midplane. With these constraints in mind we have constructed models for the mid-IR emitting disk and flared atmosphere.

[1] After this conference we have learned that the same argument is given by Pinte et al[9].

In this work we use the surface density model of Kwon et al.[8],

$$\Sigma(r) = \Sigma_0 \left(\frac{r}{R_c}\right)^{-0.22} \exp\left[1 - \left(\frac{r}{R_c}\right)^{2.22}\right], \tag{18}$$

where R_c and Σ_0 denote the typical disk size and the surface density at $r = R_c$, respectively. The former is fixed at their best fit value, $R_c = 78.9$ AU, while the latter is treated as a free parameter. We denote Σ_0 in unit of the best fit value of Kwon et al.[8], $\Sigma_{\rm Kwon} = 26.3$ g cm^{-2}.

Equation (18) does not take account of gaps since it is derived before the discovery of the gaps. We also ignore the gaps in our modeling for simplicity.[2] This is because our interest is focused on the choice of the dust model. We have removed the complexity in the dust temperature due to the gaps.

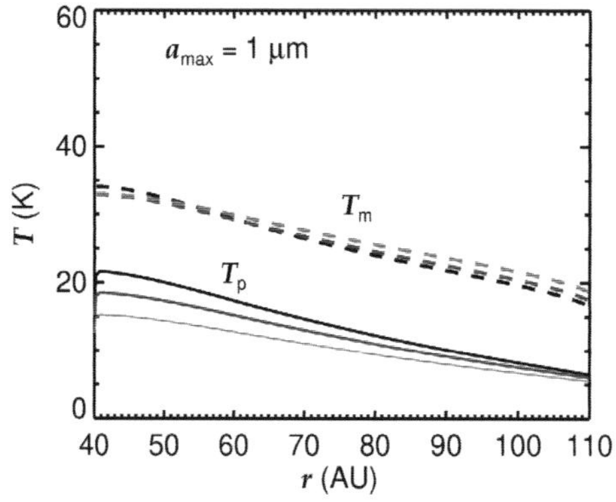

Figure 4. The solid curves denote the Planck temperature for the model opacity of $a_{\rm max} = 1$ μm, while the dashed curves do the mid plane temperature. The red, blue and black curves denote the models of $\Sigma_0/\Sigma_{\rm Kwon} = 0.4$, 0.6, and 0.8, respectively.

First we examine the model opacity in which the radius of the largest dust particle is assumed to be $a_{\rm max} = 1$ μm. The computation box covers the rectangular region of 40 AU $\leq r \leq$ 112 AU and $|z| \leq 72$ AU with resolution of $\Delta r = \Delta z = 0.6$ AU. Figure 5 shows the mid plane temperature ($T_{\rm m}$) and the Planck temperature ($T_{\rm P}$) as a function of the radial distance from the host star, r. The Planck temperature is defined to be

$$T_{\rm P} = \frac{h\nu}{k}\left[\ln\left(\frac{2h\nu^2}{c^2 I_\nu} + 1\right)\right]^{-1}, \tag{19}$$

where h and k denote the Planck and Boltzmann constants, respectively. We obtained the Planck temperature by substituting the expected intensity on the disk major axis, I_ν, into Equation (19). The red, blue and black dashed curves denote the mid plane temperature in the models of $\Sigma_0/\Sigma_{\rm Kwon} = 0.4$, 0.6, and 0.8, respectively. It is much lower than the observed one. It increases with increase in Σ_0. However, it cannot be high enough since the mid plane temperature (solid curves in Figure) is also low and decreases with increase in Σ_0. The Planck temperature coincides with the dust temperature in the limit of large optical depth. Thus we cannot expect that the observed brightness is reproduced by the model opacity of $a_{\rm max} = 1$ μm.

[2] We have included gaps in our modeling after the conference. We have relaxed the assumption that the gas to dust ratio is constant in the model.

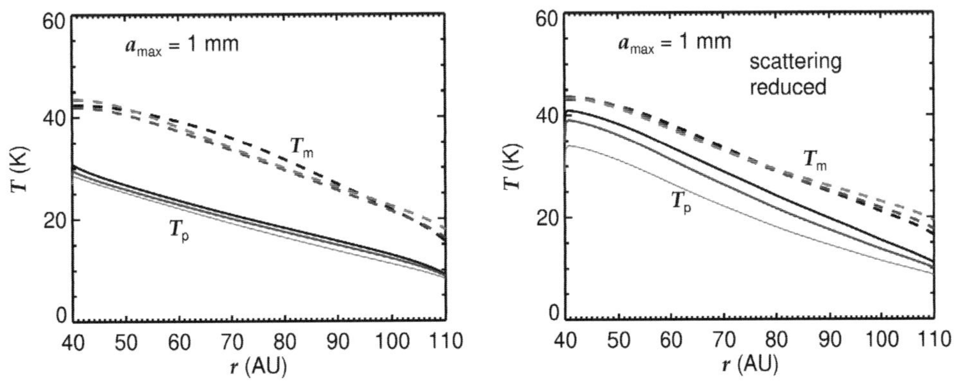

Figure 5. The left panel is the same as Fig. 4 but for the model opacity of $a_{\rm max} = 1$ mm. The red, blue and black curves denote the models of $\Sigma_0/\Sigma_{\rm Kwon} = 0.6$, 1.0, and 1.5, respectively. The right panel is the same as the left panel but for the model opacity in which the scattering opacity is reduced artificially by a factor of 10.

Next we examine the model opacity in which the radius of the largest dust particle is assumed to be $a_{\rm max} = 1$ mm. The left panel of Figure 5 is the same as Figure 4 except for $a_{\rm max} = 1$ mm and Σ_0. The surface density is taken to be $\Sigma_0/\Sigma_{\rm Kwon} = 0.6$, 1.0, and 1.5. The Planck temperature (i.e., the brightness) depends little on Σ_0 and much lower than the mid plane temperature. The large difference between $T_{\rm P}$ and $T_{\rm m}$ is due to the large scattering opacity in the mm-wave. Figure 1 shows that the scattering opacity is 10 times larger than the absorption one at $\lambda = 1$ mm.

In order to confirm this idea we reduced the scattering opacity by a factor 10, while the absorption opacity remains unchanged. The right panel of Figure 5 shows $T_{\rm P}$ and $T_{\rm m}$ in the models in which only the scattering opacity is reduced artificially. The mid plane temperature is affected little by the reduction in the scattering opacity. However the Planck temperature is increased by the reduction in the scattering opacity.

Figure 6 compares our model of $\Sigma_0/\Sigma_{\rm Kwon} = 1.5$ with the observation with ALMA. The scattering opacity is reduced by a factor of 10 artificially from the model opacity of $a_{\rm max} = 1$ mm. The computation box covers the rectangular region of 30 AU $\le r \le$ 126 AU and $|z| \le$ 96 AU with resolution of $\Delta r = \Delta z = 0.4$ AU. The black solid, dotted and dashed curves denote the Planck temperature at Bands 7, 6 and 3 shown in Zhang et al.[10], respectively. The red solid, dotted and dashed curves denote the Planck temperature in our model at 343.5, 233.0 and 101.9 GHz, respectively. They are roughly consistent with the observed ones, although the gaps are not taken into account. After the conference, we have confirmed that the gaps can be reproduced by modifying the surface density profile.

5. Summary and Discussion

We have applied our multi-color model to interesting objects, HD 142527 and HL Tau. They can evaluate the temperature distribution one the model opacity is given. We can confirm the validity by molecular line emissions as has been done in Muto et al[2]. They provide some constraints on the opacity model as shown in the modeling of HL Tau. The scattering opacity of mm-sized dust grains are much larger than the absorption one in the current model. When the scattering opacity is dominant, the Planck temperature is appreciably lower than the dust temperature even when the layer is optically thick. ALMA observations at bands 6 and 7 suggest

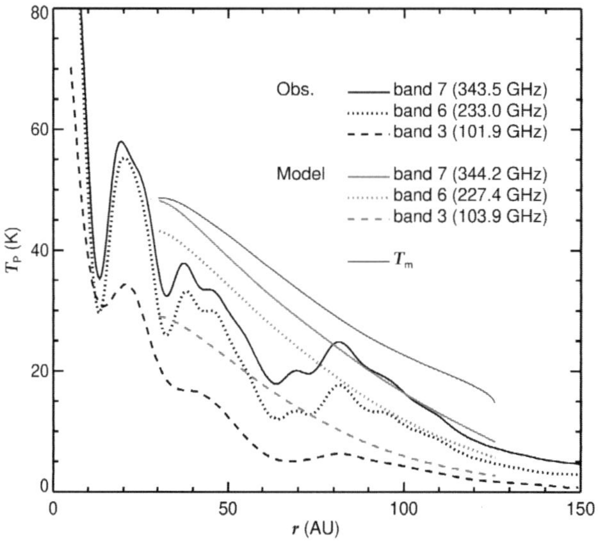

Figure 6. Comparison of our model with the observed Planck temperature at Bands 3, 6 and 7 of ALMA [10].

that the Planck temperature is saturated at ~ 55 K around $r = 30$ AU. This suggests that the dust temperature should be appreciably higher than ~ 55 K, if the dust opacity is dominant. We wonder whether the dust temperature could be so high.

TH thanks HD142527 observation team in Japan for their collaboration in modeling HD 142527. He is indebted especially to M. Momose, T. Muto, and K. Soon. This work is financially supported in part by MEXT KAKENHI Grant Number 26103702.

References

[1] González, M., Audit, E., Huynh, P. 2007, *A&A*, **464**, 429
[2] Muto, T., Tsukagoshi, T., Momose, M. et al. 2015, *PASJ* in press
[3] Partnership A., Brogan C. L., Pérez L. M. et al 2015 *ApJL*, **808**, L3
[4] Anders, E., Grevesse, N. 1989, *Geochim, Cosmochim. Acta*, **53**, 197
[5] Nomura, H., Miller, T.J. 2005, *A&A*, **438**, 923
[6] Kanno, Y.. Harada, T., Hanawa, T. 2013, *PASJ*, **65**, 72
[7] Fukagawa, M., Tsukagoshi, T., Momose, M. et al. 2013, *PASJ*, **65**, L14
[8] Kwon W., Looney L. W. and Mundy L. G. 2011 *ApJ*, **741** 3
[9] Pinte, C., Dent, W.R.F., Menard, F. et al. 2015, arXiv:1508.00584
[10] Zhang, K., Blake, G.A., and Bergin, E.A. 2015, *ApJL*, **806**, L7

A New Hydrodynamic Model for Numerical Simulation of Interacting Galaxies on Intel Xeon Phi Supercomputers

Igor Kulikov[1], Igor Chernykh[1], and Alexander Tutukov[2]

[1] Institute of Computational Mathematics and Mathematical Geophysics SB RAS, 6-th Lavrentyev avenue, Novosibirsk, Russian Federation
[2] Institute of Astronomy RAS, 48-th Pyatnitskaya Street, Moscow, Russian Federation

E-mail: kulikov@ssd.sscc.ru

Abstract. This paper presents a new hydrodynamic model of interacting galaxies based on the joint solution of multicomponent hydrodynamic equations, first moments of the collisionless Boltzmann equation and the Poisson equation for gravity. Using this model, it is possible to formulate a unified numerical method for solving hyperbolic equations. This numerical method has been implemented for hybrid supercomputers with Intel Xeon Phi accelerators. The collision of spiral and disk galaxies considering the star formation process, supernova feedback and molecular hydrogen formation is shown as a simulation result.

Introduction

The collisions between galaxies as they move in dense clusters is an important evolutionary factor because an ordinary galaxy can experience up to ten collisions with other galaxies in the Hubble time [1]. Observational and theoretical investigation of interacting galaxies is an indispensable method for studying their properties and evolution. The processes of star formation [2], AGN [3], the formation of supermassive binary black holes [4, 5], and chemodynamics [6] significantly accelerate the collision of galaxies. On the subject of research on collision of galaxies, it is worth mentioning the GALMER project of the Parisian observatory. This project consists of a database of computing experiments [7] on collisions of various types of galaxies.

The collisionless components of galaxies are usually described by means of the N-body model. This model is standard for this problem but also has some disadvantages. From the perspective of subgrid physical processes, the N-body model does not allow the total energy of a system to be maintained. As with the star formation processes (or explosion of supernova stars), we can keep only the mass of the system and the moment of an impulse. However, in this case, the thermodynamic coherence of the system is lost. That is, in the case of phase transition from gas, it is impossible to transform internal energy to particles. From the perspective of a numerical method, it is rather difficult to coordinate methods for the solution of hydrodynamic equations and particle dynamics. This is connected with the need to provide a sufficient number of particles in the case of a combination with mesh-based methods. The problem is partially solved by SPH methods, but they have shortcomings that will be described further. In the case of PIC methods, parallel realizations require many difficult algorithms of load balancing.

Advantages of SPH:	Advantages of AMR:
Robustness of algorithm	Approved methodology
Galilean invariant	Lack of artificial viscosity
Simplicity of realization	Correctness on a shock wave
Adaptation for geometry of a problem	Turbulence Reproduction
High precision of gravity	Reproduction of instability
SPH disadvantages:	AMR disadvantages:
Artificial viscosity	Complexity of realization
Radius of smoothing	Mesh effects
Oscillation on shock waves	Problem of mesh resolution
Suppression of instability	Problem of invariancy
Weak scalability	Weak scalability

Table 1. Advantages and disadvantages of SPH and AMR methods

As an alternative to the N-body model, there is a model based on the first moments of the Boltzmann equation [8]. The possibility of using this model for the description of the collisionless components of colliding galaxies [9] was experimentally proved. Sufficient conditions for the Boltzmann equation model are clustering star components, prevailing kinetics of the system and a lack of heat transfer effects.

In the last two decades, from the wide range of hydrodynamic numerical methods for the solution of nonstationary three-dimensional astrophysical problems, two main approaches have been used: the Lagrangian approach, which is generally presented by the SPH method (Smoothed Particle Hydrodynamics), and the Euler approach, which uses adaptive grids or AMR (Adaptive Mesh Refinement). In the last five years, several codes have been developed as combinations of Lagrangian and Euler approaches. The Lagrangian approach (SPH method) is used in Hydra [10], Gasoline [11], GrapeSPH [12], GADGET [13] packages. The Eulerian approach (including AMR) is used in NIRVANA [14], FLASH [15], ZEUS [16], ENZO [17], RAMSES [18], ART [19], Athena [20], Pencil Code [21], Heracles [22], Orion [23], Pluto [24], CASTRO [25] codes. The Eulerian approach with AMR was first used on hybrid supercomputers equipped with graphic accelerators in a GAMER code [26]. The advantages and disadvantages of these methods are given in table 1. BETHE-Hydro [27], AREPO [28], CHIMERA [29], GIZMO [30], and author's PEGAS/GPUPEGAS/AstroPhi [9, 31, 32] codes are based on Lagrangian and Eulerian approaches. This combination of methods solves the main problems of the approaches as described above.

In the first section, we will describe the mathematical model of the interacting galaxies. We will describe a numerical method of the solution and parallel realization in the second section. In chapter 3, the results of the computational experiment of spiral and disk galaxy collision will be shown.

1. Mathematical model

To account for chemical reactions, we will consider the equations of multicomponent single-speed gas dynamics. For the description of collisionless components, we will use the equations for the first moments of the collisionless Boltzmann equation. Our model includes star formation processes, supernova feedback, formation of molecular hydrogen and the cooling/heating function.

To describe the gas components, we will use the system of single-speed component gravitational hydrodynamics equations, which is written in Euler coordinates:

$$\frac{\partial \rho}{\partial t} + \nabla \cdot (\rho \vec{u}) = \mathcal{S} - \mathcal{D}, \qquad \frac{\partial \rho \vec{u}}{\partial t} + \nabla \cdot (\rho \vec{u} \vec{u}) = -\nabla p - \rho \nabla(\Phi) + \vec{v}\mathcal{S} - \vec{u}\mathcal{D},$$

$$\frac{\partial \rho_H}{\partial t} + \nabla \cdot (\rho_H \vec{u}) = -s_{H,H_2} + \mathcal{S}\frac{\rho_H}{\rho} - \mathcal{D}\frac{\rho_H}{\rho}, \qquad \frac{\partial \rho_{H_2}}{\partial t} + \nabla \cdot (\rho_{H_2} \vec{u}) = s_{H,H_2} + \mathcal{S}\frac{\rho_{H_2}}{\rho} - \mathcal{D}\frac{\rho_{H_2}}{\rho},$$

$$\frac{\partial \rho E}{\partial t} + \nabla \cdot (\rho E \vec{u}) = -\nabla \cdot (p\vec{v}) - (\rho \nabla(\Phi), \vec{u}) - \Lambda + \Upsilon + \varepsilon\frac{\mathcal{S}}{\rho} - \varepsilon\frac{\mathcal{D}}{\rho},$$

$$\frac{\partial \rho \varepsilon}{\partial t} + \nabla \cdot (\rho \epsilon \vec{u}) = -(\gamma - 1)\rho\epsilon \nabla \cdot \vec{u} - \Lambda + \Upsilon + \varepsilon\frac{\mathcal{S}}{\rho} - \varepsilon\frac{\mathcal{D}}{\rho}, \qquad \rho E = \frac{1}{2}\rho \vec{u}^2 + \rho\varepsilon, \qquad p = (\gamma - 1)\rho\varepsilon.$$

To describe the collisionless components, we will use the system of equations for the first moments of the Boltzmann collisionless equation, which is also written in Eulerian coordinates:

$$\frac{\partial n}{\partial t} + \nabla \cdot (n\vec{v}) = \mathcal{D} - \mathcal{S}, \qquad \frac{\partial n\vec{v}}{\partial t} + \nabla \cdot (n\vec{v}\vec{v}) = -\nabla \Pi - n\nabla(\Phi) + \vec{u}\mathcal{D} - \vec{v}\mathcal{S},$$

$$\frac{\partial \rho W}{\partial t} + \nabla \cdot (\rho W \vec{v}) = -\nabla \cdot (\Pi \vec{v}) - (n\nabla(\Phi), \vec{v}) + \varepsilon\frac{\mathcal{D}}{\rho} - \varepsilon\frac{\mathcal{S}}{\rho},$$

$$\frac{\partial \Pi_{\xi\xi}}{\partial t} + \nabla \cdot (\Pi_{\xi\xi}\vec{v}) = -2\Pi\nabla \cdot \vec{u} + \varepsilon\frac{\mathcal{D}}{3\rho} - \varepsilon\frac{\mathcal{S}}{3\rho}, \qquad \rho W = \frac{1}{2}\rho\vec{v}^2 + \frac{\Pi_{xx} + \Pi_{yy} + \Pi_{zz}}{2}.$$

The Poisson equation can be written as:

$$\Delta\Phi = 4\pi G(\rho + n),$$

where p – gas pressure, ρ_H – density of atomic hydrogen, ρ_{H_2} – density of molecular hydrogen, s_{H,H_2} – speed of formation of molecular hydrogen from atomic, $\rho = \rho_H + \rho_{H_2}$ – density of gas mixture, n – density collisionless component, $\vec{u}$ – speed gas component, $\vec{v}$ – speed collisionless component, ρE – density of total mechanical gas energy, ρW – density of total mechanical collisionless components energy, Φ – gravitational potential, ε – density of internal energy of gas, γ – adiabatic index, $\Pi_{\xi\xi} = (\Pi_{xx}, \Pi_{yy}, \Pi_{zz})$ – a diagonal tensor of dispersion of speeds collisionless components, $\mathcal{S}$ – the speed of formation of supernova stars, $\mathcal{D}$ – star formation speed, Λ – function of Compton cooling, Υ – function of heating from explosion of supernova stars.

Hydrogen molecules in intergalactic space are formed on a surface of particles and dissociated space radiation. Assuming that the density of gas is proportional to the density of particle concentration of molecular hydrogen [33] because of the good mixing of particles and gas in our model, it can be written as:

$$\frac{dn_{H_2}}{dt} = R_{gr}(T)n_H(n_H + 2n_{H_2}) - (\xi_H + \xi_{diss}(N_{H_2}, A_V))\, n_{H_2},$$

where n_{H_2} and n_H – concentration of molecular and atomic hydrogen, N_{H_2} – the column density of molecular hydrogen, speed of formation of molecular hydrogen on dust is set by function [34] of $R_{gr}(T) = 2.2 \times 10^{-18} S\sqrt{T}$ s^{-1}, where S =0.3 – efficiency of formation of molecular hydrogen on dust [35], speed of ionization of hydrogen by the space beams is set by function [36, 37] $\xi_H = 6 \times 10^{-18}$ s^{-1}, where A_v – extinction [38]. Photodissociation rate in the form of [39] $\xi_{diss}(N_{H_2}, A_V) = \xi_{diss}(0) f_{shield}(N_{H_2}) f_{dust}(A_V)$, where $\xi_{diss}(0) = 3.3 \times 1.7 \times 10^{-11} s^{-1}$ – unshielded photodissociation rate [40], $f_{dust}(A_V) = exp(-\tau_{d,1000}(A_V))$ – absorption rate on dust [39], where $\tau_{d,1000}(A_V) = 3.74 A_V = 10^{-21}\,(N_H + N_{H_2})$ – optical depth on dust particles on

Journal of Physics: Conference Series **719** (2016) 012006 doi:10.1088/1742-6596/719/1/012006

wavelength $\lambda = \text{Å}1000$, where N_H and N_{H_2} – column density. The function of the coefficient of self-shielding can be approximated [39]:

$$f_{shield}(N_{H_2}) = \frac{0.965}{(1+x/b_5)^2} + \frac{0.035}{\sqrt{1+x}} exp\left(-8.5 \times 10^{-4}\sqrt{1+x}\right),$$

where $x = N_{H_2}/5 \times 10^{10}$ m^2, $b_5 = b/10^7$ m/s, where b – the parameter of Doppler expansion. After calculation of molecular hydrogen concentration n_{H_2}, the adiabatic index [41] is defined:

$$\gamma = \frac{5n_H + 7n_{H_2}}{3n_H + 5n_{H_2}}$$

The equation for the description of hydrogen concentration dynamics has the analytical solution which was used in realization.

We will use the following necessary condition of process of star formation formulated in the form of [42]:

$$T < 10^4 K \qquad \nabla \cdot \vec{u} < 0 \qquad \rho > 1.64 \frac{M_\odot}{pc^{-3}}$$

The speed of star formation can be formulated as $\mathcal{D} = \mathcal{C}\rho^{3/2}\sqrt{\frac{32G}{3\pi}}$, where $\mathcal{C} = 0.034$ – star formation efficiency. Speed of supernova stars formation [43] can be written as $\mathcal{S} = \beta\mathcal{C}n^3/2\sqrt{\frac{32G}{3\pi}}$, where $\beta = 0.1$ – coefficient of young stars explosion. The explosion energy of one solar mass star emitted is 10^{51} erg. In this case the heat function can be written as $\Upsilon = 10^{51}\frac{M^{SN}}{M_\odot}$ erg, where M^{SN} – the mass of supernova stars in local volume.

The galactic gas which is warmed by collisions to temperature $\sim 10^4 - 10^8$ K is described by the cooling function [44] $\Lambda \simeq 10^{-22} n_H^2 cm^{-3}$ erg , where n_H – concentration of atomic hydrogen.

2. Numerical method

To solve the hydrodynamic equations, the original numerical method based on a combination of the operator splitting approach, Godunov's method, the Roe scheme and piecewise-parabolic reconstruction on a local stencil [45, 46] was used. This method unites all advantages of these methods and can be highly parallelized. To solve the Poisson equation, the FFT method is used.

At the Eulerian stage of a numerical method, the equations without the advective members and subgrid functions are solved. To approximate the spatial derivatives, the solution to a problem of the linearized Riemann's problem is used. For this purpose, the average speed and pressure (L – left cell, R – right cell) is calculated as follows:

$$\rho = \frac{\rho_L^{3/2} + \rho_R^{3/2}}{\sqrt{\rho_L} + \sqrt{\rho_R}}, \qquad p = \frac{p_L\sqrt{\rho_L} + p_R\sqrt{\rho_R}}{\sqrt{\rho_L} + \sqrt{\rho_R}}.$$

This method of averaging is associated with accurate computation of the gas vacuum border. This averaging is more accurate than the original Roe scheme [47]. We will assume that the solution is a piecewise parabolic function in the considered cells. The detailed procedure of local parabola creation can be found in classical literature [48]. This procedure also must be used in the subsequent stage (the solution of Riemann's problem) of the method.

The solution of Riemann's problem (a detailed description of Godunov's scheme can be found in [49]) for the Eulerian stage can be written as:

$$U = \frac{u_L(-\lambda t) + u_R(\lambda t)}{2} + \frac{p_L(-\lambda t) - p_R(\lambda t)}{2}\sqrt{\frac{(\sqrt{\rho_L} + \sqrt{\rho_R})^2}{\frac{\gamma_L + \gamma_R}{2}(\rho_L^{3/2} + \rho_R^{3/2})(p_L\sqrt{\rho_L} + p_R\sqrt{\rho_R})}},$$

$$P = \frac{p_L(-\lambda t) + p_R(\lambda t)}{2} + \frac{u_L(-\lambda t) - u_R(\lambda t)}{2} \sqrt{\frac{\frac{\gamma_L+\gamma_R}{2}(\rho_L^{3/2} + \rho_R^{3/2})(p_L\sqrt{\rho_L} + p_R\sqrt{\rho_R})}{(\sqrt{\rho_L} + \sqrt{\rho_R})^2}},$$

for collisionless component:

$$V = \frac{v_L(-\mu t) + v_R(\mu t)}{2} + \frac{\Pi_L(-\mu t) - \Pi_R(\mu t)}{2} \sqrt{\frac{(\sqrt{n_L} + \sqrt{n_R})^2}{3(n_L^{3/2} + n_R^{3/2})(\Pi_L\sqrt{n_L} + \Pi_R\sqrt{n_R})}},$$

$$\Pi = \frac{\Pi_L(-\mu t) + \Pi_R(\mu t)}{2} + \frac{v_L(-\mu t) - v_R(\mu t)}{2} \sqrt{\frac{3(n_L^{3/2} + n_R^{3/2})(\Pi_L\sqrt{n_L} + \Pi_R\sqrt{n_R})}{(\sqrt{n_L} + \sqrt{n_R})^2}}$$

where

$$\lambda = \sqrt{\frac{\frac{\gamma_L+\gamma_R}{2}(p_L\sqrt{\rho_L} + p_R\sqrt{\rho_R})}{\rho_L^{3/2} + \rho_R^{3/2}}}, \qquad \mu = \sqrt{\frac{3(\Pi_L\sqrt{n_L} + \Pi_R\sqrt{n_R})}{n_L^{3/2} + n_R^{3/2}}},$$

$$q_L(-\nu t) = q_i^R - \frac{\nu t}{2h}\left(\Delta q_i - q_i^6\left(1 - \frac{2\nu t}{3h}\right)\right), \qquad q_R(\nu t) = q_i^L + \frac{\nu t}{2h}\left(\Delta q_i + q_i^6\left(1 - \frac{2\nu t}{3h}\right)\right).$$

At the Lagrangian stage there is an advective transfer of hydrodynamic parameters and all equations can be written as:

$$\frac{\partial f}{\partial t} + \nabla \cdot (f\vec{v}) = 0,$$

where f can be density ρ, n, can be impulse $\rho\vec{u}, n\vec{v}$, total mechanically density $\rho E, nW$ or internal $\rho\epsilon, \Pi_{\xi\xi}$ energy of gas. A similar approach is used for the equations. This equation:

$$F = v \times \begin{cases} f_L(-\lambda t), v \geq 0 \\ f_R(\lambda t), v < 0 \end{cases},$$

is used for flux computation $F = f\vec{v}$ with $\lambda = |\vec{v}|$. where $f_L(-\lambda t)$ $f_R(\lambda t)$ – piecewise-parabolic function for f. The velocity on the interface between cells is calculated by:

$$v = \frac{v_L\sqrt{\rho_L} + v_R\sqrt{\rho_R}}{\sqrt{\rho_L} + \sqrt{\rho_R}}.$$

To create the piecewise and parabolic solution, the same procedure as in the previous stage is used. At the final stage of the hydrodynamic equations, a solution adjustment procedure is provided. In the case of a gas vacuum border, we use the procedure from [50]:

$$|\vec{v}| = \sqrt{2(E - \epsilon)}, (E - \vec{v}^2/2)/E \geq 10^{-3},$$

In the other area, an adjustment is used to guarantee non-decreasing entropy [51]:

$$|\rho\epsilon| = \left(\rho E - \frac{\rho\vec{v}^2}{2}\right), (E - \vec{v}^2/2)/E < 10^{-3},$$

This modification provides a detailed balance of energy and guarantees non-decreasing entropy.

The numerical method was verified on many problems: 1D problems of shock tube simulation, Kelvin–Helmholtz and Releigh–Taylor instabilities, Sedov blast wave problem, expansion of gas into a vacuum, Evrard's collapse and other problems. Details of the tests can be found in [9, 31, 32, 49].

A geometrical decomposition of the computational area is based on the idea of parallel realization. The uniform numerical method allows the use of the identical geometrical decomposition scheme, which facilitates simpler and effective realization. Supercomputers with the RSC PetaStream-based architecture have been used for performance and scalability tests: MVS-10P (64x Intel Xeon Phi 7120D accelerators, each of which has 16 GB of integrated RAM, JSCC RAS). We use a grid size of 5123 on each accelerator for the MVS-10P tests. In our tests, we cannot use larger grid sizes for simulation of full numerical models. Computational experiments were performed at the RSC PetaStream architecture cluster of JSCC RAS with 134x acceleration achieved in one Intel Xeon Phi accelerator. We attained 92 % efficiency with the usage of 64x Intel Xeon Phi accelerators.

3. Numerical experiment

We will simulate the collision of two galaxies as a demonstration of our code with mass $M = 10^{13}M_\odot$ and velocity $v_{cr} = 800km/s$. The first galaxy is set by self-gravitating the spherical clouds for the description of gas and collisionless components with an initial distribution of density equilibrium. The star component of the second galaxy has a spiral form. The galaxies are rotating in opposite directions with differential rotation:

$$v_\phi = \sqrt{r\frac{\partial \Phi}{\partial r}}.$$

For numerical simulation, a 512^3 computational mesh was used. This scenario of galaxy collision used the same NGC 2936/2937 interacting galaxy scenario. The figures show the results of the column density distribution of gas and star components. After the process of collision behind the front of a shock wave, there is an active increase in star formation speed, and molecular hydrogen is formed in future galaxy zones. It should be noted that the growth of the formation is observed in an area that corresponds with the spiral form.

Conclusion

In paper the new hydrodynamic model of the interacting galaxies is proposed. This model is based on the joint solution of the multicomponent gas dynamics equations, the first moments of the collisionless Boltzmann equation and Poisson's equation for gravity potential. Using of this model gave us possibility to formulate a uniform numerical method of the hyperbolic equations solution. This numerical method is effectively realized for hybrid supercomputers equipped with Intel Xeon Phi accelerators. Computational experiments were performed at the RSC PetaStream architecture cluster of JSCC RAS with 134x acceleration achieved in one Intel Xeon Phi accelerator. We attained 92 % efficiency with the usage of 64x Intel Xeon Phi accelerators. The collision problem of disk and spiral galaxies with increased star formation rate area is simulated.

Acknowledgments

This work was partially supported by Russian Foundation for Basic Research (grants 15-31-20150, 15-01-00508, and 16-07-00434) and by Grant of the President of Russian Federation for the support of young scientists number MK – 6648.2015.9. This project was partially supported by the Russian Ministry of Education and Science.

References

[1] Tutukov A., Lazareva G., Kulikov I. Gas Dynamics of a Central Collision of Two Galaxies: Merger, Disruption, Passage, and the Formation of a New Galaxy // Astronomy Reports. – 2011. – V. 55, I. 9. – P. 770-783.

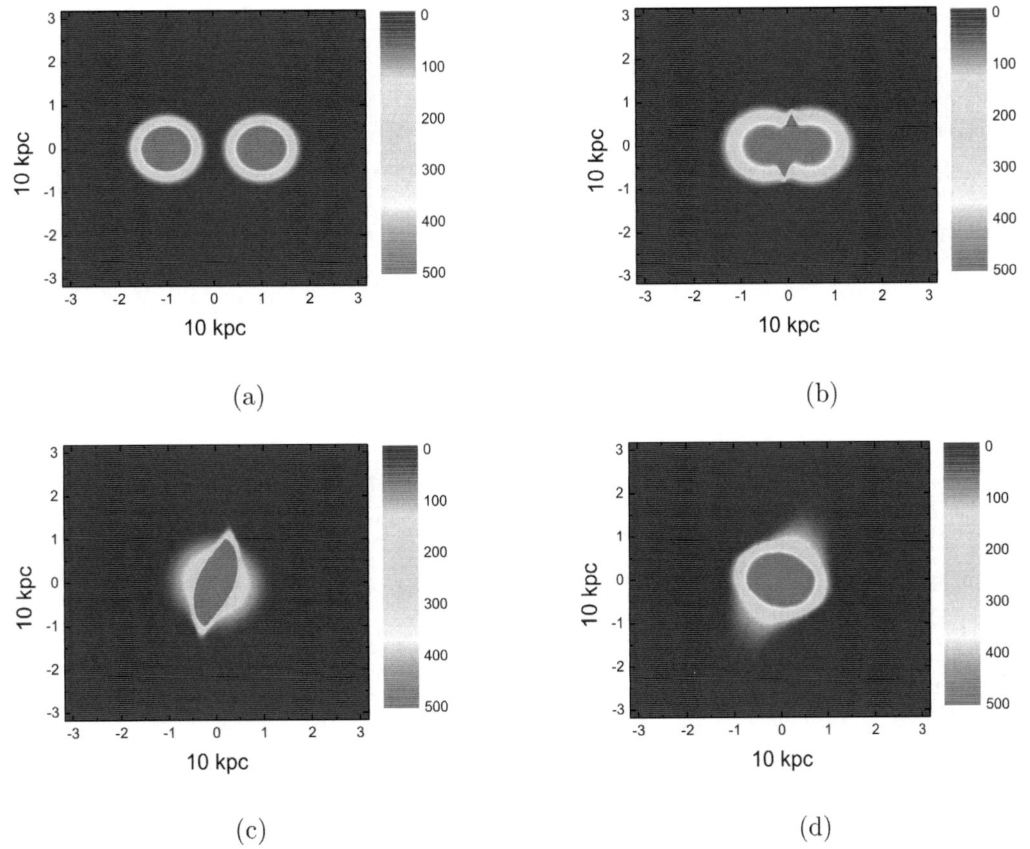

Figure 1. Column density in $M_{\odot}pc^{-2}$ of gas components in time moments $t = 0$ myr – (a), $t = 100$ myr – (b), $t = 200$ myr – (c), $t = 400$ myr – (d).

[2] Schweizer F. Merger-Induced Starbursts // Astrophysics and Space Science Library. – 2005. – V. 329. – P. 143-152.

[3] Sol Alonso M., Lambas D., Tissera P., Coldwell G. Active galactic nuclei and galaxy interactions // Monthly Notices of the Royal Astronomical Society. – 2007. – V. 375, I. 3. – P. 1017-1024.

[4] Blecha L., Loeb A., Narayan R. Double-peaked narrow-line signatures of dual supermassive black holes in galaxy merger simulations // Monthly Notices of the Royal Astronomical Society. – 2013. – V. 429, I. 3. – P. 2594-2616.

[5] Rodriguez C., Taylor G., Zavala R., Pihlstrom Y., Peck A. Hi observations of the supermassive binary black hole system in 0402+379 // The Astrophysical Journal. – 2009. – V. 697, I. 1. – P. 37-44.

[6] Combes F., Melchior A. Chemodynamical evolution of interacting galaxies // Astrophysics and Space Science. – 2002. – V. 281, I. 1-2. – P. 383-387.

[7] Chilingarian I., Di Matteo P., Combes F., Melchior A., Semelin B. The GalMer database: galaxy mergers in the virtual observatory // Astronomy & Astrophysics. – 2010. – V. 518, A61. – P. 1-14.

[8] Mitchell N., Vorobyov E., Hensler G. Collisionless Stellar Hydrodynamics as an Efficient Alternative to N-body Methods // Monthly Notices of the Royal Astronomical Society. – 2013. – V. 428. – P. 2674-2687.

[9] Kulikov I. GPUPEGAS: A New GPU-accelerated Hydrodynamic Code for Numerical Simulations of Interacting Galaxies // The Astrophysical Journal Supplements Series. – 2014. – V. 214, I. 12. – P. 1-12.

[10] Pearcea F.R., Couchman H.M.P. Hydra: a parallel adaptive grid code // New Astronomy. – 1997. – V. 2. – P. 411-427.

[11] Wadsley J.W., Stadel J., Quinn T. Gasoline: a flexible, parallel implementation of TreeSPH // New Astronomy. – 2004. – V. 9. – P. 137-158.

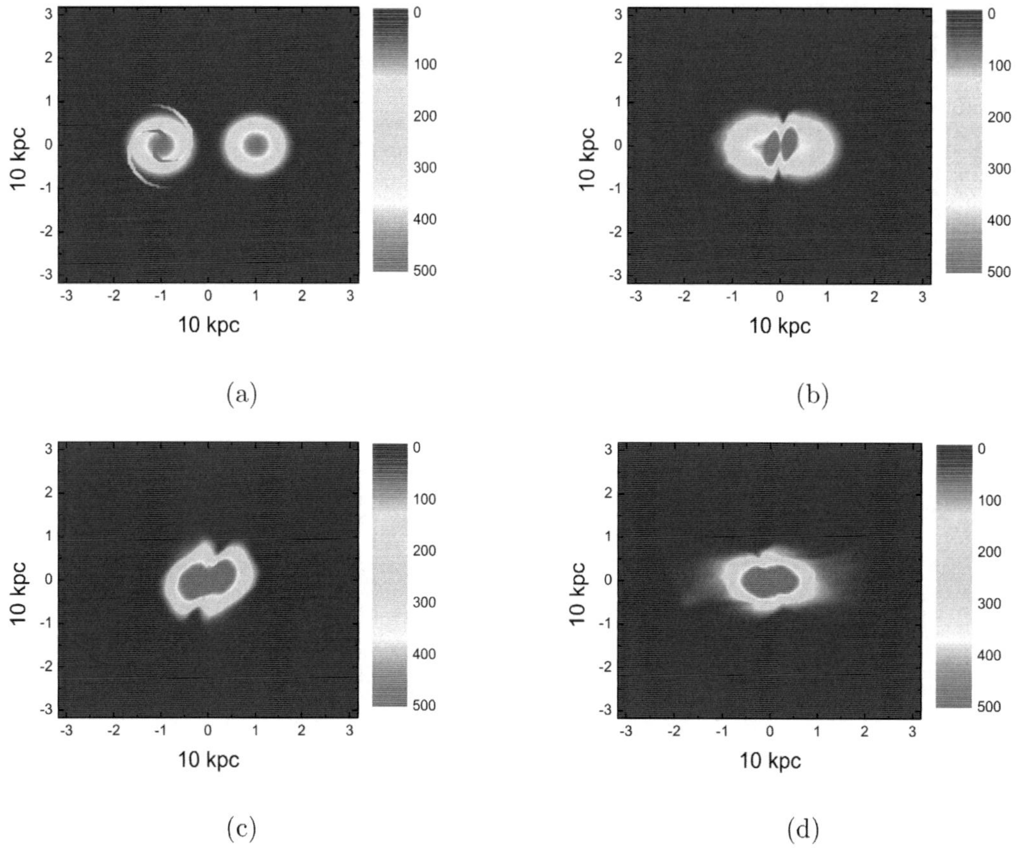

Figure 2. Column density in $M_{\odot}pc^{-2}$ of star components in time moments $t = 0$ myr – (a), $t = 100$ myr – (b), $t = 200$ myr – (c), $t = 400$ myr – (d).

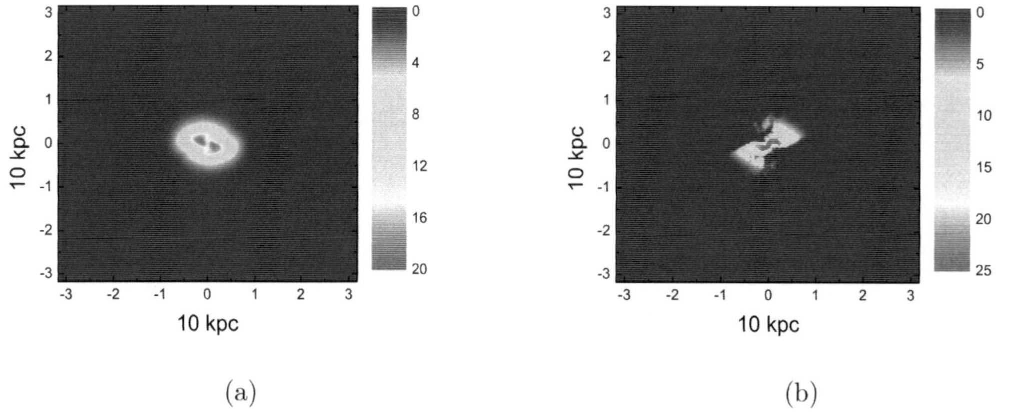

Figure 3. Column density in $M_{\odot}pc^{-2}$ of gas components in time moment $t = 400$ myr – (). The rate of star formation process $M_{\odot}pc^{-2}myr^{-1}$ at time moment $t = 400$ myr – (b).

[12] Matthias S. GRAPESPH: cosmological smoothed particle hydrodynamics simulations with the special-purpose hardware GRAPE // Monthly Notices of the Royal Astronomical Society. – 1996. – V. 278.

– P. 1005-1017.
[13] Springel V. The cosmological simulation code GADGET-2 // Monthly Notices of the Royal Astronomical Society. – 2005. – V. 364. – P. 1105-1134.
[14] Ziegler U. Self-gravitational adaptive mesh magnetohydrodynamics with the NIRVANA code // Astronomy & Astrophysics. – 2005. – V. 435. – P. 385-395.
[15] Mignone A., Plewa T., Bodo G. The Piecewise Parabolic Method for Multidimensional Relativistic Fluid Dynamics // The Astrophysical Journal. – 2005. – V. 160. – P. 199-219.
[16] Hayes J., Norman M., Fiedler R. et al. Simulating Radiating and Magnetized Flows in Multiple Dimensions with ZEUS-MP // The Astrophysical Journal Supplement Series. – 2006. – V. 165. – P. 188-228.
[17] O'Shea B., Bryan G., Bordner J., Norman M., Abel T., Harkness R., Kritsuk A. Adaptive Mesh Refinement - Theory and Applications // Lectures Notes of Computer Science Engineering. – 2005. – V. 41. – P. 341-350.
[18] Teyssier R. Cosmological hydrodynamics with adaptive mesh refinement. A new high resolution code called RAMSES // Astronomy & Astrophysics. – 2002. – V. 385. – P. 337-364.
[19] Kravtsov A., Klypin A., Hoffman Y. Constrained Simulations of the Real Universe. II. Observational Signatures of Intergalactic Gas in the Local Supercluster Region // The Astrophysical Journal. – 2002. – V. 571. – P. 563-575.
[20] Stone J. et al. Athena: A New Code for Astrophysical MHD // The Astrophysical Journal Supplement Series. – 2008. – V. 178. – P. 137-177.
[21] Brandenburg A., Dobler W. Hydromagnetic turbulence in computer simulations // Computer Physics Communications. – 2002. – V. 147. – P. 471-475.
[22] Gonzalez M., Audit E., Huynh P. HERACLES: a three-dimensional radiation hydrodynamics code // Astronomy & Astrophysics. – 2007. – V. 464. – P. 429-435.
[23] Krumholz M.R., Klein R.I., McKee C.F., Bolstad, J. Equations and Algorithms for Mixed-frame Flux-limited Diffusion Radiation Hydrodynamics // The Astrophysical Journal. – 2007. – V. 667. – P. 626-643.
[24] Mignone A. et al. PLUTO: a Numerical Code for Computational Astrophysics // The Astrophysical Journal Supplement Series. – 2007. – V. 170. – P. 228-242.
[25] Almgren A. et al. CASTRO: A New Compressible Astrophysical Solver. I. Hydrodynamics and Self-gravity // The Astrophysical Journal. – 2010. – V. 715. – P. 1221-1238.
[26] Schive H., Tsai Y., Chiueh T. GAMER: a GPU-accelerated Adaptive-Mesh-Refinement Code for Astrophysics // The Astrophysical Journal. – 2010. – V. 186. – P. 457-484.
[27] Murphy J., Burrows A. BETHE-Hydro: An Arbitrary Lagrangian-Eulerian Multidimensional Hydrodynamics Code for Astrophysical Simulations // The Astrophysical Journal Supplement Series. – 2008. – V. 179. – P. 209-241.
[28] Springel V. E pur si muove: Galilean-invariant cosmological hydrodynamical simulations on a moving mesh // Monthly Notices of the Royal Astronomical Society. – 2010. – V. 401. – P. 791-851.
[29] Bruenn S. et al. 2D and 3D core-collapse supernovae simulation results obtained with the CHIMERA code // Journal of Physics. – 2009. – V. 180. – P. 1-5.
[30] Hopkins P. A new class of accurate, mesh-free hydrodynamic simulation methods // Monthly Notices of the Royal Astronomical Society. – 2015. – V. 450, I. 1. – P. 53-110.
[31] Vshivkov V., Lazareva G., Snytnikov A., Kulikov I., Tutukov A. Hydrodynamical code for numerical simulation of the gas components of colliding galaxies // The Astrophysical Journal Supplement Series. – 2011. – V. 194, I. 47. – P. 1-12.
[32] Kulikov I.M., Chernykh I.G., Snytnikov A.V., Glinskiy B.M., Tutukov A.V. AstroPhi: A code for complex simulation of dynamics of astrophysical objects using hybrid supercomputers // Computer Physics Communications. – 2015. – V. 186. P. 71-80.
[33] Bergin E., Hartmann L., Raymond J., Ballesteros-Paredes J. Molecular Cloud Formation behind Shock Waves // The Astrophysical Journal. – 2004. – V. 612. – P. 921-939.
[34] Tielens A., Hollenbach D. Photodissociation regions. I - Basic model. II - A model for the Orion photodissociation region // The Astrophysical Journal. – 1985. – V. 291. – P. 722-754.
[35] Cazaux S., Tielens A. H_2 Formation on Grain Surfaces // The Astrophysical Journal. – 2004. – V. 604. – P. 222-237.
[36] Caselli P., Walmsley C., Terzieva R., Herbst E. The Ionization Fraction in Dense Cloud Cores // The Astrophysical Journal. – 1998. – V. 499. – P. 234-249.
[37] Bergin E., Plume R., Williams J., Myers P. The Ionization Fraction in Dense Molecular Gas. II. Massive Cores // The Astrophysical Journal. – 1999. – V. 512. – P. 724-739.
[38] Van Der Tak F., Van Dishoeck E. Limits on the cosmic-ray ionization rate toward massive young stars // Astronomy & Astrophysics. – 2000. – V. 358. – P. L79-L82.
[39] Draine B., Bertoldi F. Structure of Stationary Photodissociation Fronts // The Astrophysical Journal. –

1996. – V. 468. – P. 269-289.

[40] Draine B. Photoelectric heating of interstellar gas // The Astrophysical Journal Supplement Series. – 1978. – V. 36. – P. 595-619.

[41] Glover S., Mac Low M. Simulating the Formation of Molecular Clouds. I. Slow Formation by Gravitational Collapse from Static Initial Conditions // The Astrophysical Journal Supplement Series. – 2007. – V. 169, I. 2. – P. 239-268.

[42] Katz N., Weinberg D., Hernquist L. Cosmological simulations with TreeSPH // The Astrophysical Journal Supplement Series. – 1996. – V. 105. – P. 19-35.

[43] Springel V., Hernquist L. Cosmological smoothed particle hydrodynamics simulations: a hybrid multiphase model for star formation // Monthly Notices of the Royal Astronomical Society. – 2003. – V. 339, I. 2. – P. 289-311.

[44] Sutherland R., Dopita M. Cooling functions for low-density astrophysical plasmas // The Astrophysical Journal Supplement Series. – 1993. – V. 88. – P. 253-327.

[45] Popov M., Ustyugov S. Piecewise parabolic method on local stencil for gasdynamic simulations // Computational Mathematics and Mathematical Physics. – 2007. – V. 47, I. 12. – P. 1970-1989.

[46] Popov M., Ustyugov S. Piecewise parabolic method on a local stencil for ideal magnetohydrodynamics //Computational Mathematics and Mathematical Physics. – 2008. – V. 48, I. 3. – P. 477-499.

[47] Roe P. Approximate Riemann solvers, parameter vectors, and difference solvers // Journal of Computational Physics. – 1997. – V. 135. – P. 250-258.

[48] Collela P., Woodward P.R. The Piecewise Parabolic Method (PPM) Gas-Dynamical simulations // Journal of Computational Physics. – 1984. – V. 54. – P. 174-201.

[49] Kulikov I., Vorobyov E. Using the PPML approach for constructing a low-dissipation, operator-splitting scheme for numerical simulations of hydrodynamic flows // Journal of Computational Physics. – 2016. (submitted)

[50] Vshivkov V., Lazareva G., Snytnikov A., Kulikov I., Tutukov A. Computational methods for ill-posed problems of gravitational gasodynamics // Journal of Inverse and Ill-posed Problems. – 2011. – V. 19, I. 1. – P. 151-166.

[51] Godunov S., Kulikov I. Computation of Discontinuous Solutions of Fluid Dynamics Equations with Entropy Nondecrease Guarantee // Computational Mathematics and Mathematical Physics. – 2014. – V. 54, I. 6. – P. 1012-1024.

Journal of Physics: Conference Series **719** (2016) 012007 doi:10.1088/1742-6596/719/1/012007

The role of disk self-gravity on gap formation of the HL Tau proto-planetary disk

Shengtai Li and Hui Li

Theoretical Division, Los Alamos National Laboratory, Los Alamos, NM 87545, United States

E-mail: sli@lanl.gov

Abstract. We use extensive global hydrodynamic disk gas+dust simulations with embedded planets to model the dust ring and gap structures in the HL Tau protoplanetary disk observed with the Atacama Large Millimeter/Submillimeter Array (ALMA). Since the HL Tau is a relatively massive disk, we find the disk self-gravity (DSG) plays an important role in the gap formation induced by the planets. Our simulation results demonstrate that DSG is necessary in explaining of the dust ring and gap in HL Tau disk. The comparison of simulation results shows that the dust rings and gap structures are more evident when the fully 2D DSG (non-axisymmetric components are included) is used than if 1D axisymmetric DSG (only the axisymetric component is included) is used, or the disk self-gravity is not considered. We also find that the couple dust+gas+planet simulations are required because the gap and ring structure is different between dust and gas surface density.

1. Introduction

ALMA observations of HL Tau, a young ($\leq$ 1-2 Myr) star in Taurus, revealed bright and dark rings in the millimeter-wave continuum emission [1]. One interesting question is whether these structures are created by the embedded planets, hence yielding potentially important clues on planet formation in such disks.

Some observational features suggest however that the pattern of rings shown in HL Tau disk may be produced by planets. First, the spectral index derived from the ALMA images suggests that the dark rings are optical thin whereas the bright regions are optically thick, therefore the dark rings are real gaps in the dust distribution [1]. Second, the increase of the eccentricity of the rings at large orbital radii [1] is consistent with the fact that the orbital eccentricities of exoplanets increase with orbital radii [2, 3, 4]. Third, the dust size constrained by the polarized emission is around 150 μm [5], which means the structure of multiple rings should also exist in the gas disk because dust at this size should be well coupled with gas.

Models of disks that couple the dynamics of gas, dust and planets are crucial to interpret the observed patterns in the HL Tau system [6, 7, 8]. Various models suggest that the disk mass of HL Tau is $\sim$ 0.03-0.14 $M_\odot$ [9, 10, 11, 12], and the estimated stellar mass of HL Tau is 0.55 to 1.3 $M_\odot$ [13, 14, 15, 1]. The high disk mass might affect the disk stability and the resonance locations of potential planets [16]. A recent study by [8] used a two-fluid SPH code that includes gas+dust disk and embedded planets. Though they were able to reproduce the pattern of bright and dark rings, they adopted a disk mass of only 0.0002 $M_\odot$ within 120 AU and under predicted by about a factor 20 the observed millimeter flux density. The disk self-gravity (DSG) is not included in their simulations because it is not important with such a low disk mass.

We have run extensive global hydrodynamic disk gas+dust simulations with embedded planets, coupled with three dimensional radiative transfer calculations, to model the dust ring and gap structures in the HL Tau disk. We have performed quantitative fitting to the ALMA observations of HL Tau in terms of its millimeter flux density and spatial variations in [17]. In this paper, we use our 2D numerical simulations to demonstrate the important of the disk self-gravity in the gap formation if the proto-planet is present in HL Tau disk. We describe our approach and model set-up in Section 2 and our numerical method and code in Section 3. We then show our main results in Section 4 and discuss their implications in Section 5.

2. Numerical model and initial setup

We adopt a disk model from [11]. The disk surface density is described by

$$\Sigma_g(r) = \Sigma_0\left(\frac{r}{r_0}\right)^{0.23} \exp[-(r/r_c)^{2.23}] \ , \tag{1}$$

where $r_c = 79$ AU is a characteristic radius [18] and $r_0 = 10$ AU. The mass of the star is $M_* = 0.55M_\odot$. The locally isothermal sound speed is chosen as:

$$\left(\frac{c_s}{v_\phi}\right)(r) = 0.05\left(\frac{r}{r_0}\right)^{0.285} \ . \tag{2}$$

which implies the disk temperature varies as $T \propto T_0(r/r_0)^{-0.43}$. The whole disk is assumed to have a constant Shakura-Sunyaev viscosity parameter $\alpha = 10^{-3}$. The ratio of the scale height to the radius, $H/r(r_0) = 0.05$ (see Eq. (2), is adopted for gas after extensive experiments to fit the simulation data to the observation data. The value of H/r for dust particles is different for different particle sizes when we calculate the disk temperature and vertical distribution of the dust profile (see [17] for more details).

Although Kwon et al. [11] finds a disk mass of $0.135M_\odot$, we find that the disk with this mass quickly becomes gravitationally unstable under small perturbations. Hence we reduce the disk mass to be $M_{\rm disk} \approx 7.35 \times 10^{-2}\ M_\odot$, which leads to $\Sigma_0 = 4.83 \times 10^{-4}[M_*/(10\text{AU})^2] \approx 23.61$ g/cm^2. With these given parameters, we can estimate the Toomre's Q-parameter (see Fig. 1). Initially it has the minimum value of ~ 1.61, and we expect the disk self-gravity plays important role with this low Q-value.

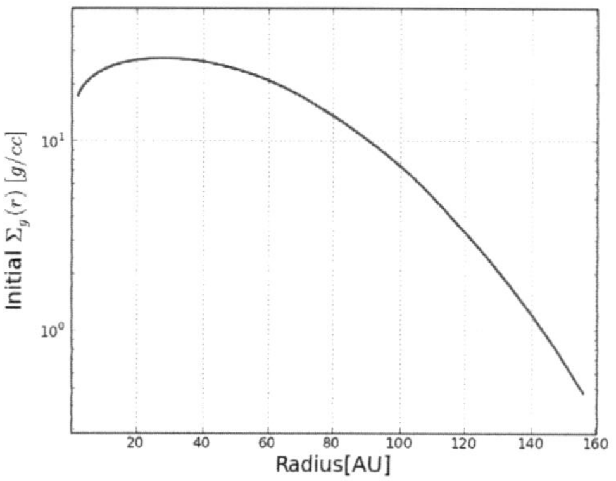

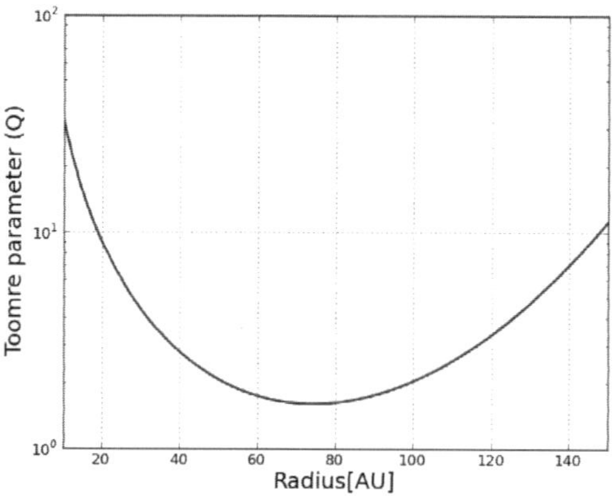

Figure 1. Initial disk surface density of the gas (Σ_g) and Toomre's Q-parameter.

We adopt dust grains with a size of 0.15 mm as best tracers of the dust emission at 1 mm. This is because single grain opacities were averaged on a grain size distribution to obtain the mean opacity used in the radiative transfer model, and the resulting dust opacity at the wavelength of 1 mm is dominated by dust grains with sizes between 0.1 and 0.2 mm.

ALMA Partnership et al. [1] identify seven pairs of dark and bright rings in the 1.0 mm image of the HL Tau disk, and among them four prominent dark rings could be considered as gaps, i.e., the dark rings D1, D2, and the adjoining D5 and D6. Based on a large body of previous literature on gap opening and formation as functions of disk gas temperature, planet mass, disk viscosity [19, 20], we have performed tens of disk+planet hydro simulations in order to determine the likely planet mass values that could roughly match the multiple gap widths and depths as shown in the ALMA observations. Based on these simulations, we adopt a set of planet mass parameters as $M_p \approx 0.35, 0.17$, and $0.26 M_{\rm Jup}$ at $13.1, 33.0$, and 68.6 AU, respectively. These three planets are fixed at their orbital locations, and planetary radial migration is turned off.

3. Numerical method

We solve the 2D disk model using LA-COMPASS code [21, 20, 22] to simulate the planet-disk interaction. The disk is treated as a 2D self-gravitating gas, whose motion is described by the Navier-Stokes equations, coupled to a similar (pressure-less) system of equations for the dust in a polar coordinate $\{r, \phi\}$ centered on the star. The dust particles are treated as pressure-less fluid. The two sets of equations for gas and dust particles are coupled via drag terms. A second-order high-resolution Godunov scheme is used to solve the equations numerically with exact Riemann solver for gas and dust. The planet motion is subject to the Newton's law with the gravitational force from the central star, other planets, and the disk material. The coupling between the disk and planet motions is carefully considered. For the small dust particles, we use a multiscale approach to accelerate the time integration. We also include the diffusion transport to the dust due to the turbulent motion of the gas. The momentum equations for the dust are modified to be consistent with the diffusion transport in the continuity equation. A fast disk self-gravity (DSG) solver is developed using the approach described in [23]. The indirect acceleration term due to the force exerted by the disk material on the star is also included together with the DSG solver without extra effort. The code is fully parallelized via message-passing interface coupled with OpenMP for multi-threads.

For numerical simulations, we have to impose the disk boundary $r \in [r_{\rm min}, r_{\rm max}]$. We adopt a value $r_{\rm min} = 2.4$AU from [11]. We use $r_{\rm max} = 156$AU. We have also tried other values of $r_{\rm min}$ and $r_{\rm max}$ and found that it does not make big difference in our simulations. A fixed boundary condition is used for gas at both inner and outer disk boundaries. For dust, we adopt an inflow (outflow) boundary condition at the inner (outer) boundary.

For the DSG calculation, we also include the disk material in the inner disk between 0.1AU and 2.4AU, and the outer disk between 156AU and 1560AU. Both the inner disk and outer disk are assumed to be fixed as the initial setup during the whole simulations. This approach avoids the sharp variation in the DSG near the $r_{\rm min}$ and $r_{\rm max}$.

For comparison, we also include the numerical results of 1D axisymmetric DSG solver [24].

4. Numerical result

The disk is discretized by 3072×768 grid in (r, ϕ) domain. We run the simulation up to 7000 orbits (measured at $r_0 = 10$AU). We find the disk surface variation becomes very small at later stage.

Fig. 2 shows the disk surface density of the gas component (Σ_g) between different disk self-gravity setup and at different times. We compare three different simulations: with fully 2D DSG, with only 1D axis-symmetric DSG, and without DSG (i.e., disk self-gravity is not considered). We plot the variation of Σ_g from 1000 orbits to 4000 orbits. It is clear that the one-dimensional axis-symmetric DSG (DSG1D in the plot) generates similar results as the case without DSG. However the fully 2D DSG generate a clearer and deeper gap at later times than the case without

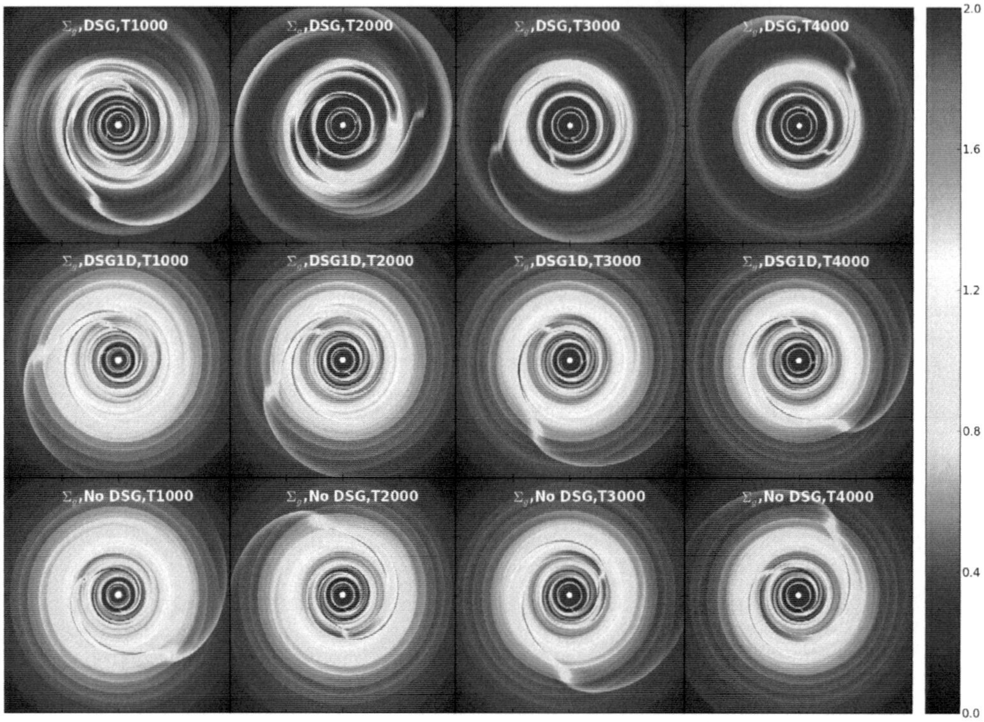

Figure 2. Disk surface density of the gas (Σ_g) at different times and between different disk self-gravity (DSG) setup. The time is represented by the orbits at 10AU radius.

DSG. This shows that the non-axisymmetric components of the disk self-gravity is important even though the disk looks axisymmetric.

An interesting feature shown in the Fig. 2 is the case at T=2000 orbits and with DSG turned on. The disk becomes slightly eccentric at this time shown by the two high density blobs near the edge of the second gap. This eccentricity eventually gets damped with the time evolution and totally disappear when the time reaches T=4000 orbits. We notice that the eccentric disk is caused mainly by both the indirect acceleration term due to the force exerted by the disk material on the star and the disk self-gravity. If we turn off this indirect acceleration term or turn off the DSG, the disk will not become eccentric. We remark that this is not purely a numerical issue as we have verified it using FARGO code and obtained similar eccentric disk. More results on this feature will be reported in the future.

Fig. 3 shows the disk surface density for the dust components. It is clear the gap become more outstanding in the dust surface density than in the gas surface density if the DSG is used. If no DSG or 1D-DSG is used, the outer ring and gap is barely visible.This shows the importance of the DSG in the gap formation, especially for the outer gap and dust ring.

Comparing Fig. 3 and Fig. 2, we find the axis-symmetric structure of the gap and dust ring is more evident in the dust surface plots than the gas surface plots. Also the spiral arms in the gas plots disappear in the dust plots. This can partially explain why the spiral arm is not visible in the observation images of ALMA.

Fig. 4 shows the azimuthal-average quantity at t = 4000 orbits. It is clear that the gap is much deeper and the gap edge is much higher if the DSG is used. We also plot the dust-to-gas ratio Σ_d/Σ_g. Initially, $\Sigma_d/\Sigma_g = 0.01$. Most simulation models that do not include dust dynamics

Journal of Physics: Conference Series **719** (2016) 012007 doi:10.1088/1742-6596/719/1/012007

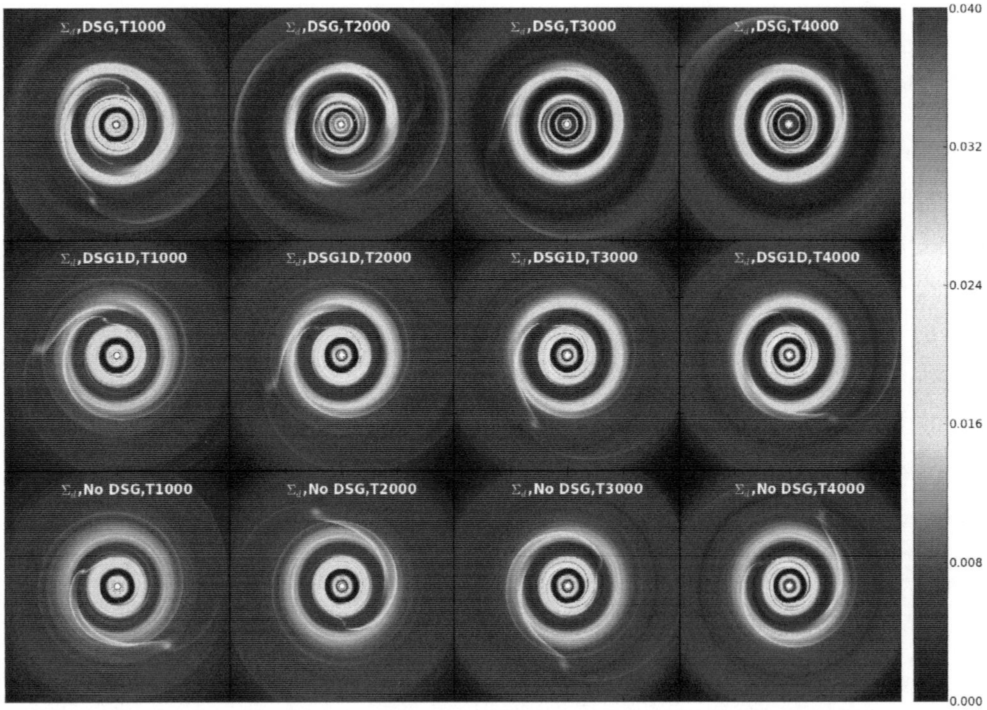

Figure 3. Disk surface density of the dust component (Σ_d) at different times and between different disk self-gravity (DSG) setup. The time is represented by the orbits at 10AU radius.

assume this ratio remains fix during the whole simulation. However, the right plot of the Fig. 4 shows this ratio can be changed dramatically in the gap region: near the bottom of the gap, $\Sigma_d/\Sigma_g \approx 0.005$; at the gap edge, $\Sigma_d/\Sigma_g \approx 0.02$; a factor of 4 variation.

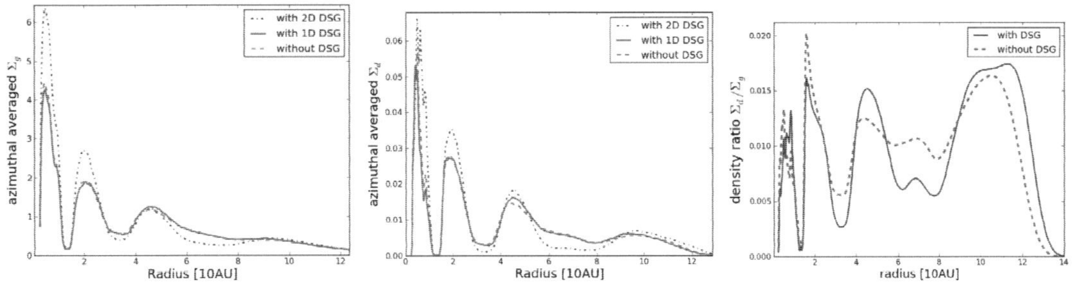

Figure 4. Azimuthal-averaged surface density for gas and dust at time T=4000 orbits. The right plot shows the azimuthal averaged dust-to-gas ratio (Σ_d/Σ_g)at T=4000 orbits

5. Conclusion

We have proposed a model to simulate the observed feature in the HL Tau disk. Our simulations show that the disk self-gravity (DSG) plays an important role in the gap formation. Without DSG, the outer gap and dust rings is barely visible and cannot be fit with the observed data.

We also show that the simplified axis-symmetric 1D DSG does not help even if it is included in the simulations.

Our simulations also show that we cannot simply use the results of gas surface density to derive the dust distribution as conventional approach does. The dust-to-gas ratio can be changed dramatically near the gap and cannot be kept as constant. The fully coupled dust+gas+planet simulation is necessary.

Acknowledgments
This research was performed under the auspices of the Department of Energy. It was supported by the Laboratory Directed Research and Development (LDRD) Program at Los Alamos.

References
[1] Partnership A, Brogan C L, Perez L and et al 2015 *ApJL* **808** L3
[2] Butler R, Wright J, Marcy G and et al 2006 *ApJ* **646** 505
[3] Shen Y and Turner E L 2008 *ApJ* **685** 553
[4] Zhang X, Li H, Li S and Lin D 2014 *ApJL* **789** L23
[5] Kataoka A, Muto T, Momose M, Tsukagoshi T and Dullemond C 2015 arXiv:1507.08902
[6] Dong R, Zhu Z and Whitney B 2014 arXiv:1411.6063
[7] Zhu Z, Stone J and Bai X 2015 *ApJ* **801** 81
[8] Dipierro G, Price D, Laibe G and et al 2015 *MNRAS* **453** L73–L77
[9] Robitaille T, Whitney B A, Indebetouw R and Wood K 2007 *ApJS* **169** 328
[10] Guilloteau S, Dutrey A, Piétu V and Boehler Y 2011 *A&A* **529** A105
[11] kwon W, Looney L and Mundy L 2011 *ApJ* **741** 3
[12] kwon W, Looney L, Mundy L and Welch W 2015 *ApJ* **808** 102
[13] Sargent A I and Beckwith S V W 1991 *ApJL* **382** L31
[14] Close L M, Roddier F, Hora J L and et al 1997 *ApJ* **489** 210
[15] White R and Hillenbrand L 2004 *ApJ* **616** 998
[16] Tamayo D, Triaud A, Menou K and Rein H 2015 *ApJ* **805** 100
[17] Jin S, Li S, Li H and et al 2016 *ApJ* **818** 76
[18] Andrews S M and et al 2009 *ApJ* **700** 1502
[19] Crida A, Morbidelli A and Masset F 2006 *ICARUS* **181** 587
[20] Li H, Li S, Lubow S H and Lin D 2009 *ApJL* **690** L52–L55
[21] Li H and et al 2005 *ApJ* **624** 1003
[22] Fu W, Li H, and S Li S L and Liang E 2014 *ApJL* **795** L39
[23] Li S, Buoni M and Li H 2009 *ApJ* **181** 244–254
[24] Huré J M 2005 *A&A* **434** 17–23

Numerical simulations of the jetted tidal disruption event Swift J1644+57

Petar Mimica

E-mail: petar.mimica@uv.es
Departamento de Astronomía y Astrofísica, Universidad de Valencia, 46100, Burjassot, Spain

Miguel A. Aloy

Departamento de Astronomía y Astrofísica, Universidad de Valencia, 46100, Burjassot, Spain

Dimitrios Giannios

Department of Physics and Astronomy, Purdue University, 525 Northwestern Avenue, West Lafayette, IN 47907, USA

Brian D. Metzger

Columbia Astrophysics Laboratory, Columbia University, New York, NY, 10027, USA

Abstract. In this work we focus on the technical details of the numerical simulations of the non-thermal transient Swift J1644+57, whose emission is probably produced by a two-component jet powered by a tidal disruption event. In this context we provide details of the coupling between the relativistic hydrodynamic simulations and the radiative transfer code. First, we consider the technical demands of one-dimensional simulations of a fast relativistic jet, and show to what extent (for the same physical parameters of the model) do the computed light curves depend on the numerical parameters of the different codes employed. In the second part we explain the difficulties of computing light curves from axisymmetric two dimensonal simulations and discuss a procedure that yields an acceptable tradeoff between the computational cost and the quality of the results.

1. Introduction

Tidal disruption events (TDE) of stars by massive black holes [1, 2, 3] provide one of the ways to detect an inactive massive black hole in the center of a galaxy. Stars whose orbit brings them too close to a black hole can be disrupted and a large fraction of their mass can be accreted. This sudden accretion event is expected to produce a luminous UV and X-ray TDE flare [4, 5]. It has recently been argued that TDEs can also produce jets [6, 7, 8]. These jets are expected to be detectable due to their non-thermal emission, and in radio they are likely to appear as transients that peak on the timescales of about a year [6, 9] after the tidal disruption.

Swift J1644+57 (SwJ1644 in the rest of the article) is a non-thermal transient detected on March 28th. 2011 [10, 11, 12]. It is thought to be prototype of a jetted TDE (jTDE). Its radio emission [13, 14, 15] is long-lived and complex. One of the first attempts to explain the radio emission of Swift J1644+57 using one-dimensional models and relativistic hydrodynamic simulations[16] assumed that the emission is coming from the shocks that form when the jet interacts with the circumnuclear medium (CNM). This model is more analogous to a GRB afterglow than to an AGN[1]. While the model was successful in explaining the first few weeks of SwJ1644 radio emission, it turns out that the long-term emission is not easily accounted for by a radio afterglow from a single ultrarelativistic, narrow jet. The reason for this is that the source rebrightened after $\gtrsim$150 days, the time at which the afterglow model predicted a steady decline[14, Figure 1]. The late time rebrightening implies that additional energy is injected in the external shock months after the burst[14]. This may happen if, for instance, the jTDE is accompanied by a powerful mildly relativistic component[18][2].

In this contribution we focus on the numerical and technical aspects of jTDE modeling. In Section 2 we describe the numerical codes used in the calculations and how they depend on each other. The Section 3 gives an overview of computational requirements and their consequences on the kind of models that we were able to produce. The discussion and summary is given in the Section 4.

2. Numerical codes

SwJ1644 is a distant object (its redsfhit is $z = 0.354$ [10]), and is thus too small to be resolved in great detail[3]. Thus, the only direct observations that we have are the total fluxes for different times and frequencies (i.e., multiwavelength light curves). Nevertheless, when simulating this source we need to construct a detailed, resolved model of the SwJ1644 jet, simulate its dynamics, compute its resolved emission, and only in the final stage integrate the resolved emission to compute the total flux to be compared to observations. In order to achieve this we use two numerical codes so that the output of the first code (*MRGENESIS*) is used as an input to the second one (*SPEV*). The former simulates the jet dynamics, while the latter computes its emission.

2.1. Relativistic hydrodynamics simulations

MRGENESIS is a code developed to solve the equations of relativistic numerical (magneto)hydrodynamics *in the laboratory frame (i.e., a frame at rest with respect to the* distant source*)*[4]. The code is massively parallelized using MPI[5] and OpenMP[6] librararies, and uses the parallel HDF5[7] library for file input and output. We employ a method of lines for our simulations in which the spatial reconstruction is a based on the third order PPM reconstruction routines[22], and the time update is performed by means of a third order TVD Runge-Kutta scheme[23]. Numerical fluxes are evaluated with the Marquina flux formula[24]. We simulate SwJ1644 in two dimensions assuming axisymmetry and discretize the computational domain using spherical coordinates. The black hole is assumed to be located at the origin and the TDE occurs at scales that are much smaller than the ones we consider in our work (we simulate neither the TDE nor the jet launching). We set the beginning of the numerical grid at $\sim 10^{16}$ cm. The jet is

[1] Interested readers can find details of the model based on the AGN jet-disc symbiosis in Section 2.1 of [17]. We note that [17] assume that the origin of the emission is *internal* to the jet, as opposed to the *external* emission that we advocate.

[2] For an overview of alternative explanations for late-time rebrightening see Section 2.2 of [18].

[3] According to our numerical models we predict that SwJ1644 may become resolvable by Very Long Baseline Inteferometry sometime in the next few years (see Figure 13 in [18]).

[4] Interested readers can find details about the original *GENESIS* in [19], its extension to relativistic magnetohydrodynamics in [20], and the current parallelized version (*MRGENESIS*) that handles ultrarelativistic magnetized outflows in [21].

[5] MPI (Message Passing Interface) is used for distributed-memory parallelism, see http://www.mpi-forum.org

[6] OpenMP (Open Multi-Processing) is used for shared-memory parallelism, see http://openmp.org.

[7] HDF5 (Hierarchical Data Format) is used to store and manage large datasets, see http://hdfgroup.org.

injected through the inner boundary in an angular interval $[0, \theta_j]$, assuming a constant Lorentz factor Γ_j, and a luminosity whose time dependence is $L_j(t) = L_{j,0} \max[1, (t/t_0)]^{-5/3}$. t_0 is the time during which the jet luminosity is constant, and the peak lumiosity $L_{j,0}$ is computed from the total jTDE energy $E_{\rm ISO}$ through $L_{j,0} = 2E_{\rm ISO}/(5t_0)$. The jet interacts with a CNM that is assumed to be at rest with a density profile $\rho_{\rm cnm} = n_{18} m_p (r/10^{18}{\rm cm})^{-k}$, where m_p is the proton mass, n_{18} is the particle number density at 10^{18} cm and k is the power-law index of the density profile. We assume that initially both the jet and the external medium are cold. During the simulation we periodically store the snapshots of the computational grid in order to obtain a sufficient coverage of the jet-CNM interaction to be able to compute the resulting emission with sufficient precision (see Section 3).

2.2. Radiative transfer

Once the relativistic hydrodynamics simulation has finished, the snapshots it produced are used by the code *SPEV* to compute the emission. The spatial distribution of the hydrodynamics variables needs to be stored with sufficient frequency because of the space-time transformations that need to be performed to compute the observable radiation in the (distant) *observer frame*. *SPEV*[25] is designed to process an arbitrary number of pre-computed sets of hydrodynamic states of the fluid (i.e., density, pressure, positions, velocities, magnetic field) and to compute the resulting non-thermal emission. The procedure can be summarized as follows:

(i) **Injection**: *SPEV* uses a set of Lagrangian particles (LPs) as a representation of the emitting volumes. Each LP has a defined position, shape and velocity. Attached to it there is a representation of the non-thermal electron (NTE) energy distribution. In *SPEV*, new LPs are created each time a snapshot is processed. The position of the shock fronts that form as a consequence of the jet-CNM interaction is detected, and LPs are injected into the shocked fluid immediately behind the front. Since the actual process of particle acceleration occurs on temporal and spatial scales far below the ones we are able to simulate (see e.g., [26] for a recent review), we use a phenomenological model to link the fluid state with NTE distribution. The model we use is a standard one [27]: the energy contained in the NTEs and in the stochastic magnetic field are assumed to be fractions ϵ_e and ϵ_B of the shocked fluid thermal energy. The injection spectrum is assumed to be a power-law in energy $n(\gamma) \propto \gamma^{-p}$, where p is the power-law index and γ is the NTE Lorentz factor. We determine the normalization and energy cutoffs of the distribution using a numerical procedure described in detail in Sections 3.2 and 3.3 of [28].

(ii) **Transport, evolution and emission**: As it processes the snapshots, *SPEV* transports the injected LPs, assuming that after they have been injected they move with the fluid. An exception are the newly injected LPs: between the snapshot in which the particle is injected and the subsequent one the two edges parallel to the shock front are assumed to move with different velocities (one with the velocity of the shocked fluid and the other one with the faster velocity of the shock front). This ensures that new LPs correctly represent the increase in the volume due to newly shocked fluid, regardless of the numerical resolution or the frequency with which snapshots are stored. Simultaneously, the NTE energy losses due to synchrotron radiation, as well as the energy gains(losses) due to adiabatic compression(expansion) are taken into account:

$$\frac{{\rm d}\gamma}{{\rm d}t} = \frac{1}{3}\frac{{\rm d}\ln\rho}{dt}\gamma - \frac{4}{3}\frac{c\sigma_T}{m_e c^2}\left(u_B + u_{\rm rad}\right)\gamma^2 \qquad (1)$$

where t, ρ, u_B and $u_{\rm rad}$ are the time, fluid density, magnetic field energy density and external radiation field energy density (all measured in the LP frame). Section 3.2 of [25] describes an efficient numerical algorithm for solving Eq. 1 for each NTE distribution in each LP.

At the same time that the particles are evolved, their non-thermal emission is computed (see Section 4 of [25] for a description of an efficient numerical algorithm for computing synchrotron emission and absorption coefficients, and Section A.2.1 of the same paper for a description of the way particles in two-dimensional axisymmetric models can be used to represent three-dimensional emitting volumes). The coefficients computed in this step are stored in a global array that is used in the final step.

(iii) **Solving the radiative transfer equation**: After all the snapshots have been processed and the emission from all LPs has been computed for all times, the time- and frequency-dependent emission can be computed. We assume that there is a virtual screen ("virtual detector", VD) located in front of the source, and the goal is to compute the intensity in each of the virtual pixels of the VD. This is done by solving the radiative transfer equation

$$\frac{\mathrm{d}I_\nu}{\mathrm{d}s} = j_\nu - \alpha_\nu I_\nu, , \tag{2}$$

where I_ν, j_ν and α_ν are the specific intensity, emissivity and absorption coefficient. s is the distance along the line of sight from the jet to a particular pixel and it relates the simulation time t to the time of observation T by $t = T + (D - s)/c$, where D is the distance from the VD to the source. In other words, for a given T and t, s is the distance of a LP from which radiation is observed at time T if that LP is emitting at a time t. We note that Eq. 2 needs to be solved in such a way that s is monotonously increasing from the jet towards the VD. Therefore, after all the snapshots have been processed, the emission and absorption coefficients need to be sorted according to their s and then Eq. 2 is solved for each pixel.

In practice the positions of LPs in steps (i) and (ii) are stored in a set of (large) intermediate files ("preprocessor" files). A postprocessor executes the radiation part of step (ii) by reading the preprocessor files and storing the coefficients in memory. After all preprocessor files have been postprocessed, the step (iii) is executed. Both preprocessor and postprocessor parts of *SPEV* are paralellized using OpenMP. The postprocessor, in particular, needs to be run on a machine with enough shared memory[8] because all information produced in step 2 needs to be held in memory for the step 3 to correctly compute the intensities.

3. Simulations of Swift J1644+57 jet

As mentioned in Section 2.1, we model SwJ1644 jet in two-dimensions. However, as will be explained in this Section, the total cost in terms of computing time, disk storage and working memory of each two-dimensional simulation is high. Therefore, we first perform a number of much less costly tests using a one-dimensional approximation to two-dimensional models. The goal of those tests is to determine the values of a number of the parameters of the numerical simulation for which the results are satisfactory. Afterward, the two-dimensional simulations are performed using the values determined in one-dimensional tests.

3.1. One-dimensional tests

Using *MRGENESIS* we simulate one-dimensional spherically symmetric jets. For the purpose of computing their emission with *SPEV* we assume that their opening angles are θ_j. We fix all the physical parameters of the jet ($E_{\rm ISO} = 4 \times 10^{54}$ erg, $\Gamma_j = 10$, $\theta_j = 0.1$ rad, $t_0 = 5 \times 10^5$ s) and of the external medium ($n_{18} = 60$ cm^{-3}, $k = 1.5$). This corresponds to the inner relativistic component of the two-component jet in [18]. The numerical parameters that will be varied are[9]:

[8] Interested readers should consult Chapter 4 of [29] for a detailed description of the optimized memory management in SPEV.

[9] We note that the results of a limited number of numerical tests have also been presented in Appendix A of [18], but here we perform an exhaustive analaysis.

- $\Delta T_{\rm snap}$: time interval between snapshots (*MRGENESIS*)
- Δr: radial resolution (*MRGENESIS*)
- $\Delta\theta$: angular resolution (*MRGENESIS*[10])
- N_b: number of NTE energy distribution bins (*SPEV*)
- n_x: resolution of the virtual detector (*SPEV*[11])

We choose as a default set of parameters $\Delta T_{\rm snap} = 8.3\times10^4$ s, $\Delta r = 5\times10^{13}$ cm, $\Delta\theta = 2\times10^{-3}$ rad, $N_b = 32$, $n_x = 1000$. We follow the jet evolution until $\sim 6.7\times10^8$ seconds. During the first 1.3×10^8 seconds the snapshots are output every $\Delta T_{\rm snap}$, and afterward the time interval increases by a factor 1.002 after every snapshot is output.

Fig. 3.1 shows the light curves for the default model evaluated at 1.4 GHz (red line) and the consequence of variations in the resolution of the hydrodynamic models, i.e., as a result of changing Δr. As can be seen, relatively small differences of the order of a percent are visible at early times and after the light curve maximum. The top right panel in Fig. 3.1 shows the result of changing angular resolution. Here the differences are somewhat larger, especially when comparing the lowest and highest resolution ($\sim 4\%$ at peak). The middle left panel in Fig. 3.1 shows how the results depend on the number of bins in which NTE energy distribution of each LP is discretized. We see that the convergence is only achieved for $N_b \gtrsim 32$. In the middle right panel we can see that the oscillations at early times are due to poor VD resolution. The angular size of the jet is very small at those times and it is only properly resolved for $n_x \sim 10^4$ for a linear pixel distribution (covering the interval $[0,\ 5\times10^{18}$ cm$]$), and $n_x \sim 10^3$ for a logarithmic one (covering the interval $[10^{12}$ cm$,\ 5\times10^{18}$ cm$]$). In the next section we will explain how we solved the problem of noise for 2D simulation light curves, especially the off-axis ones where it is not feasible to distribute the VD pixels logarithmically. Finally, the bottom left panel shows the (weak) dependence on $\Delta T_{\rm snap}$.

3.2. *Two-dimensional simulations*

In the previous section we showed how the light curves change as the five principal numerical parameters are changed. Most importantly for the 2D simulations, with the previous tests we have already obtained a very good proxy for the largest values of Δr and $\Delta\theta$ we can use in order to reduce the computational cost without compromising the accuracy of the results. In the final simulations showed in [18] we used $\Delta r = 4\times10^{14}$ cm and $\Delta\theta \simeq 2.6\times10^{-3}$ rad. In this section we discuss the technical challenges of computing the emission even from these relatively low-resolution hydrodynamic simulations.

The complete two-dimensional evolution is stored in 2412 snapshots. It was fortunate that HDF5 format permits data compression so that the compressed snapshots occupy $\simeq 460$ Gbytes. Running the preprocessor with the default parameters of Section 3.1 produces 160 (compressed) intermediate files. Each of the first 159 files contains the positions, geometry, velocity, magnetic field and the 32 bins of the NT energy distibution of 2.5×10^6 LPs, while the last file only contains $\simeq 1.9\times10^6$ LPs. This means that in total there are $\simeq 4\times10^8$ two-dimensional emitting LPs in step (ii) of the *SPEV* algorithm (see Section 2.2). Since each of the LPs has to be converted to a three-dimensional emitting volume, it may contribute to multiple VD pixels and also to multiple observing times, we soon realized that the realistic upper limit of $\lesssim$1 TByte of RAM on our

[10] When performing one-dimensional simulations we create mock two-dimensional snapshots using $\Delta\theta$ as angular cell size.

[11] In these one-dimensional tests we assume that the jet is observed on-axis (viewing angle is 0 degrees), so that its emission is axially-symmetric and we only need to compute the emission along VD x-axis. For off-axis observations a (much more costly) two-dimensional VD calculation is performed [18, Section 5.2]. We comment in detail on technical challenges of 2D simulations in next section.

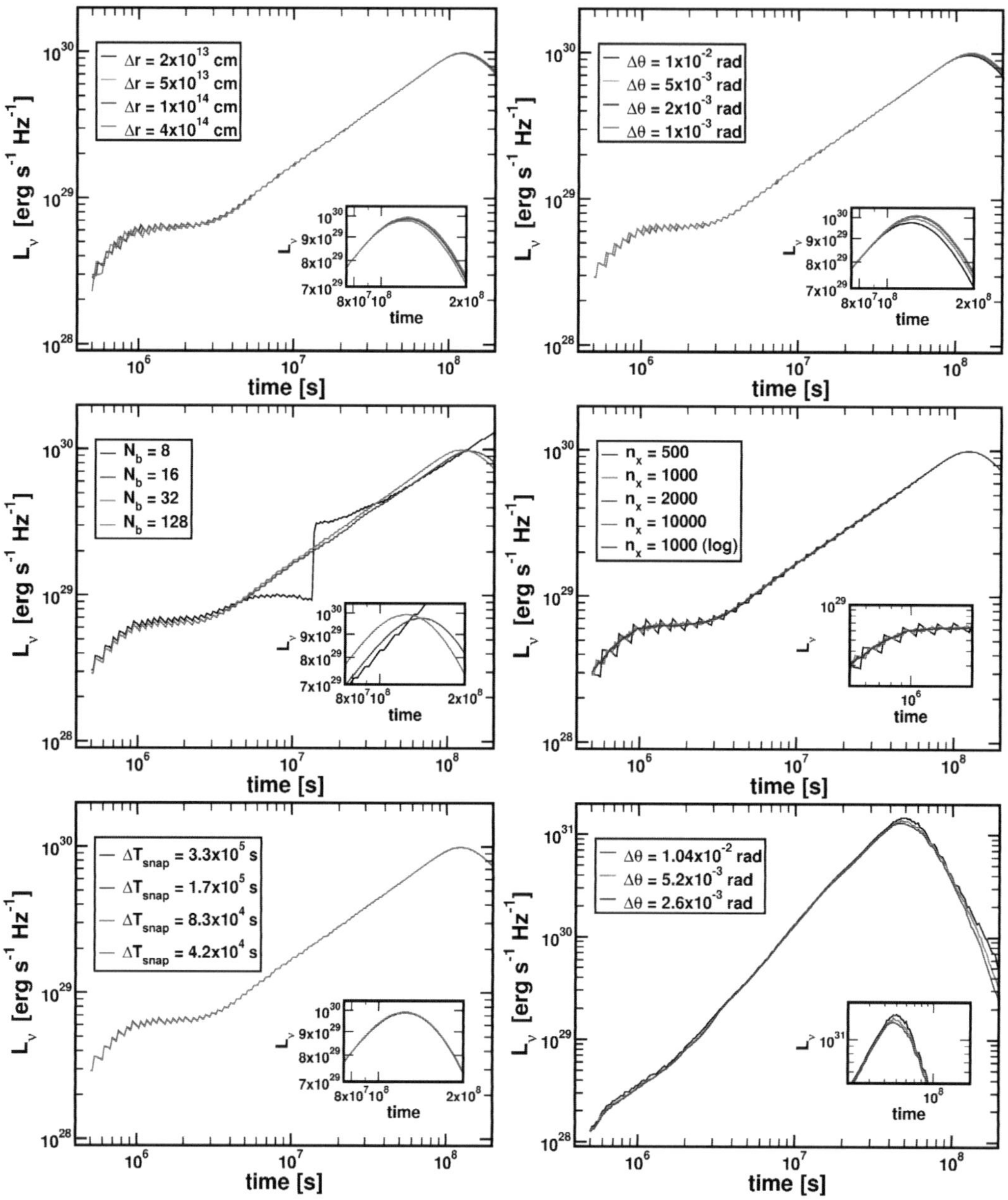

Figure 1. 1.4 GHz radio light curves of one-dimensional tests (Section 3.1). The default model is shown in red. The insets show the zoom-in around the light curve maximum. *Top left panel*: Variations in Δr. *Top right panel*: Variations in $\Delta\theta$. *Middle left panel*: Variations in the number of bins N_b. Note that $N_b = 32$ and $N_b = 128$ curves overlap. *Middle right panel*: Variations in the number of VD pixels. Note that the curves for $n_x = 10^4$ and $n_x = 10^3$ with a logarithmic distribution of pixels overlap. *Bottom left panel:* Variations in $\Delta T_{\rm snap}$. *Bottom right panel:* Emission from two dimensional two-component jet for different effective angular resolutions (see Section 3.2).

shared memory machine[12] would not be enough to compute off-axis images at full resolution. A remedy to this problem was to artificially reduce the angular resolution when preprocessing the hydrodynamic snapshots by a factor of 2 or 4 (this would roughly correspond going from the second-best to third- of fourth-best angular resolution in top left panel in Figure 3.1.). This reduces the number of emitting volumes so that even off-axis radiative transfer calculations could be performed.

3.2.1. On-axis light curves For the purposes of the numerical tests we computed the on-axis light curves (in complete analogy to the tests presented in Section 3.1). Bottom right panel in Figure 3.1 shows on-axis light curves using the full angular resolution, as well as the result of degrading it by factors of 2 and 4.

In Figure 2 we show a subset of the data that *SPEV* stores when postprocessing the intermediate files. In this particular case we show all the contributions to the emission at $T = 1.4 \times 10^6$ seconds and at 15.4 GHz. The data shown originates from LPs emitting at different times from different positions (the lower the value of y the earlier the LP had to emit for it to be observed at 1.4×10^6 seconds). In other words, what is shown does not correspond to one particular snapshot of the jet evolution. Nevertheless, we know that the edge of the emitting region corresponds to the position of the bow shock of the jet. We also see in the bottom right panel in Figure 2 that a small region behind the shock is optically thin and most of the contribution to the emission comes from this region and from the surface at which the emission becomes optically thin (the transition between light and dark blue regions). Using this type of analysis it is possible to understand, among other things, why the inverse Compton cooling of the electrons is not very efficient at low frequencies (though it may be important for higher frequency observations of SwJ1644, see [30]): the electrons cool after being accelerated by the shock, but since they are advected behind it, once they enter the opaque region (dark blue), they stop contributing to the observed emission. Therefore the observer does not see a large effect of the inverse Compton cooling [18, Section 4.1.2].

3.2.2. Off-axis light curves Here we comment on the procedure we use to reduce the numerical noise that shows at early times (Figure 3.1). As was commented in Section 3.1, the jet is small at early times, so that at a typical resolution of 1000 points per spatial direction it is very poorly resolved by the VD. Due to memory limitations we were not able to use VD resolution of 10000^2. However, the resolutions of 1000^2 pixels were feasible. Therefore, we adopted the following procedure:

(i) Pre-determine the jet viewing angle $\theta_{\rm obs}$ and the times T_i ($i = 1, .. n_t$) at which it is to be observed.
(ii) Compute the radio images using 1000×1000 VD for each T_i in a single calculation.
(iii) For each T_i, determine the rectangle that contains 99.99% of the emission. Then expand the rectangle by a safety margin.
(iv) For each T_i, independently re-compute the radio image by limiting the VD to the rectangle computed in the previous step, but using the resolution 1000×1000.

The above procedure enables us to compute the emission from early times without numerical artifacts[13]. The tradeoff is that we need to postprocess the intermediate files multiple times: once

[12] We use the machine *LluisVives* hosted at the University of Valencia. It has 30 Xeon 7500 hexacore CPUs clocked at 2.67 GHz and almost 1 TBytes of RAM. For more information see http://www.uv.es/siuv/cas/zcalculo/calculouv/des_vives.wiki (in Spanish).
[13] Figures 12-14 in [18] were computed using this procedure. Note that in Figure 14 in [18] the jet size in the upper panels is $\simeq 3-5$ times smaller than in the lower panels, but all are computed using the same VD resolution.

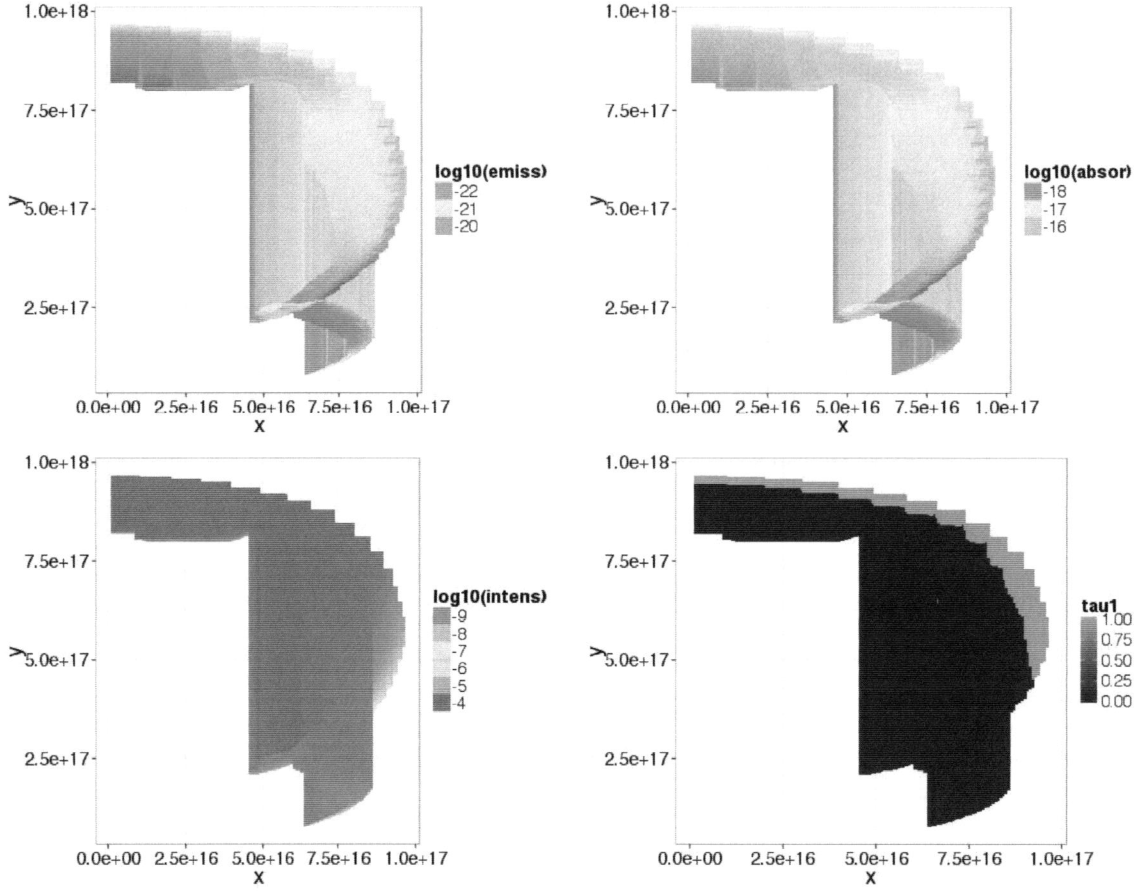

Figure 2. Observer frame simultaneous emission, absorption, intensity and optical depth that contribute to the emission at $T = 1.4 \times 10^6$ seconds at 15.4 GHz. *Top left*: synchrotron emissivity in erg s^{-1} cm^{-3} Hz^{-1}. *Top right*: synchrotron self-absorption in cm^{-1}. *Bottom left*: specific intensity in erg s^{-1} cm^{-2} Hz^{-1}. The intensity is the result of integrating Equation 2 from bottom to top for each vertical line. *Bottom right*: optical depth measured from the observer (integral of the absorption from top to bottom along vertical lines). The light blue colored region denotes the part of the emitting volume that is transparent, while the emission in the dark blue region is absorbed. As explained in Section 3.2.1, the information displayed here is a subset of the data stored in the global array used by *SPEV* to compute the intensity in each pixel (see Section 2.2). The luminosity is obtained by integrating the intensity along vertical lines in the top right panel (taking into account that the emission is axially symmetric).

in step (ii) and then additional n_t times in step (iv). The computing time does not necessarily scale as n_t since, for each T_i, the emission and absorption coefficients are only computed from those emitting volumes that contribute to the emission, while the rest (typically corresponding to most of the volume of the model) are ignored. This is made efficient by indexing the intermediate files. For a given observing angle the calculations at 8 different observing frequencies last $\simeq 3-5$ days.

4. Summary

In this contribution we have discussed the technical details of the one and two dimensional simulations of the SwJ1644 jet. In Section 2 we explain in detail how the relativistic hydrodynamic simulations are coupled to the radiative transfer code to produce the synthetic observations (light curves) that can be directly compared to the observations. In Section 3.1 we show that many one-dimensional tests can be used to probe the sensitivity of the computed light curves on the resolution of the hydrodynamic simulations, on the frequency of stored snapshots, on the resolution of the discretization in energy space and of the virtual detector. The results obtained helped to determine the resolution of the two dimensional simulation so that its computational cost is acceptable, but also so that the final light curves are as free of numerical artifacts as possible. This procedure has been successful in enabling us to produce a viable two-component jet model of the SwJ1644 jet.

Acknowledgements

PM and MAA acknowledge the support from the European Research Council (grant CAMAP-259276), and the partial support of grants AYA2013-40979-P, PROMETEO-II-2014-069 and SAF2013-49284-EXP . DG acknowledges support from the NASA grant NNX13AP13G. BDM acknowledges support from the NSF grant AST-1410950 and the Alfred P. Sloan Foundation. We thankfully acknowledge the computer resources, technical expertise and assistance provided by the "Centre de Càlcul de la Universitat de València" through the use of *Lluís Vives* cluster and *Tirant*, the local node of the Spanish Supercomputation Network. We use the open source software packages *Grace*[14] and *ggplot2*[15] to produce the graphics in this paper.

References

[1] Lacy J H, Townes C H and Hollenbach D J *ApJ* **262, 120**
[2] Rees M J 1988 *Nature* **333**, 523
[3] Evans C R and Kochanek C S 1989 *ApJ* **346**, 13
[4] Strubbe L E and Quataert E 2009 *MNRAS* **400**, 2070
[5] Lodato G and Rossi E M 2011 *MNRAS* **410** 359
[6] Giannios D and Metzger B D 2011 *MNRAS* **416**, 20102
[7] Krolik J H and Piran T 2012 *ApJ* **749**, 92
[8] Tchekhovskoy A, Metzger B D, Giannios D and Kelley L Z 2014 *MNRAS* **437**, 2744
[9] Metzger B D, Williams P K G and Berger E 2015 *ApJ* **806**, 224
[10] Bloom J S *et al.* 2011 *Science* **333**, 203
[11] Burrows D N *et al.* 2011 *Nature* **476**, 421
[12] Levan A J *et al.* 2011 *Science* **333**, 109
[13] Zauderer B A *et al.* 2011 *Nature* **476**, 425
[14] Berger E *et al.* 2012 *ApJ* **748**, 36
[15] Zauderer B A *et al.* 2013 *ApJ* **767**, 152
[16] Metzger B D, Giannios D and Mimica P 2012 *MNRAS* **420**, 3528
[17] van Velzen S, Körding E and Falcke H 2011 *MNRAS* **417**, L51
[18] Mimica P, Giannios D, Metzger B D and Aloy M A 2015 *MNRAS* **450**, 2824
[19] Aloy M A, Ibáñez J M, Martí J M and Müller E 1999 *ApJS* **122**, 151

[14] http://plasma-gate.weizmann.ac.il/Grace
[15] http://ggplot2.org

[20] Leismann T *et al.* 2005 *A&A* **436**, 503
[21] Mimica P, Giannios D and Aloy M A 2009 *A&A* **494**, 879
[22] Colella P and Woodward P R 1984 *JCP* **54**, 174
[23] Shu C W and Osher S J 1984 *JCP* **77**, 439
[24] Donat R, Font J A, Ibáñez J M and Marquina A 1988 *JCP* **146**, 58
[25] Mimica P *et al.* 2009 *ApJ* **696**, 1142
[26] Sironi S, Keshet U and Lemoine M 2015 *Space Science Reviews* **191**, 519
[27] Sari R, Piran T and Narayan R 1998 *ApJ* **49**, L17
[28] Mimica P and Aloy M A 2012 *MNRAS* **421**, 2635
[29] Aloy C., 2013, masters thesis, Univ. Polit. de Valencia., https://riunet.upv.es/handle/10251/35350
[30] Kumar P, Barniol Duran R, Bošnjak Ž and Piran T 2013 *MNRAS* **445**, 3919

Termination of the MRI via parasitic instabilities in core-collapse supernovae: influence of numerical methods

T. Rembiasz[1], M. Obergaulinger[1], P. Cerdá-Durán[1], M.Á. Aloy[1] and E. Müller[2]

[1] Departamento de Astronomía y Astrofísica, Universidad de Valencia, C/ Dr. Moliner 50, 46100 Burjassot, Spain

[2] Max-Planck-Institut für Astrophysik, Karl-Schwarzschild-Str. 1, 85748 Garching, Germany

E-mail: martin.obergaulinger@uv.es

Abstract. We study the influence of numerical methods and grid resolution on the termination of the magnetorotational instability (MRI) by means of parasitic instabilities in three-dimensional shearing-disc simulations reproducing typical conditions found in core-collapse supernovae. Whether or not the MRI is able to amplify weak magnetic fields in this context strongly depends, among other factors, on the amplitude at which its growth terminates. The qualitative results of our study do not depend on the numerical scheme. In all our models, MRI termination is caused by Kelvin-Helmholtz instabilities, consistent with theoretical predictions. Quantitatively, however, there are differences, but numerical convergence can be achieved even at relatively low grid resolutions if high-order reconstruction methods are used.

1. Introduction

The iron core of a massive star (zero-age main-sequence mass $M_{\rm ZAMS} \gtrsim 8\,M_{\odot}$) collapses within a few hundred milliseconds from a radius of a few thousand kilometer and a central density of the order of $\rho_{\rm c} \sim 10^{10}\,{\rm g\,cm^{-3}}$ to a proto-neutron star (PNS) with a radius of $\sim 50\,{\rm km}$ and a central density above nuclear matter saturation density, $\rho_{\rm nuc} \sim 2\times10^{14}\,{\rm g\,cm^{-3}}$. A hydrodynamic shock wave forms when the core bounces because of the stiffness of nuclear matter. This prompt shock stalls because of severe energy losses due to photo-disintegration of heavy nuclei into free nucleons, and turns into an accretion shock initially located at a radius $\sim 100\,{\rm km}$, through which matter continues to accrete onto the PNS. Most of the gravitational binding energy liberated during the collapse is radiated away in the form of neutrinos. A small fraction of these neutrinos inteact with matter in the post-shock layers depositing energy, which revive the stalled shock and eventually cause a core-collapse supernova explosion (CCSN). A failure to revive the shock before accretion increases the mass of the PNS beyond the maximum mass that can be supported against gravity, will lead to a secondary collapse to a black hole. Hence, the fate of the core hinges on the efficiency of the energy transfer to the post-shock matter, which may be enhanced by various hydrodynamic instabilities (convection and the standing accretion shock instability, SASI). The exact conditions for the neutrino-driven explosion mechanism to launch an explosion of the star compatible with the observations (in particular concering the explosion energy) is currently a matter of intense investigation (for a review see, e.g., [1], [2], [3], [5]).

If the pre-collapse core rotates rapidly, part of its rotational energy might be tapped to aid the neutrino-driven mechanism. The efficiency of this effect is greatly enhanced in strongly magnetised cores. However, the presence of both rapid rotation and strong fields is unlikely except for possibly a very small set of progenitors. Therefore, assuming a rapidly rotating core, the prospects for rotationally aided or even rotationally driven explosions greatly increase if strong magnetic stresses can be generated after collapse, i.e. if a weak seed field is efficiently amplified during the time before either an explosion is launched by other mechanisms or the PNS collapses to a black hole.

Angular momentum conservation during collapse causes the core to speed up and strong differential rotation is generated at the edge of the homologously collapsing inner core, even for initially rigidly rotating iron cores. This opens additional channels for the amplification of seed fields besides flux conservation in the (compressive) accretion flow and a possible turbulent dynamo driven by hydrodynamic instabilities that do not rely on rotation. Winding of a pre-existing poloidal field component creates a toroidal component that grows linearly with time (e.g. [6]). Possibly more important still, a significant fraction of the core fulfils the instability criterion for the magneto-rotational instability (MRI) [7]. In its most basic form, MRI requires a negative radial gradient of the angular velocity, $\partial_r \Omega < 0$, and a weak background field to increase field perturbations on an exponential timescale. The growth time of the instability is comparable to the rotational period of the core that can be much shorter than the post-collapse evolution timescale ($\sim 100\,\mathrm{ms}$).

The MRI has been intensively studied in the context of accretion discs, where it is considered to provide the most important source of turbulence and enhanced angular-momentum transport [8]. In CCSN, indications for its activity have been found in global axisymmetric simulations [9, 10, 11, 12]. Furthermore, it has been investigated in local shearing box simulations [13, 15]. Both types of models suggest that the MRI can indeed grow under conditions typical for PNS. So far, these studies leave, however, several questions open, most importantly what is the amplitude of the turbulent stress generated by the MRI. In a new set of simulations, we focus on the transition from the initial phase of exponential growth of the MRI to turbulence. In the former phase, the MRI takes the form of laminar *channel modes*, pairs of radial inflows and outflows containing magnetic fields of opposite polarity. They contain shear layers as well as current sheets, which can become unstable against Kelvin-Helmholtz (KH) instabilities or tearing-modes (TM). The growth rate of these secondary, *parasitic* instabilities is proportional to the radial component of the velocity or magnetic field, respectively. Since these components grow exponentially at a constant rate, γ_{MRI}, the growth rate of the parasitic instabilities increases exponentially with time and, consequently, their amplitude grows super-exponentially. When the amplitudes of the parasites are as large as the amplitude of the MRI itself, they terminate the MRI growth, initiating the saturation phase. This parasitic termination scenario was described by [16] and analysed in detail by [17], who determined the growth rate of the KH instabilities and TM, and identified which of the two parasites should dominate given the viscosity and resistivity of the gas.

To test the predictions of [17] under conditions representative for the outer layers of a differentially rotating PNS, we performed a set of local three-dimensional simulations of the MRI in the shearing-disc approach ([18, 13]) varying the viscosity and resistivity. For a detailed discussion of our simulations, see [19]. Here, we will focus on the impact that different numerical methods have on the results.

2. Setup and numerical methods

We solve the equations of visco-resistive MHD in the presence of an external gravitational field and use the same initial and boundary conditions as in [19]. The simulation domain covers 1 km to 4 km in each direction of a cylindrical box (r, z, ϕ) centered at a radius of $r_0 = 15.5\,\mathrm{km}$. The

name	reco.	reso.	box	$\gamma_{\rm MRI}$ [$\rm s^{-1}$]	$\mathcal{M}_{r\phi}^{\rm term}$ [10^{30} $\rm G^2$]
PLM-8	PLM	8	s	926	2.2
PLM-10	PLM	10	s	959	2.3
PLM-16	PLM	16	s	1089	1.9
PLM-20	PLM	20	l	1116	1.9
PLM-34	PLM	34	s	1123	1.8
MP5-8	MP5	8	s	1093	1.1
MP5-10	MP5	10	s	1104	1.4
MP5-16	MP5	16	s	1127	1.1
MP5-20	MP5	20	s	1134	0.85
MP5-34	MP5	34	s	1127	1.0
MP9-8	MP9	8	s	1104	0.79
MP9-10	MP9	10	s	1122	1.3
MP9-16	MP9	16	s	1130	1.1
MP9-20	MP9	20	l	1126	1.1
MP9-25	MP9	25	l	1127	1.0
MP9-34	MP9	34	s	1127	0.93
MP9-67	MP9	67	l	1127	0.73
MP9-134	MP9	134	s	1128	0.73

Table 1. List of models. The first column displays the name of the model, which is a combination of the reconstruction method ($2^{\rm nd}$ column) and the grid resolution, measured in grid cells per MRI wavelength ($3^{\rm rd}$ column). In the $4^{\rm th}$ column, letters indicate the size of the simulation domain: 's' (small) and 'l' (large) stand for box sizes of $1\,\rm km \times 1\,km \times 0.333\,km$ and $1\,\rm km \times 4\,km \times 1\,km$, respectively. The last two columns give the growth rate of the MRI and the termination value of the Maxwell stress (defined in Eq. 2), respectively.

initial profiles of gravitational potential, density, and pressure are derived from the post-bounce state of the global simulations of [9]. In addition, we assume an angular velocity with a simple power law profile,

$$\Omega = \Omega_0 \left(\frac{r}{r_0}\right)^{-q}, \tag{1}$$

where $q = 1.25$, and $\Omega_0 = 1824\,\rm s^{-1}$. This profile is hydrodynamically stable.

If we add a uniform magnetic field in z-direction, the MRI will grow from small perturbations. The geometry and strength of the CCSN cores our models are supposed to represent depends on the pre-collapse fields as well as the evolution during and immediately after collapse and, as a result, are quite uncertain. Stellar-evolution models support a combination of poloidal and toroidal components [4]. Both kinds of fields are susceptible to the MRI. We focus here on the case of a poloidal, in our case vertical, field because the MRI grows fastest for this field component. By focusing on a uniform field, we furthermore neglect possible small-scale structures, which might yield small corrections to the geometry, but not the growth rate of the MRI.

We impose periodic boundary conditions in z and ϕ direction, and shearing-disc boundaries in radial direction, i.e. boundaries that are periodic in the deviation of the hydrodynamic variables from the initial (background) state.

The parameters of our simulations are based on the models of [19]. The theoretical growth rate of the MRI is $\gamma_{\rm MRI} = q\Omega_0/2 \approx 1140\,\rm s^{-1}$. We use an initial magnetic field of $b_0^z = 4.6 \times 10^{13}\,\rm G$, corresponding to $\lambda_{\rm MRI} \approx 333\,\rm m$. We neglect viscosity and use a uniform resistivity $\eta = 4.45 \times 10^8\,\rm cm\,s^{-1}$. The most important parameter that, as we will show below,

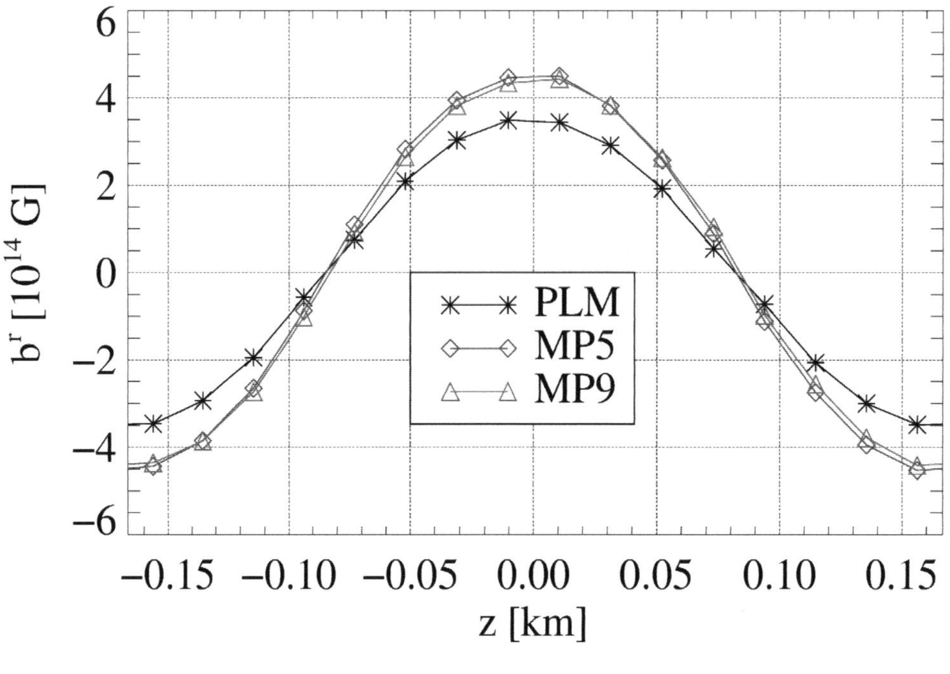

Figure 1. Vertical profiles of the radially averaged radial magnetic field at $t = 10\,\mathrm{ms}$ for models PLM-16, MP5-16, and MP9-16.

causes differences between the simulations is the grid resolution, which we express in terms of the number of grid cells per channel width. We present our set of models in Tab. 1. We point out that our simulation results are scalable to different values of r_0, Ω_0, and b_0^z (see, [19]).

Our numerical code is based on a dimensionally unsplit finite-volume discretisation of the MHD equations and the constrained-transport scheme to ensure a divergence-free evolution of the magnetic field. The spatial reconstruction can be done using the following methods:

- piecewise-constant reconstruction of 1^{st}order;
- TVD piecewise-linear (PLM) reconstruction (2^{nd}order) with several slope limiters;
- 4th-order WENO reconstruction;
- monotonicity-preserving reconstruction (MP; [20]) of 3rd, 5th, 7th, or 9th order.

In [19], we used the most accurate of these schemes, viz. 9th-order MP reconstruction. Here, we will make a comparison to two methods of lower order: 5th-order MP, and PLM with the MC slope limiter. The other ingredients of the code are the same as in [19]: a multi-stage (MUSTA) version of the HLLD Riemann solver and a 3rd-order Runge-Kutta time integrator.

3. Results

For the conditions in supernova cores, [19] found, as predicted by [17], that the MRI growth is terminated by KH modes, unless special conditions such as an insufficient size of the numerical domain somewhat artificially restrict their growth. This qualitative finding also holds for less accurate reconstruction schemes.

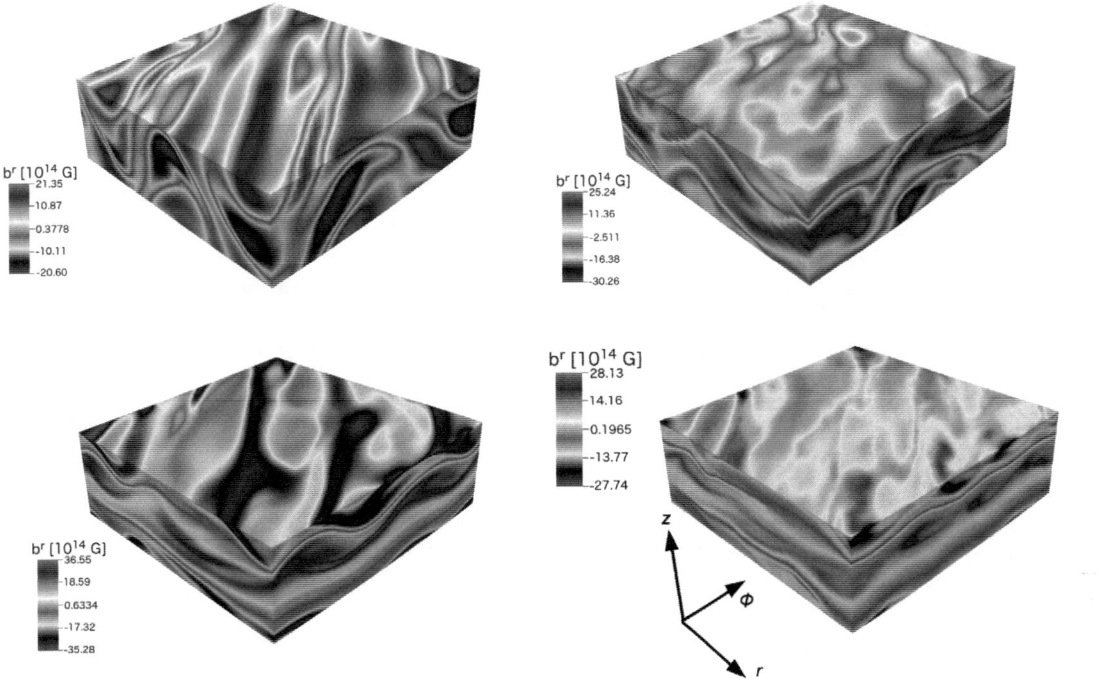

Figure 2. Visualisation of the radial component of the magnetic field close to the moment of termination for models (top left to bottom right) PLM-16 (at $t = 12\,\mathrm{ms}$), MP9-16 (at $t = 11.5\,\mathrm{ms}$), PLM-34 (at $t = 11.5\,\mathrm{ms}$), and MP9-34 (at $t = 11\,\mathrm{ms}$), respectively. Note that not all snapshots are taken at the same evolutionary time, but they correspond to the same evolutionary instant, namely the time at which (for each model) saturation of the MRI growth sets in.

Irrespective of the reconstruction method, MRI channel modes start to grow from the weak perturbations introduced at $t = 0$. Even for low grid resolution and low-order schemes, the vertical profiles that we find in the simulations approximate very well sine functions, as we show for the radial component of the magnetic field of models PLM-16, MP5-16, and MP9-16 in Fig. 1. The numerical growth rates of the MRI depend on the resolution and the reconstruction scheme. Under-resolved simulations show in general a slower MRI growth, but as the grid resolution increases, all methods exhibit numerical convergence to the theoretical value. The good agreement between numerical schemes shown in Fig. 1 holds well throughout the exponential growth phase. Differences show up when the growth is terminated, leading to a spread in the value of the termination amplitude (see below).

The channel modes constitute the background in which parasitic instabilities grow. We compare visualisations of the radial component of the magnetic field for models with PLM and MP9 reconstruction schemes and two resolutions in Fig. 2. We note that the snapshots show a very dynamical phase of the evolution, during which a slight time difference corresponds to a very pronounced difference in the state of the flow. Therefore (and because of the influence of random perturbations), a point-wise comparison between the four panels cannot be made. Nevertheless, we conclude that the parasites have the same structure as the one found in [19], and thus can be identified as KH instabilities. In particular, patterns such as the vortices visible in the r–z-planes of the lower two panels provide evidence for KH modes. The rather different appearance of the two upper panels (with a resolution of 16 zones per channel) is not caused by a difference in the parasites, but is a consequence of the fact that the parasites have grown somewhat more than in the two bottom panels.

Analogously to MRI channels, KH modes grow at a reduced rate if the numerical resolution is insufficient. This means that, for the same MRI amplitude, the KH parasites grow slower. Consequently, the time at which the parasites are strong enough to terminate the MRI growth is delayed for a coarse grid. Furthermore, KH parasite modes develop structures smaller than the channel mode. Therefore, insufficient resolution affects them more than the MRI, and parasitic modes can be suppressed on a grid where the MRI growth rate is hardly reduced. Thus, non-converged simulations tend to *overestimate* the termination amplitude of the MRI as measured by, e.g. the maximum value of the volume-averaged Maxwell stress component,

$$\mathcal{M}_{r\phi} = \frac{|\int b^r b^\phi \mathrm{d}V|}{V}, \tag{2}$$

where V is the total grid volume. This tendency is ameliorated for more accurate numerical schemes.

We summarise the results described above in Fig. 3. The three top panels show the time evolution of $\mathcal{M}_{r\phi}$ for simulations with resolutions of 8, 16, and 34 zones per channel. Except at the lowest resolution, for which the PLM run exhibits a low growth rate, the exponential MRI growth is captured well. Comparing the growth rates for all simulations (bottom left panel), we can quantify the differences between the reconstruction methods: PLM requires roughly 34 zones per channel to attain a growth rate $\gamma_{\mathrm{MRI}} > 1120\,\mathrm{s}^{-1}$, whereas the MP schemes achieve that value already at 16 (MP5) and 10 (MP9) zones, respectively.

In the top panels, MRI termination corresponds to the maximum value of $\mathcal{M}_{r\phi}(t)$. The bottom right panel compares this maximum for all models. Because of the suppression of the growth of the parasites at low resolutions, the termination amplitude increases with resolution until a sufficiently large number of zones per channel width is reached. Then it settles to a roughly constant value. Please note that the runs with 8 zones per channel represent a special case for which other effects beside parasitic KH modes affect termination such as the diffusion of magnetic field and momentum across the current and shear layers, respectively. Furthermore, model MP5-20 deviates from the trend, possibly indicating the range of stochastic variations of the termination amplitude. We find considerable differences between PLM and the two MP schemes, with the former yielding an almost twice as high $\mathcal{M}_{r\phi}^{\mathrm{term}}$ as the two other at the same resolution. Restrictions in computing time allowed us to obtain numerical convergence only for the MP9 models, which approach a value of $\mathcal{M}_{r\phi}^{\mathrm{term}} = 0.73 \times 10^{30}\,\mathrm{G}^2$, if we use at least 67 zones per channel. We expect, however, the other models to behave similarly for a sufficiently high grid resolution.

4. Summary and conclusions

In an extension of the studies of [13] and [19], we investigated the influence of numerical effects on the early evolution of the MRI in conditions similar to those expected in core-collapse supernovae. We performed a series of three-dimensional MHD simulations in the shearing-disc framework, fixing the physical conditions (rotational profile, density, initial field) in such a way that we could observe the growth of MRI channel modes and their disruption by parasitic Kelvin-Helmholtz modes and varying the grid resolution and the reconstruction method. For the latter, we used a 2nd order piecewise linear reconstruction with the MC slope limiter, as well as monotonicity-preserving methods of 5th and 9th order accuracy.

The main results of this study are the following:

- The physical nature of the parasitic instability terminating the MRI does not depend on the numerical settings. The structure of the flow and the field show the signatures of KH modes, which are consistent with the theoretical predictions of [17].

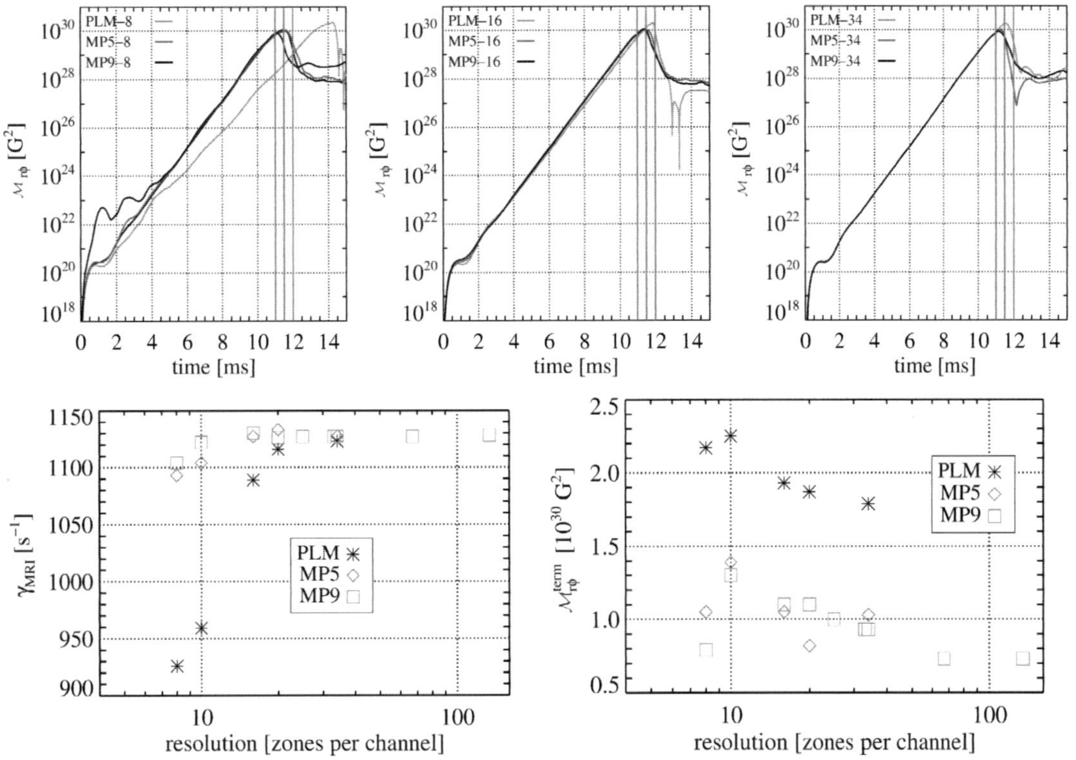

Figure 3. Top panels: time evolution of $\mathcal{M}_{r\phi}$ for models with grid resolutions of 8, 16, and 34 zones per channel (left to right). Bottom panels: dependence of the growth rate (left) and the termination amplitude (right) of the MRI on the grid resolution.

- All methods capture the exponential growth of the MRI well at low resolution of the order of 10 zones per channel, with MP schemes performing better than PLM.
- The termination value of the Maxwell stress exhibits convergence with grid resolution. Under-resolved simulations overestimate its value because numerical viscosity reduces the growth rate of the parasites. Again, MP schemes are superior to PLM. In general, a much higher resolution is required to obtain the correct parasitic growth rate than that of the MRI itself, because KH modes grow on much smaller length scale, viz. the extent of the shear layer between the channel flows.

From these results we conclude that the most important findings concerning the MRI termination are robust w.r.t. the effects of numerical schemes, which affirms our previous studies. Our initial field strengths, and thus the MRI wave lengths, most likely exceed those that can be expected in most supernova cores (the post-collapse fields of [14] would, in a rapidly rotating core, correspond to $\lambda_{\rm MRI} \sim 0.01 \ldots 10\,{\rm m}$). This means that for realistic initial fields, the constraint to resolve the KH instability in the shear layer can translate into a very severe restriction on the grid resolution beyond what can be currently afforded in global models. Simulations that cannot meet this resolution requirement run the risk of producing too high amplitudes of the MRI at termination, which may lead to unphysically strong magnetic stresses and consequently tend to overestimate the influence of magnetic fields on the dynamics. Although the use of high-order methods, such as our MP5 and MP9 schemes, does not overcome this limitation, at least it helps to alleviate it. We estimate that a three-dimensional simulation with a high-order method following the evolution for a time scale of $\sim 100\,\mathrm{ms}$ and resolving channels of $1\,\mathrm{m}$ length scale in a PNS of radius $30\,\mathrm{km}$ requires $\sim 10^8$ grid cells and $\sim 10^{14}$ CPU hours. Even if adaptive mesh refinement (AMR) techniques were employed, the simulations would still be beyond current possibilities.

Acknowledgments

TR acknowledges support from The International Max Planck Research School on Astrophysics at the Ludwig Maximilian University Munich, EM & TM acknowledge support from the Max-Planck-Princeton Center for Plasma Physics, and MA, PCD, TR and MO acknowledge support from the European Research Council (grant CAMAP-259276). We also acknowledge support from grants AYA2013-40979-P and PROMETEOII/2014-069. The computations have been performed at the Leibniz Supercomputing Center of the Bavarian Academy of Sciences and Humanities (LRZ), the Max Planck Computing & Data Facility (MPCDF), and at the Servei d'Informàtica of the University of Valencia.

References

[1] Mezzacappa2005Mezzacappa, A. et al. 2005, Annual Review of Nuclear and Particle Systems, vol. 55, 467
[2] Janka, H.-T. 2012, Annual Review of Nuclear and Particle Science, 62, 407
[3] Kotake, K. et al. 2012, Advances in Astronomy, vol. 2012, id. 428757
[4] Braithwaite, J., Nordlund, Å., 2006, A&A, 450, 1077
[5] Burrows, A. 2013, Reviews of Modern Physics, 85, 245
[6] Meier, D. L., Epstein, R. I., Arnett, W. D., & Schramm, D. N. 1976, ApJ, 204, 869
[7] Akiyama, S., Wheeler, J. C., Meier, D. L., & Lichtenstadt, I. 2003, ApJ, 584, 954
[8] Balbus, S. A. & Hawley, J. F. 1998, Reviews of Modern Physics, 70, 1
[9] Obergaulinger, M., Aloy, M. A., Dimmelmeier, H., & Müller, E. 2006, A&A, 457, 209
[10] Cerdá-Durán, P., Font, J. A., Antón, L., Müller, E. 2008, A&A, 492, 937
[11] Sawai, H., Yamada, S., & Suzuki, H. 2013, ApJL, 770, L19
[12] Sawai, H., & Yamada, S. 2015, arXiv:1504.03035
[13] Obergaulinger, M., Cerdá-Durán, P., Müller, E., & Aloy, M. A. 2009, A&A, 498, 241
[14] Obergaulinger, M., Janka, H.-Th., Aloy, M.Á, 2014, MNRAS, 445, 3169
[15] Masada, Y., Takiwaki, T., Kotake, K., & Sano, T. 2012, ApJ, 759, 110
[16] Goodman, J. & Xu, G. 1994, ApJ, 432, 213
[17] Pessah, M. E. 2010, ApJ, 716, 1012
[18] Klahr, H. H. & Bodenheimer, P. 2003, ApJ, 582, 869
[19] Rembiasz, T., Obergaulinger, M., Cerdá-Durán, P., Müller, E., & Aloy, M. A., submitted
[20] Suresh, A. & Huynh, H. 1997, J. Comput. Phys., 136, 83

Journal of Physics: Conference Series **719** (2016) 012010 doi:10.1088/1742-6596/719/1/012010

Gas dynamic modeling of the CME propagation through the envelope of a hot Jupiter-type exoplanet

A.A. Cherenkov, P.V. Kaygorodov, D.V. Bisikalo

Institute of Astronomy of the Russian Academy of Sciences, 48 Pyatnitskaya St. 119017, Moscow, Russia

E-mail: bisikalo@inasan.ru

Abstract. We propose a 3D gasdynamic numerical model for the study of the interaction between the extended envelopes of hot Jupiters, overfilling their Roche lobes, and non-stationary stellar wind. In the model we use a Roe-Osher numerical scheme with Eindfeldt entropy fix. To test the model we have simulated a flow structure, forming due to the interaction between the extended quasi-stationary envelope of the hot Jupiter planet HD 209458b and the bow shock formed ahead of a propagating coronal mass ejection (CME). We have adopted the solar CME parameters in our computations and taken into account the fact that the planet is located close to its host star. The simulation results show that the bow shock of the CME partially destroys the stream, starting from the L_1 point of the quasi-closed planet's envelope. A bow shock, existing ahead of the planet in its orbital motion when the stellar wind is undisturbed, almost disappears when the CME shock passes through the system.

1. Introduction

The Hubble Space Telescope observations of hot Jupiter HD 209458b [1, 4] have shown that within the transit the absorption in the Ly-α reaches 9%–15% while the planet's is able to absorb only 1.8% of the star's light. Later the same effect have been observed in the C, O and Si lines [5–7]. The results of these observations have driven the scientists to suppose that the planet is surrounded by an extended gaseous envelope. Similar results have been obtained for the HD 189733b and WASP-12b planets. In addition, the observations of WASP-12b, taken in 2009, showed that the transit onset occurs at different time moments in different wavelength ranges, which means that a bulk of relatively dense gas exists ahead of the planet at a distance of 4-5 planet's radii [8].

Hot Jupiters have a number of outstanding features, caused by the proximity to the host star. For example, the proximity of the planet to the host star may result in the effect of gas outflowing from the planet's atmosphere to the star, which is well known in the physics of close binary stars. Besides, the short distance between the planet and the star gives a high value of the planet's orbital velocity. If the planet's orbital velocity exceeds the local sound speed a bow-shock forms ahead of the planet. These effects substantially change the form of interaction of the planet and the stellar wind.

A system consisting of a star and a hot Jupiter may be regarded as a close binary system with extremely low mass ratio. We can consider the force field in a binary system with the components that are a star and a hot Jupiter. We can also assume that the orbits of the components are circular and their proper rotations are synchronized with the orbital motion of the planet. Under these assumptions we can use the Roche approximation, as it described in the Model section below. The Roche potential has five libration points, called Lagrangian points, where its gradient is equal to zero. The equipotential, going through the inner Lagrangian point L_1, encloses two contiguous volumes, known as critical surfaces or Roche lobes. For the Hot Jupiters the L_1 point locates close enough to the planet to allow significant mass transfer in the system [3].

In the papers [9, 10] the authors proposed a gasdynamic model for the study of interaction between exoplanet's atmospheres and stellar wind. The reported computation results have shown that for typical hot Jupiters, orbiting their host stars with supersonic velocities, a bow shock, forming ahead of the planet, strongly influences the final solution. In particular, in [9] it has been shown that the type of the exoplanet's envelope depends on the position of the head-on collision point (a point where the wind's dynamic pressure is equal to the pressure of the planet's atmosphere). The envelopes of planets, where head-on collision points are located inside the Roche lobe, are almost spherical, only slightly distorted by the gravitational influence of the star and stellar wind. If the head-on collision point is located outside the Roche lobe, the planet's atmosphere starts to outflow through the vicinity of the L_1 and L_2 points, which results in the formation of an extended asymmetric envelope. The latter envelopes may be related to two different types. If the dynamic pressure of the stellar wind is strong enough to stop the outflow through the inner Lagrangian point L_1, a quasi-closed stationary envelope can form around the planet (see [10]). If the wind cannot stop the outflow from L_1, an open envelope forms in the system. The estimates of mass-loss rates made in [11] have shown that a quasi-closed envelope loses have mass loss rate comparable to that of a "classical" closed envelope, which means that quasi-closed envelopes may be stationary.

Previously, we simulated the dynamics of hot Jupiter atmospheres under an assumption that the stellar wind is stationary. Nonetheless, it is known that solar-type stars demonstrate flare activity, including coronal mass ejections (CME), which may drastically change the parameters of stellar wind. The first attempts to simulate the interaction between a hot Jupiter envelope and non-stationary stellar wind has been described in [12] where the main attention has been payed to the physical aspects of the problem. In this paper we give a detailed description of the proposed numerical model that allows one to study all the stages of CME development. In particular, we study the propagation of a shock wave, induced by CME, through hot Jupiter envelopes.

2. The model

Hot Jupiters are located close to their host stars. Therefore, when modeling these objects, one should take into account all the forces acting in the binary star/planet system. In our paper we use the Roche approach, within which the star and planet are represented by point masses, moving in circular orbits around the system's center of mass. This approach is fairly appropriate, since the mass of the gaseous envelopes under study is much less than the planet's mass.

The flow is described by a system of 3D gasdynamic equations that take into account the gravitational field of the binary system. The system of equations is closed by the equation of state of ideal monatomic gas. In this model we neglect non-adiabatic processes of radiative heating and cooling, and radiative transfer processes. The resulting equation system is following:

$$\begin{cases} \dfrac{\partial \rho}{\partial t} + \mathrm{div}\,(\rho \mathbf{v}) = 0 \\ \dfrac{\partial \rho \mathbf{v}}{\partial t} + \mathrm{div}\,(\rho \mathbf{v} \otimes \mathbf{v}) + \mathrm{grad}\, P = -\rho\, \mathrm{grad}\, \Phi - 2\left[\mathbf{\Omega} \times \mathbf{v}\right] \rho \\ \dfrac{\partial \rho(\varepsilon + |\mathbf{v}|^2/2)}{\partial t} + \mathrm{div}\, \rho \mathbf{v}(\varepsilon + P/\rho + |\mathbf{v}|^2/2) = -\rho \mathbf{v}\, \mathrm{grad}\, \Phi \\ P = (\gamma - 1)\rho\varepsilon \\ \Phi = -\left(\dfrac{GM_*}{\sqrt{x^2+y^2+z^2}} + \dfrac{GM_{pl}}{\sqrt{(x-A)^2+y^2+z^2}}\right) - \\ \quad -\dfrac{1}{2}\Omega^2\left[\left(x - A\dfrac{M_{pl}}{M_* + M_{pl}}\right)^2 + y^2\right] \end{cases} \tag{1}$$

where ρ is the density, $\mathbf{v} = (u, v, w)$ is the velocity vector, P is the pressure, ε is the specific internal energy, Φ is the Roche potential, $\gamma = 5/3$ is the adiabatic index, Ω is the angular velocity of the star-planet system's rotation, G is the gravitational constant, M_* and M_{pl} are the masses of the star and planet respectively, A is the orbital radius of the planet.

In our computations we neglected the magnetic field of the planet, since the estimates, made in [13] for the typical hot Jupiter HD 209458b, show that its magnetic moment is only ~ 0.1 of the Jupiter's magnetic moment, which is in agreement with theoretical estimates made for magnetic fields of tidaly-locked hot Jupiters. A field of this intensity may stop the plasma of stellar wind at a distance of only ~ 3 planet radii, which means that one should take it into account only when studying an envelope, whose size is smaller than the size of the Roche lobe. However, the observed parameters of the atmosphere reported in [9] show that the planet may have a much bigger envelope, and one may neglect magnetic field when modeling this envelope. Assuming that ratio between magnetic pressure and dynamical pressure in the wind is the same as on the Earth orbit (less then 0.1 during the CME propogation, [15]) we can neglect effects caused by magnetic fields in our study.

We solved the system of gasdynamic equations in the Cartesian coordinate frame (X, Y, Z) with the origin in the star's center. The coordinate frame rotates along with the star/planet system with the angular velocity Ω. The planet coordinates in this frame are $(A, 0, 0)$.

All the physical quantities are associated with cell centers of a Cartesian grid that has the dimension of $(468 \times 468 \times 178)$ in X, Y and Z dimensions, respectively. The physical size of the computational domain is $(40 \times 40 \times 10)\, R_{pl}$, where R_{pl} is the photometric radius of the planet. The computational grid was locally refined near the exoplanet. Hence, the size of the cells at the photometric radius of the planet $(\Delta x, \Delta y \simeq 0.04\, R_{pl})$ is smaller than the smallest scale height of the simulated atmosphere. The scales of the major flow elements, located near the boundaries of the computational domain (at a distance of $r \simeq 20\, R_{pl}$ from the planet's center) is significantly larger than the size of the cells in this region $(\Delta x, \Delta y \leq 0.25\, R_{pl})$.

In this work we investigate the dynamics of the outer atmosphere located far beyond the planet's photometric radius. Therefore, we excluded the inner region of the planet from the computations and set the solid wall boundary conditions at the distance of $r = 0.5\, R_{pl}$ from the planet's center. In this way, we avoid the region with the high gradients of gasdynamic quantities. In the boundary cells we set the normal component of the velocity equal to zero.

To damp waves that may propagate inside the planet due to the interaction between the atmosphere and supersonic stellar wind and cause undesired oscillations of the solution we added artificial viscosity in the cells located below the photometric radius $(r < R_{pl})$. At each step within this region we reduced the gas velocities and maintained the constant temperature. Thus the sphere with the radius of $r < R_{pl}$ may be considered as a "soft" boundary that

Phase	1	2	3	4
Phase duration (hours)	-	8.5	13	21.5
(P_{orb})	-	0.1	0.15	0.25
N (cm^{-3})	10^4	$4 \cdot 10^4$	$6 \cdot 10^3$	10^5
T (K)	$7.3 \cdot 10^5$	$3.65 \cdot 10^6$	$5.84 \cdot 10^5$	$2.19 \cdot 10^5$
V (km/sec)	100	133	144	111

Table 1. Average parameters of the stellar wind at a moment, when an exoplanet with the orbital parameters of HD 209458b passes through CME

effectively absorbs waves.

We adapted the solution with a quasi-stationary envelope reported in [9] as the initial conditions. This solution includes a stream starting from the vicinity of the L_1 point. At a certain distance from the L_1 point the dynamical pressure of the wind becomes equal to that of the stream, which stops the stream. As a result, a portion of the stream material returns toward the planet and form a quasi-stationary circulating envelope. In [9] we studied the interaction of the atmosphere of a hot Jupiter planet with the stellar wind having constant parameters that correspond to the parameters of the solar wind at the distance equal to the radius of the planet's orbit. The wind was set as the boundary conditions at the outer boundary of the computational domain.

In [9, 10] the authors assumed that the parameters of the stellar wind are constant in time. However, real stellar winds are not stationary. This can be observed, for instance, in the Sun that demonstrates solar wind disturbances caused, in particular, by powerful coronal mass ejections (CME). In accordance with [14], solar CMEs are characterized by the following average quantities: the mass ejected into the interplanetary medium is $\sim 10^{16}$ g, the ejection energy is $\sim 10^{31}$ erg, the velocity of the ejected gas varies from 400 to 1000 km/s, which means that it moves with supersonic velocities and a shock wave should form in the medium. In addition, in a CME the abundance of He ions is usually $10-15\%$ higher than in the undisturbed solar wind. It is important to note that the frequency of CMEs is very high even for the Sun. It varies from 0.5 to 5 CMEs per day, depending on the stage of the solar activity cycle [14]. This means that an exoplanet orbiting a relatively quiescent star experiences CMEs fairly often (not less than 2 CMEs per month for Earth). These CMEs should significantly change the flow structure in the planet's gaseous envelope.

During a CME propogation the parameters of stellar wind may significantly vary on short timescales. If a host star of a hot Jupiter is similar to the Sun, one may assume the parameters of its CMEs to similar to those of the Sun. In [15] the authors report the parameters of the solar wind in the vicinity of Earth, measured by ACE, WIND, and SOHO on May 1-4, 1998, when Earth passed through a CME. In this work the reported structure of the CME is complicated and consists of three major elements: a shock wave, an early CME, and a late CME. In the shock wave the density and temperature jump by approximately 2 times, and the velocity grows from ~ 450 to ~ 600 km/s. The early CME has a lower density and temperature, and its parameters do not differ much from those of the undisturbed solar wind, except the gas velocity that is equal to the velocity behind the shock. In the late CME the temperature is almost equal to that of the undisturbed wind, the velocity monotonously decreases, and the density grows by up to 5 times. It reaches the maximum in approximately the middle of the late CME and demonstrates strong variations, which means a significantly non-uniform structure of the CME.

The parameters of CME in an orbit of a hot Jupiter should significantly differ from those observed in Earth's orbit. If the stellar wind density is inversely proportional to the squared distance from the star, and the velocity and temperature dependence on the distance is described by Parker's model [16], then the average values at four CME stages (undisturbed wind, shock wave, early and late CMEs) should correspond to those given in Table 1. We should note that the duration of the CME phases in the orbit of the hot Jupiter should also differ from the respective phases in Earth's orbit, since the size of CME gasdynamic elements grows while the CME propagates. Unfortunately, we cannot take this effect into account, since we should have conducted a detailed MHD simulation, which is beyond the scope of this work.

In our simulations we use a TVD Roe-Osher numerical scheme with the Einfeldt entropy fix. This explicit scheme of a higher-order of approximation possesses low numerical viscosity in the regions of smooth solution and does not smear out shock waves. This scheme is considered in detail in [17]. It is based on the Roe scheme designed to solve the equations of gasdynamics. To increase the order of spatial approximation the Osher TVD correction is applied [18]. To avoid unphysical discontinuities, occurring in the unmodified Roe scheme, we apply the Einfeldt entropy fix [19].

We use a parallel gasdynamical code, parallelized using MPI library. The computations have been performed with 324 processors of the Kurchatov Institute Supercomputer.

3. Modeling results

To test the proposed model we adapted the simulation results described in [9]. These results have been obtained for the quasi-closed envelope of the hot Jupiter HD 209458b. In the top-left panel of Fig. 1 the density distribution that existed in 30 minutes since the CME shock wave started propagating through the system is shown. As one can see, at this moment the flow structure is still almost undisturbed, though the CME shock has already started to sweep out the rarified material and interact with the bow shock located ahead of the stream issuing from the L_1 point. In the top-right panel of Fig. 1 one can see a more developed stage of the process (one hour since the start). By this moment a new more powerful bow shock has formed ahead of the stream, and the density jump in this wave is noticeably stronger. A part of the CME shock, located below the stream in Fig. 1 starts to dissolve after having met a turbulent stream forming behind the stream. We also should mention the noticeable regions of the interaction between the shock waves, where the density is significantly increased. One also can see secondary shock waves propagating from these regions. The middle-left panel demonstrates the flow structure that occurs in 1.5 hours since the CME shock have started propagating. In addition to the previously found features one also can see that the stream structure has changed. This change is caused by with the compression of the stream by the shock wave. In fact the stream now has a pipe-like structure with dense walls and rarified interiors. In the middle-right panel one can clearly see the pipe-like shape of the stream. By this moment new shock waves has formed ahead of the planet and stream. In addition, a weaker attached shock wave has formed. It starts near the head-on collision point. Behind this wave, in the rarifaction wave, one can see the "walls" of the pipe-like decay, which results in the formation of gas filaments. In the bottom-left panel one can see the time moment of 2.5 hours since the process started. By this moment the stream has totally collapsed and its rarified core has disappeared. It continues to decay, though remains visible as a whole and still starts near the L_1 point. The CME shock is well seen in the upper part of the figure (ahead of the planet in its orbital motion), but is almost absent in the bottom part, since it has decayed while propagating through the nonuniform and strongly turbulent trail behind the planet. The outflow from the vicinity of the L_2 still exists, though becoming pipe-like and also starting to decay. Finally, in the bottom-right panel we show the moment when the planet has passed through the CME shock (8 hours after onset). By this moment the stream debris are still at the same place, but they have no connection to the L_1 point. The wind pressure is high enough to prevent the Roche lobe overflow. The outflow from the L_2 point has a shape of separate filaments swept out by the wind. The planet's envelope has a ragged shape where shock waves dissipated. As a result we see only weak secondary attached shock waves in different parts of the solution, while the bow shock ahead of the planet is absent. The shock wave, located ahead of the whole envelope still exists, but it is out of the region, shown in Fig. 1.

4. Conclusions

We have developed a numerical model that allows studying the influence of non-stationary stellar wind on the gaseous envelopes of hot Jupiter exoplanets. In the model we use the Roe-Osher numerical scheme with the Einfeldt entropy fix. This scheme belongs to the TVD family, possesses low numerical viscosity and does not smear shock waves. This allows us to consider the interaction between the shock waves, induced by CMEs, and the flow structure of the planet's gaseous envelope.

To test the model we simulated the process of CME shock wave propagation through an extended quasi-stationary envelope of the hot Jupiter HD 209458b, forming due to the gas outflow through the vicinity of L_1 and L_2 points. The simulation result confirms the efficiency of the proposed model. The results show that the CME shock, propagating through an extended quasi-stationary envelope, prevents the formation of the stream in the vicinity of the L_1. The previously existing stream decays and losses most of its material by the moment when the planet will have passed through the CME shock. The material issuing from the L_2 point is almost totally swept out by the CME shock. The CME shock wave also destroys the bow shock, forming ahead of the planet.

Acknowledgements

This work is supported by the Russian Scientific Foundation (project 15-12-30038). The results of the work were obtained using computational resources of MCC NRC "Kurchatov Institute"

References

[1] Vidal-Madjar A, Lecavelier des Etangs A, Désert J M, Ballester G E, Ferlet R, Hébrard G and Mayor M 2003 Nature **422** 143–146
[2] Bisikalo D V, Kaigorodov P V, Ionov D E and Shematovich V I 2013 *Astronomy Reports* **57** 715–725 (*Preprint* 1311.4441)
[3] Bisikalo D V, Kaigorodov P V, Arakcheev A S 2015 *Living Together: Planets, Host Stars and Binaries, ASP Conf. Ser.* **496** 337
[4] Ben-Jaffel L 2007 ApJ **671** L61–L64 (*Preprint* 0711.1432)
[5] Vidal-Madjar A, Désert J M, Lecavelier des Etangs A, Hébrard G, Ballester G E, Ehrenreich D, Ferlet R, McConnell J C, Mayor M and Parkinson C D 2004 ApJ **604** L69–L72 (*Preprint* arXiv:astro-ph/0401457)
[6] Ben-Jaffel L and Sona Hosseini S 2010 ApJ **709** 1284–1296 (*Preprint* 0912.1409)
[7] Linsky J L, Yang H, France K, Froning C S, Green J C, Stocke J T and Osterman S N 2010 ApJ **717** 1291–1299 (*Preprint* 1005.1633)
[8] Fossati L, Haswell C A, Froning C S, Hebb L, Holmes S, Kolb U, Helling C, Carter A, Wheatley P, Collier Cameron A, Loeillet B, Pollacco D, Street R, Stempels H C, Simpson E, Udry S, Joshi Y C, West R G, Skillen I and Wilson D 2010 ApJ **714** L222–L227 (*Preprint* 1005.3656)
[9] Bisikalo D V, Kaigorodov P V, Ionov D E and Shematovich V I 2013 *Astronomy Reports* **57** 715–725 (*Preprint* 1311.4441)
[10] Bisikalo D, Kaygorodov P, Ionov D, Shematovich V, Lammer H and Fossati L 2013 ApJ **764** 19 (*Preprint* 1212.2779)
[11] Cherenkov A A, Bisikalo D V and Kaigorodov P V 2014 *Astronomy Reports* **58** 679–687
[12] Bisikalo D V and Cherenkov A A 2016 *Astronomy Reports* **60** in press
[13] Kislyakova K G, Holmström M, Lammer H, Odert P and Khodachenko M L 2014 *Science* **346** 981–984 (*Preprint* 1411.6875)
[14] Howard R A, Sheeley Jr N R, Michels D J and Koomen M J 1985 J. Geophys. Res. **90** 8173–8191
[15] Farrell W M, Halekas J S, Killen R M, Delory G T, Gross N, Bleacher L V, Krauss-Varben D, Travnicek P, Hurley D, Stubbs T J, Zimmerman M I and Jackson T L 2012 *Journal of Geophysical Research (Planets)* **117** E00K04
[16] Withbroe G L 1988 ApJ **325** 442–467

[17] Boyarchuk A A, Bisikalo D V, Kuznetsov O A and Chechetkin V M 2002 *Mass transfer in close binary stars* (London: Taylor & Frances)
[18] Chakravarthy Sukumar R and Osher S *AIAA Paper* chap NEW CLASS OF HIGH ACCURACY TVD SCHEMES FOR HYPERBOLIC CONSERVATION LAWS.
[19] Einfeldt B 1988 *SIAM Journal on Numerical Analysis* **25** 294–318

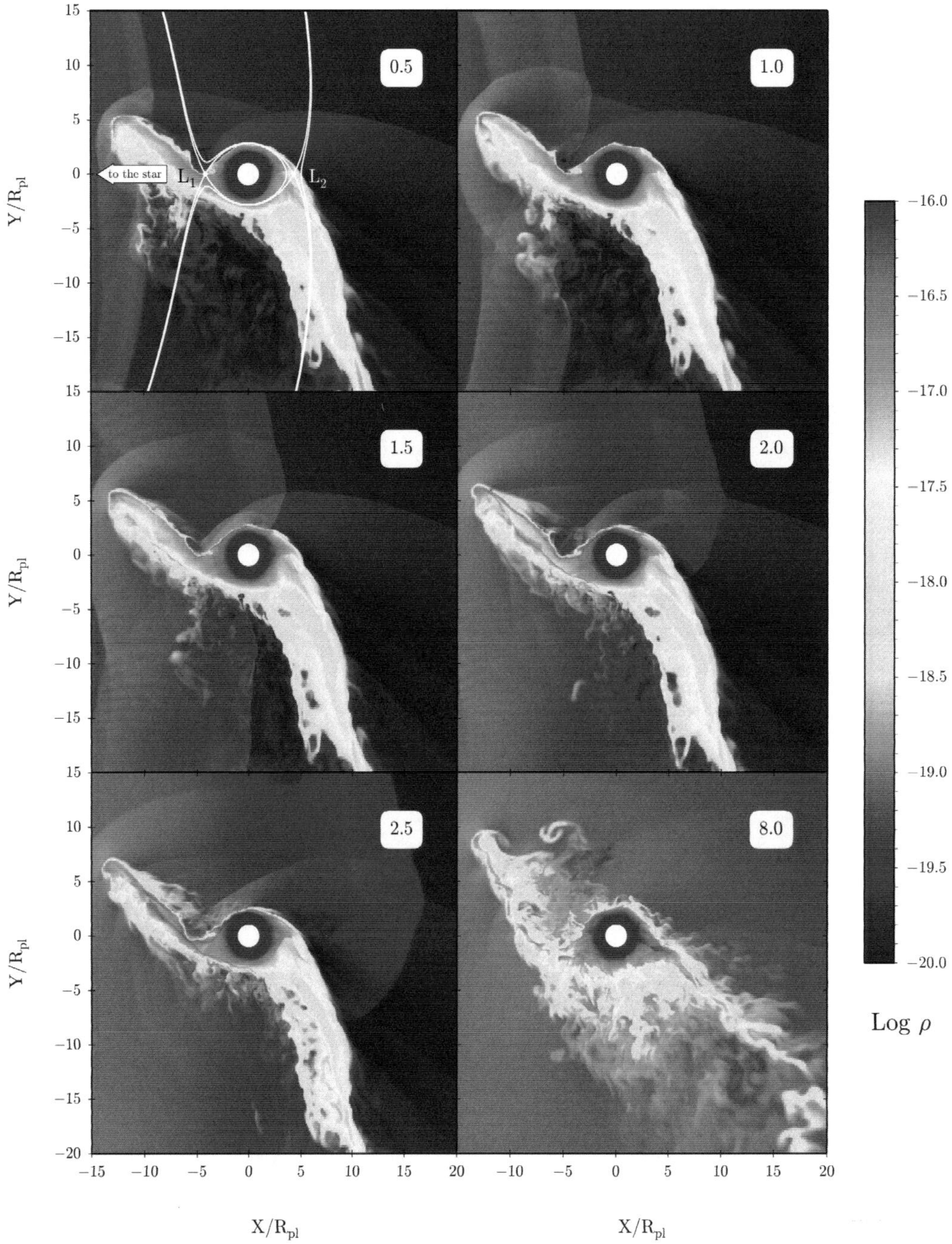

Figure 1. The density logarithm distribution in the equatorial plane of the system for 6 moments during the propagation of the shock wave of CME. The parameters of the atmosphere and the initial state correspond to the model with the temperature of 7500 K described in [9]. On the first diagram white solid lines depict the isolines of the Roche lobe. We also show the positions of the L_1 and L_2 points. White arrow shows direction to the star.

Where should MMS look for electron diffusion regions?

G. Lapenta[1], M. Goldman[2], D. Newman[2], S. Markidis[3]

[1] Center for mathematical Plasma Astrophysics, University of Leuven, KU Leuven, Belgium
[2] Department of Physics, University of Colorado, Boulder, USA
[3] HPCVIZ, KTH, Stockholm, Sweden

E-mail: giovanni.lapenta@wis.kuleuven.be

Abstract. A great possible achievement for the MMS mission would be crossing electron diffusion regions (EDR). EDR are regions in proximity of reconnection sites where electrons decouple from field lines, breaking the frozen in condition. Decades of research on reconnection have produced a widely shared map of where EDRs are. We expect reconnection to take place around a so called x-point formed by the intersection of the separatrices dividing inflowing from outflowing plasma. The EDR forms around this x-point as a small electron scale box nested inside a larger ion diffusion region. But this point of view is based on a 2D mentality. We have recently proposed that once the problem is considered in full 3D, secondary reconnection events can form [Lapenta et al., Nature Physics, 11, 690, 2015] in the outflow regions even far downstream from the primary reconnection site. We revisit here this new idea confirming that even using additional indicators of reconnection and even considering longer periods and wider distances the conclusion remains true: secondary reconnection sites form downstream of a reconnection outflow causing a sort of chain reaction of cascading reconnection sites. If we are right, MMS will have an interesting journey even when not crossing necessarily the primary site. The chances are greatly increased that even if missing a primary site during an orbit, MMS could stumble instead on one of these secondary sites.

1. Introduction

The Magnetospheric Multiscale (MMS) mission launched successfully on March 12, 2015 at 10:44pm. At the time of writing, MMS data is already arriving and being analysed, revolutionising our field with a quantum jump in quality and resolution. The goals of the mission as stated on its web site aim squarely at the core of reconnection physics: *What determines when reconnection starts and how fast it proceeds? What is the structure of the diffusion region? How do the plasmas and magnetic fields disconnect and reconnect in the diffusion regions? What role do the electrons play in facilitating reconnection? What is the role of turbulence in the reconnection process? How does reconnection lead to the acceleration of particles to high energies?* [From: mms.gsfc.nasa.gov].

To answer these questions, MMS must first find a reconnection site and pass through a region where the processes discussed above are happening. A typical configuration of a reconnection site is shown in Fig. 1. Reconnection has a central fulcrum, called x-point where field lines advect to from above and below, break and reconnect to eventually advect outwards left and right [1, 2]. The electrons and ions follow this process as well but become separated in the course of reconnection. The ions separate first in the so-called ion diffusion region (green area)

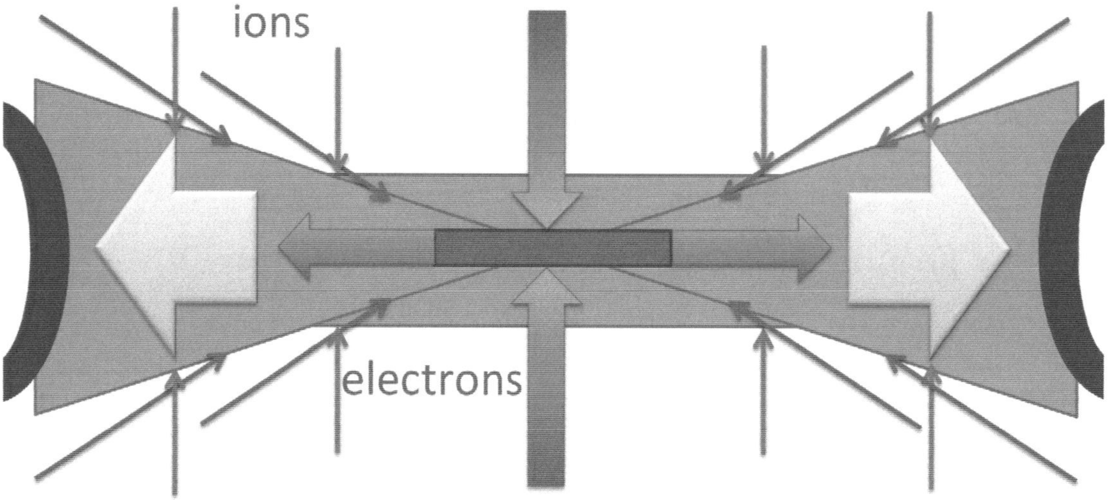

Figure 1. Cartoon of the electron and ion flow patterns in kinetic reconnection. Only a minority of electrons and ions pass near the x-point. Around this point, an electron diffusion region forms with electrons arriving from the inflow region and being ejected as two narrow jets into the outflow (large blue arrows). The vast majority of electrons (blue arrows) and ions (green arrows) pass through the separatrices where their flow speed receives a large increase over a narrow transition layer. The area within the separatrix (in green) is filled with reconnected downstream plasma and the electrons and ions cross at the separatrices. At the exit of the reconnection region the outflow meets its environment forming two fronts (darker green).

where their speed separates from that of the field lines and of the electrons that remain frozen in. The electrons eventually also break their link with the field lines in the electron diffusion region (EDR, in blue). The electrons emerge from the electron diffusion region forming jet that can be deflected by the presence of guide fields [3]. The separatrices play a critical role as most electrons and ions do not move into the reconnection exhaust passing via the central diffusion regions, but rather cross via the separatrices [4].

The central EDR is small, of the electron skin depth scale [2][1]. A crowning achievement for MMS would be to cross a few of these regions and observe their inner structure. Where should we look for them? Where will MMS have the best chances of finding one? Based on Fig. 1 the EDR should be located around the x-point where the separatrices converge.

Figure 1 is a cartoon based on 2D understanding of reconnection. This picture is still valid in 3D only if sufficient symmetry is assumed in the out of page direction for the x-point to become what is defined as a x-line. This concept assumes that the conditions around the Earth are locally relatively azimuthal symmetric, that is the conditions do not change rapidly in the dawn-dusk direction going around the Earth on the ecliptic plane.

However, we know from satellite data that if the large scale structure of the Earth magnetosphere has some regions of slow variation in the azimuthal direction, there are in fact processes that introduce rapid and local changes in the azimuthal direction [5, 6]. Recently we have proposed that one among these instabilities can significantly alter our understanding of where EDR are likely to be found [7]: drift-interchange modes downstream of regions of reconnection outflow can lead to secondary reconnection sites.

The region where the outflow from reconnection come into contact with the pre-existing

[1] See expecially the chapter by Shay and Drake

plasma environment forms fronts [8]. The snow plower analogy is often used to describe the process: like a snow plower pushing the show to the edge of a road forms two crests of snow piling up higher than further afield, similarly the outflow from reconnection tends to form a piling up of vertical magnetic field and of plasma species density. We need to recall that the outflow is generated by the Lorentz force of the curved field lines formed by reconnection [3]. The Maxwell stress pushes the lines to try to straighten them. This force accelerates the plasma outward, left and right of the central x-line, against the environment. This motion encounters resistance from the ambient plasma: the divergence of the pressure tensor balances that of the Maxwell stress tensor. At the front we essentially have magnetic tension pushing against plasma pressure. The force caused by the Maxwell stress tensor is equal and opposite to the divergence of the pressure tensor.

These are the perfect conditions to give rise to the type of instabilities foreseen by Lord Rayleigh and Sir Geoffrey Ingram Taylor: the curvature of the magnetic field lines is co-directed with the pressure gradient, the conditions are met for the interchange mode [9, 10]. However, in plasma physics, interchange modes are much more complex and diversified than in regular fluids [11]. In the conditions typical of the magnetosphere, a fluid treatment is not sufficient and we need to seek a fully kinetic answer. The answer in this case is that the interchange mode acquires a drift nature with the growth rate better described by the standard theory of the lower hybrid drift instability that by fluid models [12, 13]. The dual nature of interchange and drift mode of the present instability is not unique of these conditions and has been observed in tokamak experimental investigations [14] in regions where unfavourable curvature plays a significant role in determining the properties of drift modes. We refer to this instability with the same nomenclature used in fusion research: *drift-interchange*.

We then have a kinetic variant of the Rayleigh-Taylor (RT) mode developing along the advancing fronts. The front will become rippled [6, 15, 16] with the dominant wavelength determined by kinetic theory [15, 12]. The ripples grow and tend to merge in a inverse cascade that produces fewer and bigger finger-like structures characteristic also of common fluids [17, 15].

The question we want to answer here is what happens in the region affected by these fingers?

Figure 2 shows one front with a fully developed system of RT-type ripples. The magnetic field is intertwined with the ripples and the electrons remain frozen in the large fields of the fronts. The interaction among the ripples become then topological relevant and can lead to secondary reconnection sites. Where two interacting ripples happen to push together oppositely directed fields, a new reconnection site can be formed.

In the remainder of the present work we will show evidence for the formation and interaction of the ripples. We will show that, in some regions, nulls form where field lines of opposite polarity are pushed together. We will prove that secondary reconnection develops there and we will show that under common definitions of electron slippage and breaking of the frozen in condition [18], secondary EDR are formed. The results presented here are new but based on the same simulations reported in Ref. [7]. We consider here longer times where the instabilities have further developed and expand the list of diagnostics used to make a positive identification of reconnection and of EDR.

2. Drift-interchange destabilization of fronts formed by reconnection outflows

We consider an initial Harris equilibrium with density:

$$n = n_0 \operatorname{sech}^2(2y/d_i) + n_0/10$$

in a domain of $L_x = 40d_i$, $L_y = 15d_i$ and $L_z = 10d_i$ with d_i defined by n_0. The initial field includes the Harris reversal plus a guide field:

$$\mathbf{B} = B_0\hat{\mathbf{x}} + B_0\hat{\mathbf{z}}/10$$

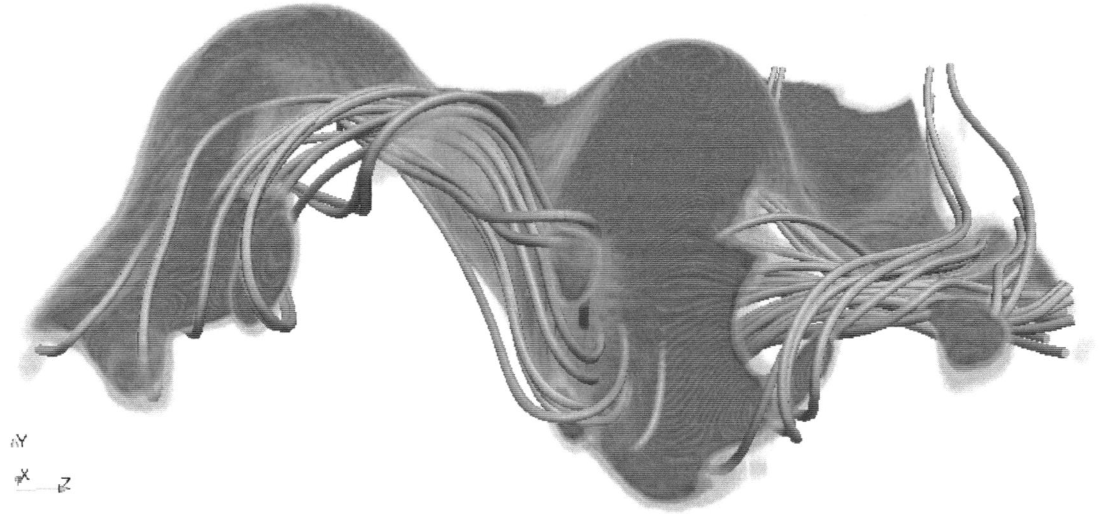

Figure 2. A drift-interchange unstable front generated by the interaction of a reconnection outflow from a reconnection site. The orange surface is a false color representation of the plasma density via volume rendering. The plasma sheet becomes warped and kinked. The magnetic field lines are shown in grey and are intertwined with the ripples in the front. Data from the simulation reported in [7] at cycle 16,000 ($\omega_{ci}t = 19.4$). The point of view is from the central x-line of the primary reconnection site looking towards the positive x-axis and therefore viewing only half of the system in perspective, as indicated by the orientation axis.

Reconnection is initiated uniformly along the z direction via a localized perturbation exponentially decreasing away from the central x-line positioned at $x = L_x/2, y = L_y/2$ (for details of the perturbation, see [19]) and is followed using the iPic3D code [20]. Open boundary conditions are used in x and y to allow reconnection to progress for long times without interference from the boundary. Plasma is allowed to enter from an outside infinite reservoir along the y direction and to exit from the x direction. But the z direction is assumed periodic.

Figure 3 shows the evolution of the ion density and magnetic field on the neutral plane defined by the condition that B_x is minimum. At each x, z coordinate we locate the position y where the amplitude of the magnetic field component $|B_x|$ reaches its minimum and project the false colour representation of the density and magnetic field intensity on this plane to the xz-plane of Fig. 3. The neutral plane is initially located at $y = L_y/2$ but changes its position as the instabilities establish initially and develops in response to the changes induced by reconnection. Following the evolution on the neutral plane has the advantage of providing a 2D view of the most critical 3D plane, simplifying the presentation on a 2D page.

Figure 3 shows the formation from the initial straight x-line in the center of two fronts where the magnetic field piles up (right panels). Following the panels from top to bottom, we can follow the evolution of the fronts in time. Ahead of the fronts, a snow plow effect accumulates plasma density (left panels). As time progresses, the fronts move away from the central line. An instability is seen forming at the fronts leading to their rippling. The instability at the fronts extends also downstream and the plasma between the fronts and the edge of the box is perturbed. The front instability affects then not just the front proper but a vast region downstream of the fronts. It is in this vast region and not just at the fronts where we expect to see secondary reconnection.

The nature of this instability is that of a drift mode [15, 12]: the density at the front has a

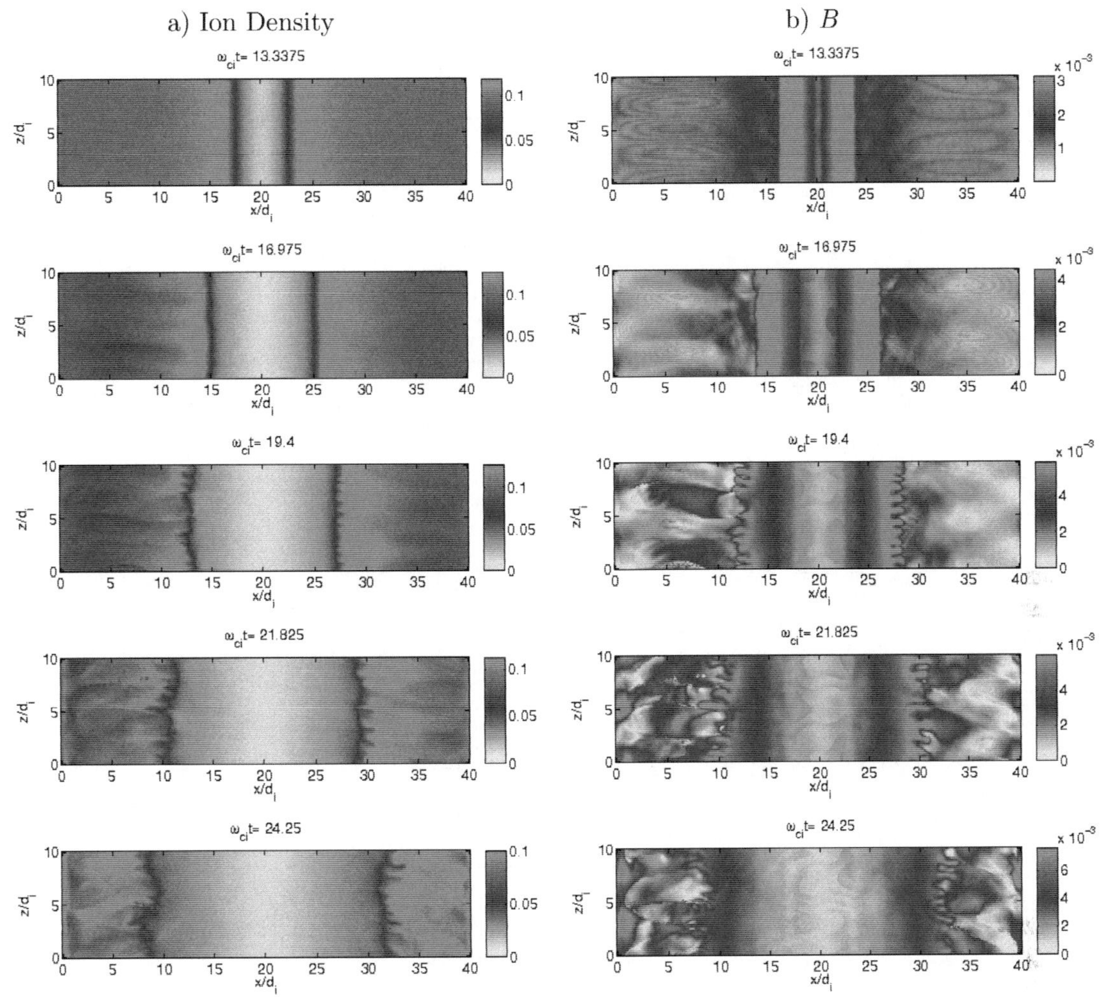

Figure 3. False color representation of the density (left) and magnetic field intensity (right) on the neutral plane (defined as the plane where $|B_x|$ reaches its minimum). Different subsequent times are shown marked by the labels. All quantities are in code units.

strong gradient in a direction normal to the field, the typical condition for drift modes [12]. We reported elsewhere the analysis of the mode spatial and temporal spectrum [7]. The temporal fluctuations at the front peak at the lower hybrid frequency ω_{lh} and the wavenumber of the perturbation along z is in the $(\rho_e\rho_i)^{1/2}$ range. The instability is then reasonably described by the linear theory of the lower hybrid drift instability [12]. Further, theoretical investigation will be required to fully ascertain the role of curvature in these drift-interchange modes [14].

3. Energy exchanges associated with drift-interchange unstable fronts

A first most obvious consequence of the front instability described above is the exchange of large amounts of energy. Figure 4 reports the energy exchange between fields and particles. The energy exchange is not frame independent. The left panels report the analysis in the simulation frame and the right panels that in the local frame moving with the local center of mass (CM) speed. The latter measure has the advantage of removing the energy expended in merely pushing the fronts outward, a work done by the Maxwell stress tensor against the pressure tensor.

Journal of Physics: Conference Series **719** (2016) 012011 doi:10.1088/1742-6596/719/1/012011

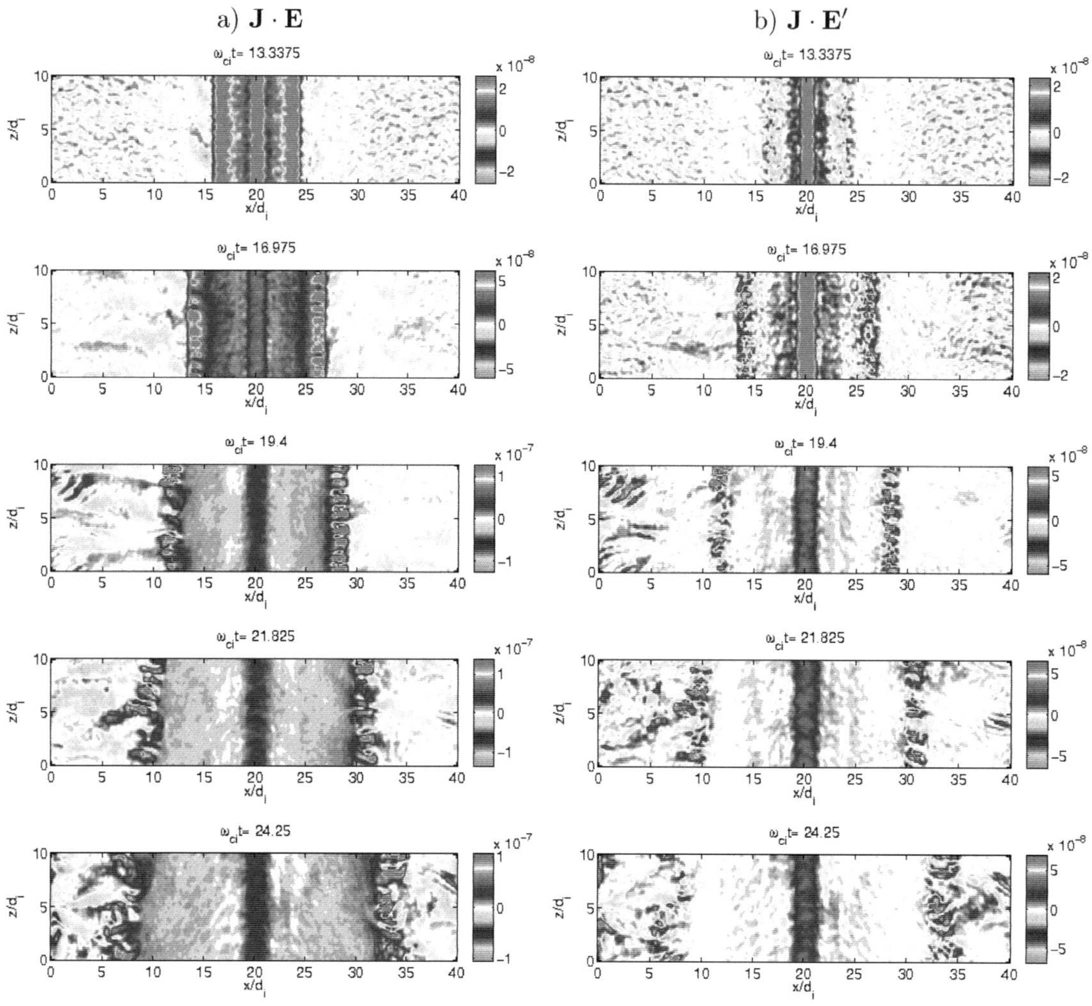

Figure 4. False color representation of the energy exchange term $\mathbf{J} \cdot \mathbf{E}$ in the simulation frame (left) and in the frame $\mathbf{J} \cdot \mathbf{E}' = \mathbf{J} \cdot (\mathbf{E} + \mathbf{u}_{CM} \times \mathbf{B})$ moving with the local center of mass speed $\mathbf{u}_{CM} = (\mathbf{u}_e m_e + \mathbf{u}_i m_i)/(m_i + m_e)$ (right). The results are shown as a xz projection of the neutral plane defined for each x, z by the y location where $|B_x|$ reaches its minimum. Different subsequent times are shown marked by the labels. All quantities are in code units.

Considering the energy exchange in the simulation frame, the fronts are overwhelmingly the dominant loci of energy exchange. However this energy is mostly expended in moving the fronts. When the local CM frame is considered, the main energy exchange $\mathbf{J} \cdot \mathbf{E}'$ is observed at the central x-line.

Even in the $\mathbf{J} \cdot \mathbf{E}'$ measure, the fronts remain host to intense energy exchanges. Following the ripples of the front, one can notice how energy exchange alternates from positive to negative alternating generator regions (where the particle energy is transferred to the fields) to load regions (where electromagnetic energy goes to the particles). Recently, observational evidence has accumulated supporting such strong and at the same time oscillating energy exchanges at the fronts [21, 22]. Based on observations from the Cluster[21] and THEMIS [22] missions, $\mathbf{J} \cdot \mathbf{E}$ has been estimated locally, observing large values in the fronts, with a tendency for load regions to be found near the reconnection site and for a more equal distribution of load and generator regions further downstream from reconnection [21]. This evidence is obviously in agreement with an interpretation based on the drift-interchange mode discussed here [23].

The energy exchange measured in the frame moving with the center of mass speed is essentially equivalent [18] to the dissipation measure recently proposed by Zenitani and collaborators [24]. The virtue of this measure is to identify true local dissipation from energy linked to simple bulk motions. Figure 4 identifies that the fronts themselves as well as many other site downstream of the fronts are regions of intense energy exchanges. Not only the region of the fronts (located in the range $x/d_i \in [5, 8]$ and $x/d_i \in [32, 35]$ at the final time) but also further downstream intense energy exchanges are observed. The present simulation uses open boundary conditions and a relatively large (for 3D runs) box along x, allowing us to follow a significant distance. Future work will be needed on even larger boxes to discover how far downstream these perturbations reach.

Are these regions of intense $\mathbf{J} \cdot \mathbf{E}'$ secondary reconnection sites? And are they surrounded by secondary electron diffusion regions? The next section will show that indeed this is the case with several new reconnection sites generated by the perturbations introduced by the front instability.

4. Evidence for secondary reconnection downstream of drift-interchange unstable reconnection outflows

Recently, Goldman et al. [18] analysed several measures of reconnection that identify topological changes and electron diffusion regions. We have recently applied such measures to show that secondary reconnection sites develop in the reconnection outflow [7].

We extend here the analysis, by considering additional diagnostics (while referring the readers to the previous study for the others [7]) and considering a later time than earlier reported. Figure 5 reports three among the several measures we analysed. All agree to give one clear answer: the regions of intense energy exchange identified above in the reconnection outflow are often also reconnection regions. Future research will investigate whether all regions of energy exchange are also reconnection sites or if energy can be exchanged via other processes also (e.g. shocks, wave-particle interaction). The three measures selected are:

- The electron agyrotropy, defined via the two non zero eigenvalues of the auxiliary pressure tensor computed from the perpendicular velocities [25]:

$$A\O = 2\frac{\lambda_+ - \lambda_-}{\lambda_+ + \lambda_-}$$

  defined positive by definition, $\lambda_+ > \lambda_-$. Electron agyrotropy is taken to measure the extension of the EDR because agyrotropy leads to non ideal terms in the Ohm's law and violation of the frozen in condition.
- The topological measure of violation of conservation of moving magnetic field lines $|\mathbf{B} \times \nabla \times \mathbf{E}_{||}| / B^2$ [26]. A direct measure of reconnection in its topological consequences.
- Perpendicular slippage, $\delta_\perp u_e/c = |\mathbf{u}_{e\perp} - \mathbf{E} \times \mathbf{B}/B^2| /c$, a direct measure that the electrons are not frozen in.

As can be observed the same regions that were identified by the $\mathbf{J} \cdot \mathbf{E}'$ are in most cases also identified by the direct measures of reconnection and break of the frozen in condition. Recalling that we observed above the rippled fronts to be at $x/d_i \in [5, 8]$ and at $x/d_i \in [32, 35]$ at the final time, the observed secondary reconnection sites are not just at the front but also further downstream. For example, prominent is a site at $x/d_i \approx 2.5, z/d_i \approx 2.5$, well downstream of the front.

The secondary reconnection sites are not limited to the neutral plane. At later times the current sheet becomes severely warped and its geometry is complex, see Fig. 2. The analysis

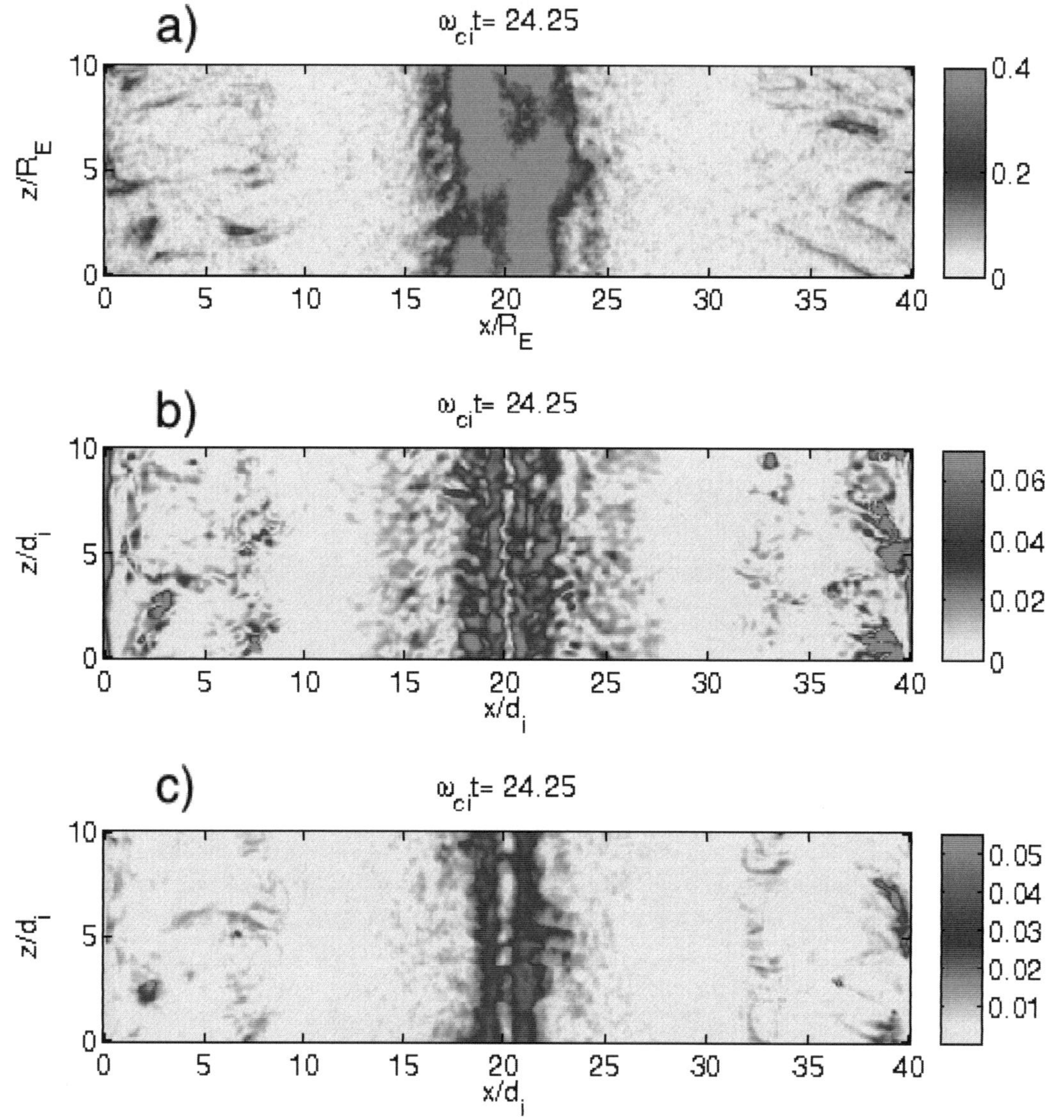

Figure 5. Three reconnection indicators on the neutral plane at the end of the run. From top to bottom: a) agyrotropy $A\O$ of the electron pressure tensor [25]; b) topological indicator $\left|\mathbf{B} \times \nabla \times \mathbf{E}_{||}\right| / B^2$, defined in [26]; c) perpendicular velocity slippage, $\delta_\perp u_e/c = \left|\mathbf{u}_{e\perp} - \mathbf{E} \times \mathbf{B}/B^2\right|/c$.

of reconnection indicators needs to be extended to full 3D visualization. Figure 6 shows a 3D volume rendering of half of the domain ($x < L_x/2$). Two quantities are shown, the energy exchange rate in CM frame and the measure of smallness of the magnetic field intensity (defined as B_0/B). As can be observed the indicators are far from limited to a plane and become volume filling occupying a large potion of the region between the separatrices.

The ultimate proof of secondary reconnection is finding field lines whose topology is incompatible with reconnection happening only at the central x-line. Figure 6 reports in green field lines reconnected only once by the x-line. Two sets are shown. One freshly reconnected and still near the x-line and one already advected far downstream and located just before reaching a secondary reconnection site.

The newly reconnected field lines that have reconnected a second time in the secondary reconnection site are shown in violet. These field lines are not connected to the end wall in x and are in fact spending their entire life inside the exhaust, meandering around. These lines are not initially there before reconnection start (see Ref. [7] for a detailed analysis) and owe their existence to secondary reconnection.

5. Conclusions

In conclusion, we observe that the region of interaction between a reconnection exhaust and the surrounding environment becomes host to a drift-interchange instability that severely perturbs a large area of plasma downstream of the the interaction front. The area in between the separatrices becomes host of secondary reconnection sites, each with its own EDR and topological changes.

This finding is of great present interest for the MMS mission, increasing significantly the chances that EDR can be encountered by chance during the orbit of the four spacecraft. An interesting question will be to determine how a secondary site can be distinguished empirically from a primary site. In our simulations the primary site is a 2D-like reconnection x-line while the secondary site is of the 3D null type. But in reality the transition towards turbulent behavior might spur a cascade of reconnection site where the distinction between primary and secondary becomes meaningless.

Acknowledgments

The present work is supported by the NASA MMS Grant No. NNX08AO84G, by the Onderzoekfonds KU Leuven (Research Fund KU Leuven), by the Interuniversity Attraction Poles Programme of the Belgian Science Policy Office (IAP P7/08 CHARM) and by the DEEP-ER project of the European Commission. The simulations were conducted on NASA (NAS and NCCS) supercomputers, on DOE-NERSC supercomputers, at the VSC Flemish supercomputing centre and on the facilities provided by 4 consecutive PRACE research infrastructure Tier-0 grants.

References

[1] Biskamp D 2000 *Magnetic Reconnection in Plasmas* (Cambridge University Press, UK)
[2] Birn J and Priest E R 2007 *Reconnection of magnetic fields: magnetohydrodynamics and collisionless theory and observations* (Cambridge University Press)
[3] Goldman M, Lapenta G, Newman D, Markidis S and Che H 2011 *Physical review letters* **107** 135001
[4] Lapenta G, Markidis S, Divin A, Newman D and Goldman M 2014 *Journal of Plasma Physics* 1–39
[5] Sergeev V, Runov A, Baumjohann W, Nakamura R, Zhang T, Volwerk M, Balogh A, Reme H, Sauvaud J, André M *et al.* 2003 *Geophysical research letters* **30**
[6] Guzdar P, Hassam A, Swisdak M and Sitnov M 2010 *Geophysical Research Letters* **37** L20102
[7] Lapenta G, Markidis S, Goldman M V and Newman D L 2015 *Nature Physics* **11** 690–695
[8] Sitnov M, Swisdak M and Divin A 2009 *Journal of Geophysical Research: Space Physics (1978–2012)* **114**
[9] Kruskal M and Schwarzschild M 1954 *Proceedings of the Royal Society of London A: Mathematical, Physical and Engineering Sciences* vol 223 (The Royal Society) pp 348–360
[10] Nakamura M, Matsumoto H and Fujimoto M 2002 *Geophysical research letters* **29** 88–1
[11] Pritchett P and Coroniti F 2010 *Journal of Geophysical Research: Space Physics (1978–2012)* **115**
[12] Divin A, Khotyaintsev Y V, Vaivads A, André M, Markidis S and Lapenta G 2015 *Journal of Geophysical Research: Space Physics* **120** 2675–2690
[13] Divin A, Khotyaintsev Y V, Vaivads A and André M 2015 *Journal of Geophysical Research: Space Physics*
[14] Poli F M, Brunner S, Diallo A, Fasoli A, Furno I, Labit B, Mller S H, Plyushchev G and Podest M 2006 *Physics of Plasmas* **13** 102104
[15] Vapirev A, Lapenta G, Divin A, Markidis S, Henri P, Goldman M and Newman D 2013 *Journal of Geophysical Research: Space Physics* **118** 1435–1449

[16] Lapenta G and Bettarini L 2011 *Geophysical Research Letters* **38** DOI: 10.1029/2011GL047742
[17] Lapenta G and Bettarini L 2011 *EPL (Europhysics Letters)* **93** 65001 (*Preprint* 1102.4791)
[18] Goldman M, Newman D and Lapenta G 2015 *Space Sci. Rev.* **154** DOI: 10.1007/s11214–015–0154–y
[19] Lapenta G, Markidis S, Divin A, Goldman M and Newman D 2010 *Physics of Plasmas* **17** 082106 (pages 11)
[20] Markidis S, Lapenta G and Rizwan-uddin 2010 *Mathematics and Computers and Simulation* **80** 1509–1519
[21] Hamrin M, Marghitu O, Norqvist P, Buchert S, André M, Klecker B, Kistler L M and Dandouras I 2011 *Journal of Geophysical Research: Space Physics (1978–2012)* **116**
[22] Angelopoulos V, Runov A, Zhou X Z, Turner D, Kiehas S, Li S S and Shinohara I 2013 *science* **341** 1478–1482
[23] Lapenta G, Goldman M, Newman D, Markidis S and Divin A 2014 *Physics of Plasmas (1994-present)* **21** 055702
[24] Zenitani S, Hesse M, Klimas A and Kuznetsova M 2011 *Physical review letters* **106** 195003
[25] Scudder J and Daughton W 2008 *Journal of Geophysical Research: Space Physics (1978–2012)* **113**
[26] Hesse M and Schindler K 1988 *Journal of Geophysical Research: Space Physics (1978–2012)* **93** 5559–5567

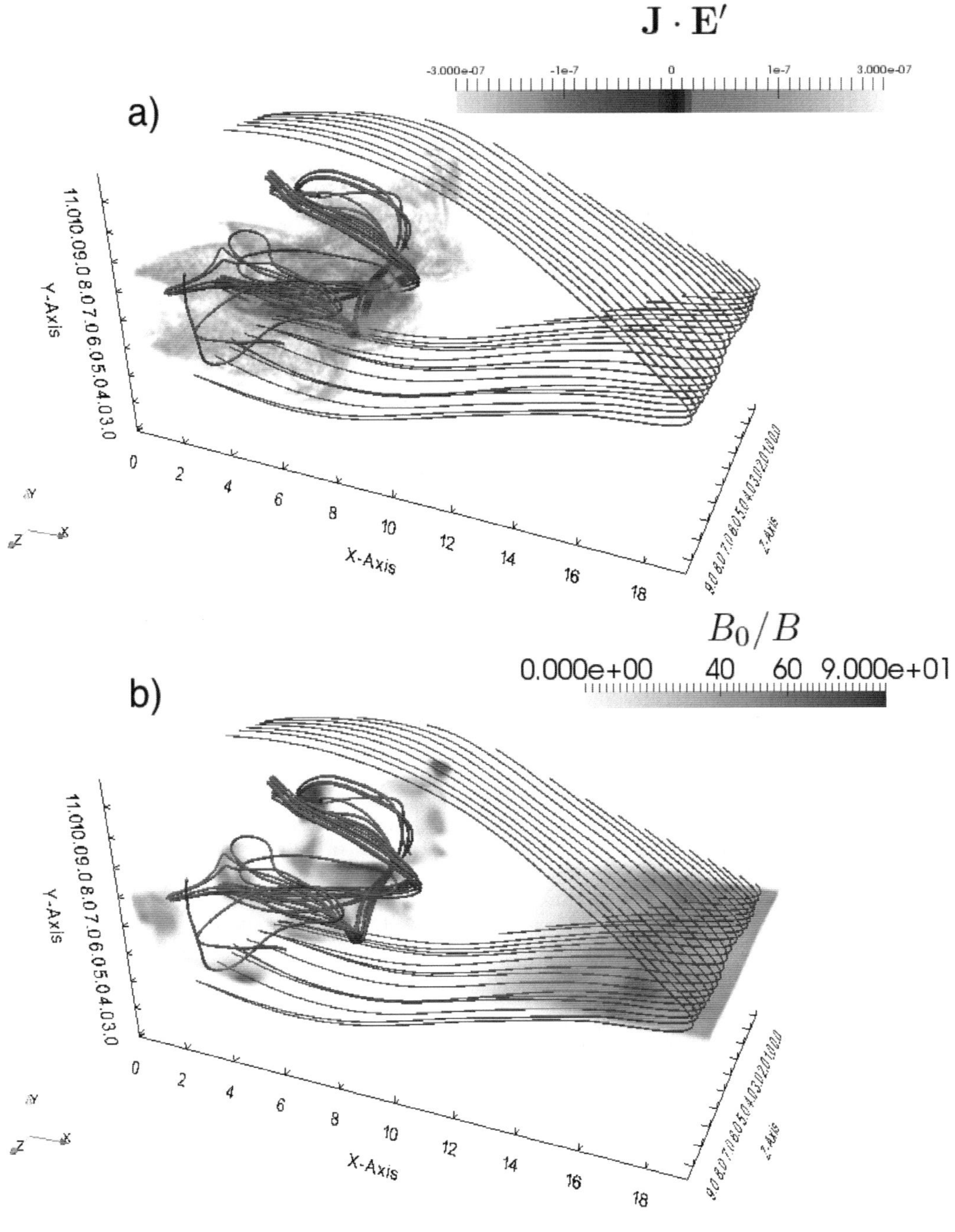

Figure 6. Three dimensional volume rendering of the energy exchange rate $\mathbf{J} \cdot \mathbf{E}'$ measure in the CM frame (a) and of the inverse of the magnetic field intensity B_0/B (b), a measure of local magnetic field vanishing. Selected field lines are show. A set of field lines just reconnected near the x-line is shown in green. Another set of field lines reconnected once at the x-line but already advected far down stream are shown also in green in proximity of a secondary reconnection site. A final set of field lines that have reconnected a second time at the secondary reconnection site is shown in violet.

An Empirically Driven Time-Dependent Model of the Solar Wind

Jon A. Linker[1], Ronald M. Caplan[1], Cooper Downs[1], Roberto Lionello[1], Pete Riley[1], Zoran Mikic[1], Carl J. Henney[2] Charles N. Arge[2], Tae Kim[3] and Nikolai Pogorelov[3]

[1]Predictive Science Inc., San Diego, CA 92121
[2]Space Vehicles Directorate, Air Force Research Laboratory, Kirtland AFB, NM 87117
[3]Department of Space Science, University of Alabama, Huntsville, AL35899

E-mail:
linkerj@predsci.com,caplanr@predsci.com,cdowns@predsci.com,lionel@predsci.com, pete@predsci.com,mikicz@predsci.com, cjhenney@gmail.com, afrl.rvb.pa@kirtland.af.mil,tkk0023@uah.edu,np0002@uah.edu

Abstract. We describe the development and application of a time-dependent model of the solar wind. The model is empirically driven, starting from magnetic maps created with the Air Force Data Assimilative Photospheric flux Transport (ADAPT) model at a daily cadence. Potential field solutions are used to model the coronal magnetic field, and an empirical specification is used to develop boundary conditions for an MHD model of the solar wind. The time-dependent MHD simulation shows classic features of stream structure in the interplanetary medium that are seen in steady-state models; it also shows time evolutionary features that do not appear in a steady-state approach. The model results compare reasonably well with 1 AU OMNI observations. Data gaps when SOLIS magnetograms were unavailable hinder the model performance. The reasonable comparisons with observations suggest that this modeling approach is suitable for driving long term models of the outer heliosphere. Improvements to the ingestion of magnetograms in flux transport models will be necessary to apply this approach in a time-dependent space weather model.

1. Introduction

The outer atmosphere of the Sun, the solar corona, is a plasma that expands to become the supersonic solar wind, which envelopes the planets and forms the heliosphere. The structure and dynamics of the solar wind are strongly influenced by the solar magnetic field, including the locations of fast and slow solar wind and the position of the heliospheric current sheet (HCS). The magnetized solar wind is also the primary medium by which solar activity is transmitted to the Earth and beyond, in the form of coronal mass ejections (CMEs), which evolve and propagate in the wind, and energetic particles, which are transported along the magnetic field. Realistic models of the solar wind (in the sense of having sufficient fidelity to be compared with observations) are thus important not only for understanding how the large-scale heliosphere arises from coronal structure and dynamics that we observe remotely, but also for the prediction of solar wind impacts on space weather at Earth.

The key observational input to coronal and solar wind models are maps of the global photospheric radial magnetic field (B_r). Global solar maps are developed from full-disk magnetograms of the line-of-sight (LOS) magnetic field (in turn derived from measurements of the Zeeman splitting of spectral lines) and are available from ground and space-based observatories. These observations are presently available only from the Earth view, so the maps are built up over the course of a solar rotation. Often referred to as "synoptic" maps, they are really diachronic maps [1]. These maps serve as an essential boundary condition for coronal models, be they potential field or full magnetohydrodynamic (MHD).

MHD models with a lower boundary at the solar surface can in principle predict both the structure of the solar corona and the solar wind. However, the processes that heat the solar corona and accelerate the solar wind occur on scales far below that of a global model, are far from completely understood, and indeed may not be realizable in single fluid MHD. An active area of research is to develop global MHD solutions that parameterize/model these processes to fully describe the detailed properties of the corona (as observed in EUV and X-ray emission, for example) and also predict the properties of the solar wind [2, 3, 4].

Empirical models have been shown to be able to predict solar wind structure with reasonable success from a magnetic field model based on observed photospheric fields [5, 6, 7, 8]. At the present time, the U.S. National Centers for Environmental Prediction compute potential field source surface (PFSS) [9] and potential field current sheet (PFCS) [10] models for the magnetic field based on photospheric magnetic maps, and then use the Wang-Sheeley-Arge (WSA) model [8] to empirically specify the boundary conditions for the Enlil MHD model [11]. The MHD equations are integrated to steady state. These solar wind solutions are used to predict the arrival of fast solar wind streams and as the background state for "cone model" CME simulations. Models of the ambient corona and solar wind are typically integrated to steady state, in part because of the lack of a description of the time-evolving photospheric magnetic field. However, the processes by which the magnetic flux on the Sun evolves have been studied for many years. Flux transport models [12, 13, 14, 15] incorporate these processes (primarily differential rotation, meridional flow, supergranular diffusion, and random flux emergence) and have been successful in predicting the evolution of photospheric magnetic fields. These models can assimilate magnetograms from available observatories to produce a continuous approximation of the state of the photospheric magnetic field, as a sequence of synchronic maps. The Air Force Data Assimilative Photospheric flux Transport (ADAPT) model [16] is a unique example in that it includes rigorous data assimilation and ensemble modeling techniques [17].

In this paper, we demonstrate a time-dependent, empirically driven MHD model of the solar wind. We describe how sequences of ADAPT maps can be used to create time-dependent boundary conditions for an evolving solar wind model. By construction, our simulation does not include the effects of CMEs, but focuses on modeling the ambient wind through which CMEs propagate. A goal of this work is to develop a time-dependent model of the solar wind for the 5-year period from 9/27/2003 to 9/27/2008, to be used to drive simulations of the outer heliosphere using the MS-FLUKSS code [18]. We show our initial results for the time period 9/27/2003-9/27/2004, and compare them with OMNI observations.

2. Model Description

2.1. Empirically Driven Models

The ambient solar corona (and in turn the solar wind) restructures in response to changes in the photospheric magnetic field. To capture the evolution of the Sun's magnetic field, we use maps of the photospheric magnetic field derived from the ADAPT model. For the time period of interest, ADAPT was run using daily full disk magnetograms obtained from National Solar Observatory's Synoptic Optical Long-term Investigations of the Sun (NSO SOLIS) Vector Spectromagnetograph (VSM). The maps were provided on a 180 × 360 uniform latitude and longitude mesh. We smoothed the maps to remove abrupt pixel to pixel variations. Eventually, we would like to compute time-dependent MHD solutions of the solar corona driven by ADAPT maps [19], but the calculation of such solutions over years of time would require a large computational investment. In the place of an MHD model for the magnetic field, we first compute a PFSS model from 1 to 2 solar radii (R_S) for daily ADAPT maps, followed by a PFCS model to compute the field from 2-30R_S. This last step is required to obtain a field where $|B_r|$ shows little variation in latitude, consistent with observations [20]; such configurations arise naturally in MHD models. The PFSS and PFCS models are computed numerically on nonuniform $101 \times 181 \times 361$ (r, θ, ϕ) spherical meshes using a preconditioned conjugate gradient method. We note that during 2004, there were a number of significant data gaps for SOLIS magnetograms, related to testing the VSM and moving it to Kitt Peak. During these time periods, ADAPT continued to produce daily synchronic maps, but there are obviously quality problems as the model will drift away from observations during these time periods. We will return to this issue in section 3.3.

Empirically driven MHD models of the solar wind can circumvent the requirement of a detailed coronal model. Reference [5], using PFSS models, first showed an inverse correlation between coronal flux tube expansion (known as the expansion factor) and solar wind speed. This technique was later implemented using PFSS models based on daily updated observatory maps [6]. Reference [7] developed an empirical technique using polytropic MHD models as the the underlying field model and a mapping based on the distance to the coronal hole boundary (DCHB); this empirical method was then used to form the boundary conditions for solar wind models with an inner radius beyond the Alfven and sonic critical points ($30R_S$). Reference [8] augmented the expansion factor technique used in [6] with a factor based on the DCHB; this is the present WSA model. Recently, reference [21] compared the three empirical techniques and showed that the techniques of [8] and [7] produce similar results that are superior to use of the expansion factor alone. The WSA and DCHB models produce similar results because the slow wind produced in the WSA model arises predominantly from the DCHB term. We use the DCHB model to specify the velocity as a 4π steradian map at $30R_S$; the density n is then computed from the empirical observation of roughly constant momentum flux [22], and the temperature T from thermal pressure balance. The details are described by reference [7]. Figure 1 summarizes our empirically driven model.

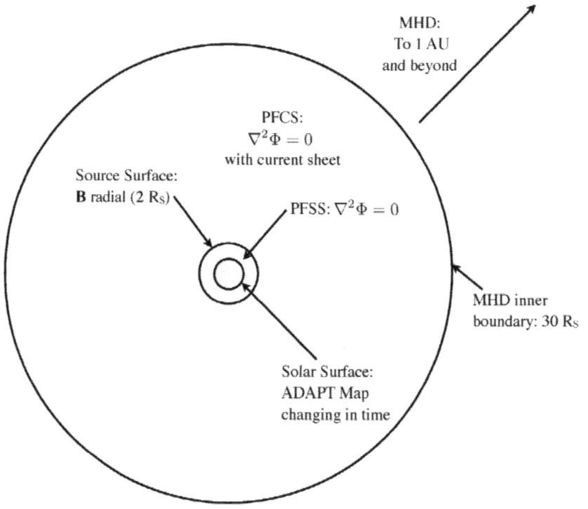

Figure 1. Diagram showing the components of the empirically driven time-dependent model. Sequences of ADAPT maps are supplied at the solar boundary. These are used to calculate sequences of PFSS and PFCS models, that in turn provide the radial magnetic field at the inner boundary of the heliospheric MHD model. The DCHB prescription (see text) is used to provide radial velocity, density, and temperature at the MHD inner boundary.

2.2. MHD model

The empirical procedure described above provides us with a sequence of boundary maps at $30R_S$ of radial magnetic field, density, temperature, and radial velocity, in the rotating frame of the Sun, at a cadence of 1 per day, to drive our heliospheric MHD solutions. The solutions are

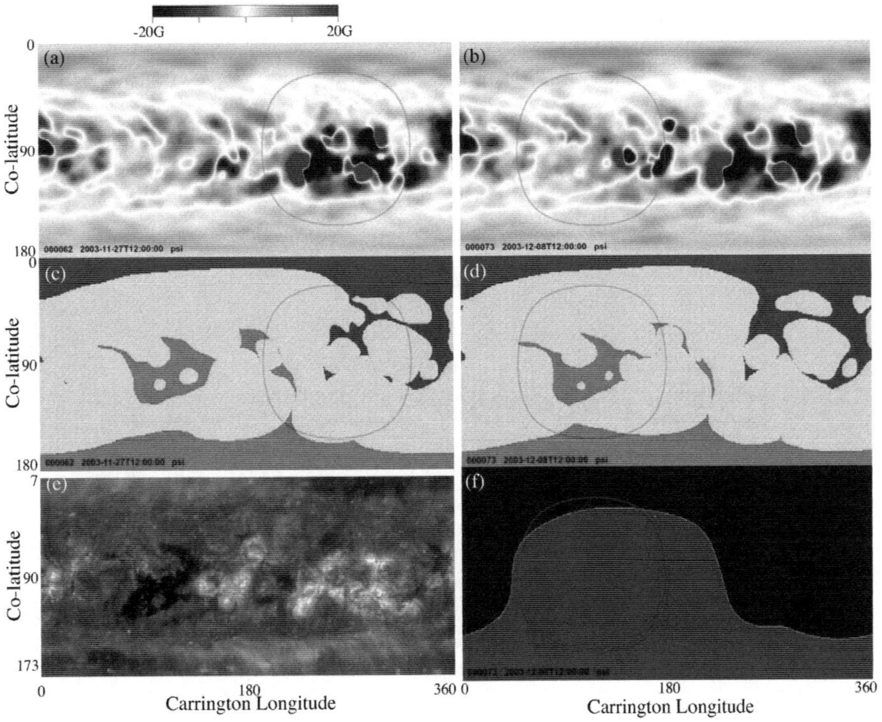

Figure 2. Boundary conditions and results for PFSS models. (a) ADAPT magnetic map for 11/27/2003 at 12:00UT. The approximate location of the assimilation window for new magnetic data is shown as a green oval. (b) The same as (a) but for 12/08/2003 at 12:00UT. (c) Signed coronal hole boundaries (blue for inward magnetic flux, and red for outward) for the PFSS model on 11/27/2003. (d) The same as (c) for 12/08/2003. (e) Synoptic EIT 195Å image for this time period. (f) Position of the HCS (boundary between blue and red) at 12/08/2003.

computed on a nonuniform $265 \times 181 \times 361$ spherical mesh extending from $30 - 230R_S$ with the Magnetohydrodynamic Algorithm outside a Sphere (MAS) code (Earth is at $\sim 215R_S$ or 1 AU). MAS has been used extensively in models of coronal structure [23, 24, 2, 25], coronal dynamics [26, 27, 28] and coronal mass ejections [29, 30]. MAS solves the time-dependent resistive MHD equations on a nonuniform spherical coordinate grid using a semi-implicit algorithm [31]. The code is massively parallel and scales to thousands of cores. Among its many advantageous properties, the formulation exactly preserves $\nabla \cdot \mathbf{B} = 0$. MAS is the primary MHD model in CORHEL (Corona-Heliosphere), a suite of models [32] for describing the solar corona and inner heliosphere.

Our method for coupling coronal and heliospheric solutions (including the set of equations solved) is described by [30]. Here we make a few brief remarks. As the inner boundary for the calculation is chosen beyond the critical points, all of the MHD values should be specified, subject to the constraint $\nabla \cdot \mathbf{B} = 0$. The heliospheric calculations are performed in the inertial frame, so the supplied boundary values must be transformed from the rotating frame to the inertial frame. In practice this means the boundary values move in the longitudinal direction with the Sun's rotation rate; this is accomplished by linearly interpolating the values to their new position on the grid at each time step. The boundary values also change in time as the PFSS/PFCS solutions change. As the times steps in the calculation ($\sim$2 minutes) are much smaller than 1 day, values are interpolated in time between the values specified for each PFSS/PFCS. The

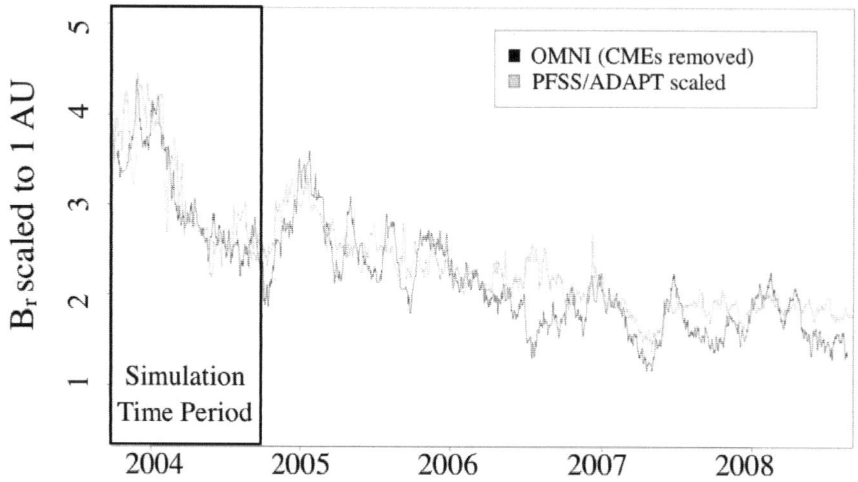

Figure 3. The magnitude of the radial magnetic field at 1 AU from OMNI data (black) and the average global radial magnetic field at 1 AU predicted from PFSS models using ADAPT maps (green). The ADAPT maps were scaled by a factor of 1.5 (see text). ADAPT captures the general trend of the interplanetary magnetic field during this time period, when there was a large-scale decrease in the field as part of the approach to and during the recent unusual solar minimum.

magnetic field is advanced via specification of the tangential (to the boundary) electric field; in practice this is accomplished by a Helmholtz decomposition of the electric field into two potentials. As described by [30], this has the advantage of exactly preserving the supplied B_r. At the outer boundary, all the quantities are computed with the help of characteristic equations.

3. Results

3.1. PFSS solutions

Figure 2 shows the PFSS solutions for two time snapshots: 11/27/2003 at 12:00UT (Figures 2(a)&(c))and 12/08/2003 at 12:00UT (Figures 2(b),(d)&(f)). Figures 2(a)-(b) show the ADAPT maps used as boundary conditions. The maps are scaled from -20 to 20 G to better illustrate the structure in the weaker field regions. Figure 2(c)-(d) shows the signed open/closed regions (proxies for coronal hole boundaries), with blue indicating inward-pointing open field regions ($-B_r$), red indicating outward-pointing open field regions ($+B_r$), and gray indicating closed field regions. It is the structure of the open/closed regions that determines the parameters that drive the heliospheric model. Figure 2(e) shows a SOHO Extreme Ultraviolet Imaging Telescope (EIT) 195Å "synoptic" map [33] for this time period (Carrington rotation 2010; 11/19-12/17/2003). The EIT map is built up over the solar rotation from measurements near disk center, similar to a magnetic synoptic map (available at http://sun.stanford.edu/synop/EIT). The map is filled between 7°-173° co-latitude. The dark regions in emission are observed coronal holes. The open field regions in the model correspond roughly to the EIT map, although the shapes are somewhat different and in general the modeled coronal holes are somewhat larger than those observed; this could in part be due to obscuration of coronal holes by nearby active region loops. Figure 2(f) shows the HCS location predicted by the model at this time.

We can compare the heliospheric open magnetic flux predicted by the model with observations of B_r at 1 AU from OMNI (http://omniweb.gsfc.nasa.gov). We calculate an equivalent field strength at 1 AU by computing the magnitude of the open flux in the model, dividing by the

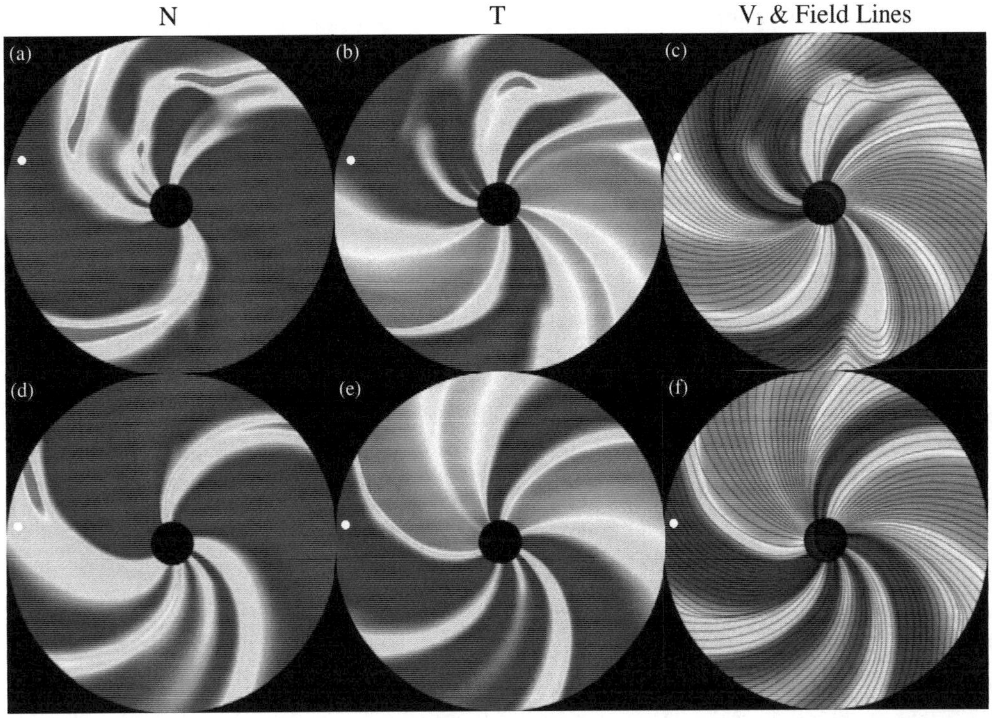

Figure 4. Time-dependent solar wind simulation results in the heliographic equator. Red indicates the largest values and blue the smallest. The Earth's location is shown as white disk in each image. Figures (a)-(c) correspond to 11/27/2003 15:07UT while Figures (d)-(f) are 8 days later at 12/08/2003 15:07UT. Figures (a) and (d) show plasma density, (b) and (e) show plasma temperature. Figures (c) and (f) show the radial velocity and magnetic field lines traced from the outer boundary of the calculation inward. The field lines are traced in 3D, so lines that appear to end actually pass beneath the displayed surface.

surface area at the upper boundary, and scaling to 1 AU using the expected r^{-2} fall off. We have found that coronal models, whether simple (PFSS) or complex (e.g. thermodynamic MHD models), tend to underestimate the interplanetary magnetic field (IMF) when using typical observatory maps [34, 1]; the reasons for this are unresolved. To account for this underestimate, we uniformly multiply the ADAPT values by 1.5. During 2003-2008, the IMF exhibited an unprecedented drop in IMF field strength [35], associated with the approach to the unusual solar minimum of 2008-2009. Figure 3 shows that the model with ADAPT maps captures this trend rather well, particularly in the time period of the simulation presented here (2003-2004).

3.2. Time-dependent Solution

The MHD solution is initialized by taking the first ADAPT map on 9/27/2003, computing PFSS and PFCS models, and empirically prescribing the boundary values at $30R_S$ as described in section 2.1. The initial magnetic field in the heliosphere is a potential field consistent with the supplied B_r. We advance the solution for 200 hours of real time assuming the boundaries are not changing (in the rotating frame) to obtain a steady-state solar wind with a Parker spiral magnetic field. We then proceed to compute the time-dependent solution until 9/27/2004, modifying boundary values as described in section 2.2.

We illustrate the behavior of the time-dependent solution in Figure 4, which shows the

Journal of Physics: Conference Series **719** (2016) 012012 doi:10.1088/1742-6596/719/1/012012

solution at 11/27/2003 15:07UT and 12/08/2003 15:07UT in the heliographic plane. The approximate location of the Earth (actually slightly out of this plane) is shown as a white disk. The density (n) has been scaled by r^2 and the temperature (T) by r in the images for easier visualization of the quantities. The figure shows classic features of interplanetary solutions, with corotating fast and slow solar wind streams (Figures 4(c)&(f)). The density n in Figures 4(a)&(d) is enhanced at streamer interface regions; these regions are at the leading edge of the high-speed wind where the interaction between fast and slow wind occurs. Temperature (Figures 4(b)&(e)) is enhanced at the streamer interface and within the fast streams. Superimposed on the radial velocity (v_r) in Figures 4(c)&(f) are magnetic field lines traced from $230R_S$ inward. The field lines are traced in three dimensions so lines that appear to end actually pass beneath the velocity surface. Figure 4(f) shows the expected spiral magnetic field lines.

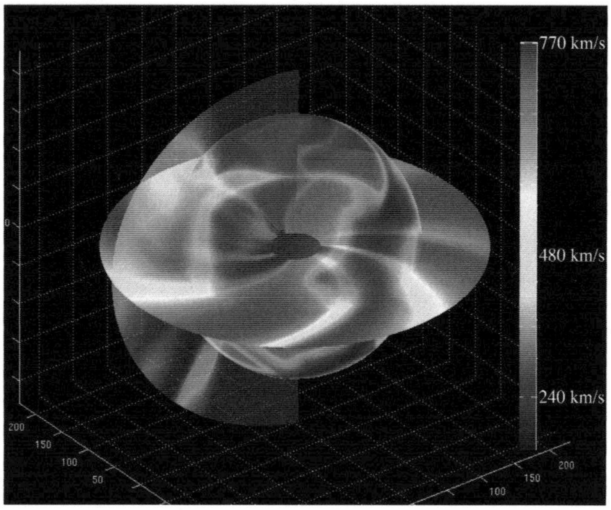

Figure 5. 3D view of the radial velocity at 11/27/2003 15:07UT. Values of velocity are shown on a surface in the heliographic plane, on a semi-transparent sphere at $r = 150R_S$, and on a semi-transparent longitudinal cut plane at the location of the Earth.

The solution also exhibits features that are not found in a steady-state description, particularly in the 11/27/2003 frames. Figures 4(a)-(c) show the formation of a new fast stream (near the top of each figure) that has not yet propagated out to 1 AU. This feature is associated with a restructuring of the magnetic field; note the presence of disconnected field lines (i.e., field lines that do not trace back to the Sun but are connected at both ends to the outer boundary) in the upper and lower parts of Figure 4(c). These disconnections occur at the HCS and appear to come about because, at times, the HCS at the lower boundary can change significantly on the order of a day. If we were using a true time-dependent coronal model to drive the heliospheric solutions, these reconfigurations might be topologically different; closed field lines might expand outward and reconnect with surrounding open field lines in so-called interchange reconnection [36].

Figure 5 shows a three-dimensional view of the v_r at 11/27/2003 15:07UT. The opaque plane cutting through the center of the plot shows v_r in the heliographic plane, the same as in Figure 4(c). Near the north pole of the semi-transparent sphere and in the longitudinal cut plane, high speed wind is seen to dominate; however, there is a mix of slow and fast wind at all latitudes, arising from the complex coronal structure present at this time period.

3.3. Comparison with Observations

To test the simulation results against observations, we collect data from the simulation at the L1 point every 6 hours to form a time series. As the model is only realistic for large scale structures, we compare it with 6 hour running averages of B_r, v_r, n, and T from OMNI, also plotted every 6 hours. The results from the entire year of the simulation are shown in Figures 6-7. Also shown as green boxes are SOLIS data gaps, defined as periods where SOLIS magnetograms were unavailable for three or more days (the box is placed on the third and subsequent days). CMEs

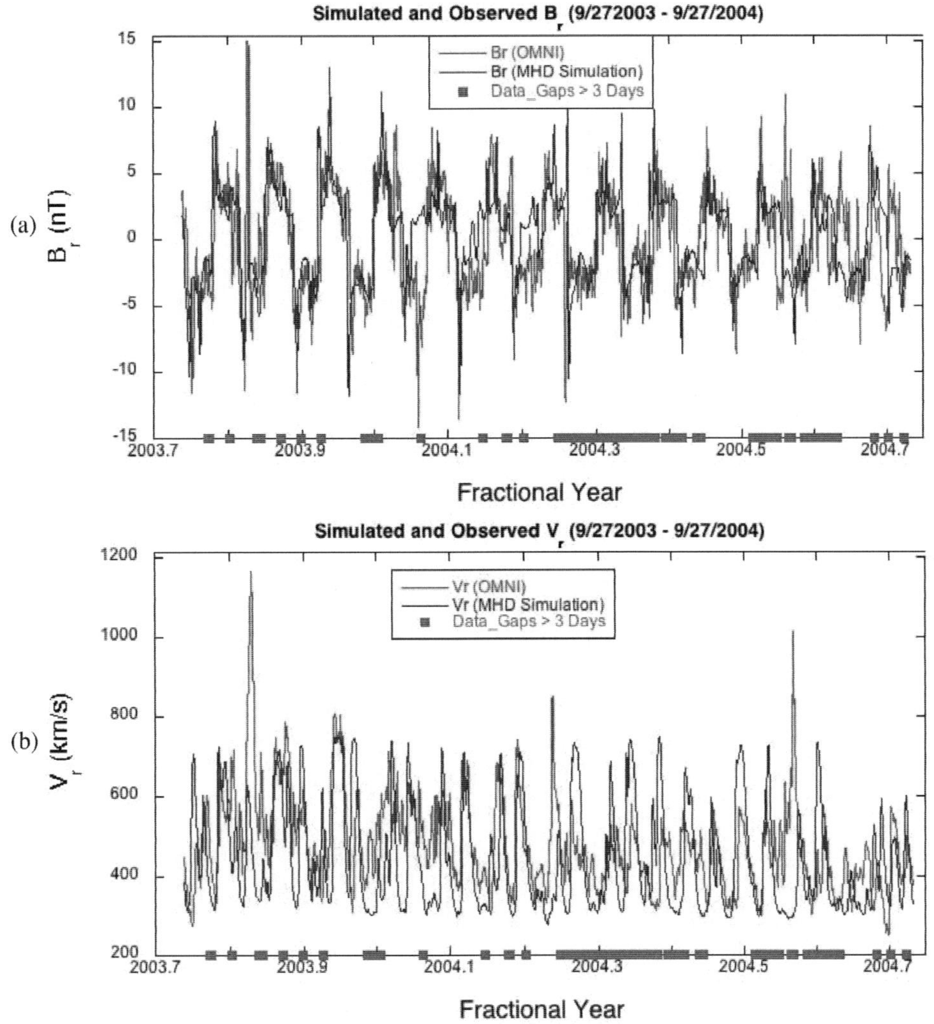

Figure 6. Comparison of the simulation results with OMNI observations for 9/27/2003-9/27/2004. Time is given along the x axis as fractional year. Data are plotted every 6 hours. Time periods when SOLIS magnetograms were unavailable for the ADAPT model for more than 3 days are shown as green boxes at the bottom of each plot. (a) The radial magnetic field (B_r) in nT. (b) The radial velocity (v_r) in km/s.

have not been removed from the data set, the largest of which are noticeable as velocity spikes in Figure 6(b).

Figure 6(a) shows that the simulation reproduces the overall sector structure and magnitude of the heliospheric magnetic field quite well. The location of fast streams also often show a good correspondence with the observed streams (Figure 6(b)), particularly in 2003 and early 2004. Figure 7(a) shows that the model uniformly predicts much higher densities at the stream interfaces than are observed. This discrepancy appears to arise in the boundary specification, where the momentum flux is assumed to be constant. The lower speed wind in the vicinity

(a)

(b)

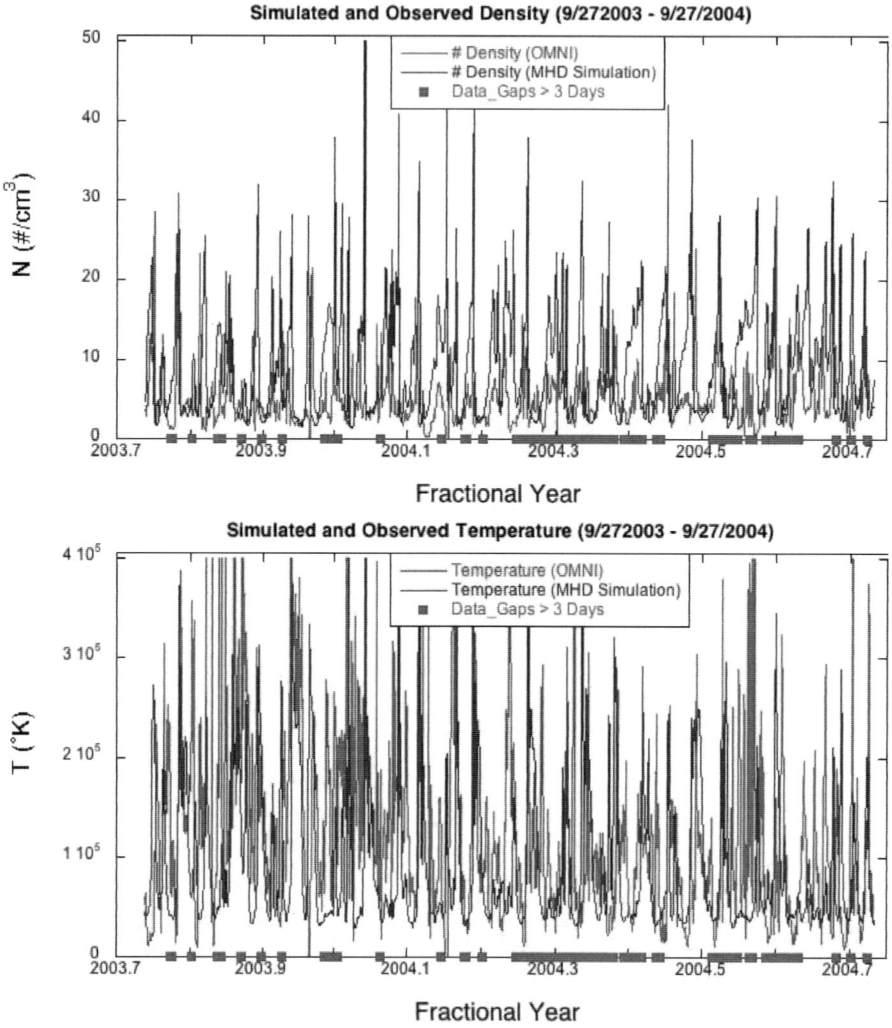

Figure 7. Comparison of the simulation results with OMNI observations for 9/27/2003-9/27/2004, in the same format as Figure 6. (a) The number density n in cm^{-3} (b) The temperature T in Kelvin.

of the HCS makes the density much higher there, indicating that this aspect of the empirical specification can perhaps be improved. The plasma temperature in the fast streams is roughly similar to the observations (Figure 7(b)), but there are temperature spikes in the OMNI data that are not associated with fast speed wind and are not reproduced by the model.

It is perhaps surprising that B_r shows reasonable correspondence with the OMNI data even during some of the time periods of long SOLIS data gaps, such as 2004.25 to 2004.45 (3/31-6/01/2004), when the VSM was operational for only 7 out of 63 days. However, while the location of the fast streams during this period were not entirely dissimilar from the OMNI data, the model predicted much faster speeds than were measured. During another period of long data gaps (2004.51-2004.61, 7/6-8/10/2004, VSM operational for 12 out of 36 days) the model

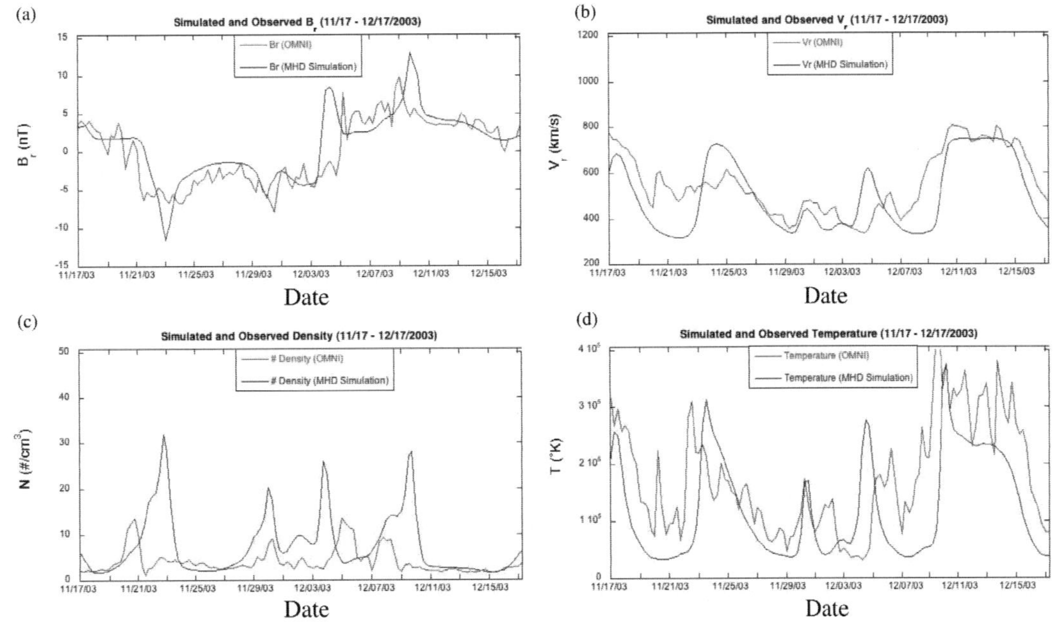

Figure 8. Comparison of the simulation results with OMNI observations for 11/17-12/17/2003. Data are plotted every 6 hours. This period had few SOLIS data gaps. (a) The radial magnetic field (B_r) in nT . (b) The radial velocity (v_r) in km/s. (c) The number density n in cm^{-3} (d) The temperature T in Kelvin.

appears to drift away from reality for all quantities.

For a more detailed comparison, we examine the same quantities in Figures 6-7 for a 30 day period from 11/17-12/17/2003 in Figure 8. This time period had only a few SOLIS data gaps. The magnitude and sign of B_r from the model correspond well with the observations, as do v_r and T. As we saw in Figure 7(a), the model predicts densities at the stream interfaces that are much higher than observed. While the solution corresponds quite well with the observations in a global sense, the predicted arrival of the HCS crossings and the fast wind streams can differ from the measurements by as much as a day. The measurements (even with 6 hour averages) also show more structure than is in the model.

4. Summary & Discussion

We have developed a time-dependent model of the solar wind using ADAPT maps as input. For the year of time simulated here (9/27/2003-9/27/2004), the model compares reasonably well with OMNI measurements. It is perhaps surprising that the model performed well in late 2003, because the Sun was very active at that time, with some of the largest CMEs ever recorded (and are not included in the model). Despite this activity, the model comparisons indicate that the large scale heliospheric structure was reproduced during this time period, indicating that this is a valid approach for driving models of the outer heliosphere over long time intervals. Data gaps for SOLIS magnetograms certainly hinder the performance of the model; there are fewer of these in subsequent years. By circumventing the calculation of a full coronal model, the computational requirements for modeling the inner heliosphere are greatly reduced. The one year simulation required about 20 hours of wall clock time on 576 processors.

It is interesting to speculate on the usefulness of this approach for space weather models.

While it does not appear thus far that the results are markedly better than solutions computed for individual Carrington rotations, the time-dependent approach is more efficient than computing sequences of steady-state models with an artificial relaxation time for each solution. A higher time cadence of maps (as is possible now with GONG or HMI) may improve the results and be essential for space weather applications. However there are important data issues to address if this approach is to be used in a more practical or operational setting.

In their present state, ADAPT maps (or those from any flux transport model) differ significantly from the true evolution of the Sun's photospheric magnetic field. When the model is run in near real time, active regions appear in an unrealistic manner; first one polarity of the active region appears in the assimilation window, followed by the opposite polarity appearing later. Our procedures automatically balance the total positive and negative magnetic flux in every map, but this is done globally on the map. Balancing the flux to compensate for a poorly observed AR, followed by rebalancing when the full region comes into view, can lead to artificial changes to open/closed boundaries and the HCS. This likely drives some artificial evolution in our present model. This problem could be avoided by automatically identifying these instances and delaying the appearance of the active region until the full region is observed, but such procedures have yet to be developed.

A second difficulty is that data are only assimilated on the Earth-facing side of the Sun. Active regions that emerged several days earlier appear in the map only when they rotate into view; this artificially forces the evolution to occur in the narrow assimilation window, which sweeps across the Sun at the rate of the solar rotation. This issue could be mitigated for scientific studies by retrospective analysis to estimate where and when the active regions first emerged and incorporating them in the maps prior to the first Earth-based observation. This could in principle be done most effectively during the STEREO era. However, for space weather applications, this limitation may only be diminished by obtaining magnetograph measurements from a view away from the Sun-Earth line.

Acknowledgments

Work at Predictive Science was supported by AFOSR, NASA via a subcontract from the University of Alabama, Huntsville, the LWS Focus team on the Slow Solar Wind, the LWS Focus team on Prediction of the Interplanetary Magnetic Field Vector Bz at 1AU, and NSF. Computational resources were provided by the NSF supported Texas Advanced Computing Center (TACC) in Austin and the NASA Advanced Supercomputing Division (NAS) at Ames Research Center. The ADAPT maps are produced collaboratively between AFRL and NSO/NISP. Work at the University of Alabama Huntsville was supported by NASA grants NNX12AB30G and NNX14AF41G, and NSF grant AGS-1358386.

References

[1] Riley P, Ben-Nun M, Linker J A, Mikic Z, Svalgaard L, Harvey J, Bertello L, Hoeksema T, Liu Y and Ulrich R 2014 Sol. Phys. **289** 769–792
[2] Lionello R, Linker J A and Mikić Z 2009 ApJ **690** 902–912
[3] Lionello R, Velli M, Downs C, Linker J A and Mikić Z 2014 ApJ **796** 111 (*Preprint* 1410.1789)
[4] van der Holst B, Sokolov I V, Meng X, Jin M, Manchester IV W B, Tóth G and Gombosi T I 2014 ApJ **782** 81 (*Preprint* 1311.4093)
[5] Wang Y M and Sheeley Jr N R 1990 ApJ **355** 726–732
[6] Arge C N and Pizzo V J 2000 J. Geophys. Res. **105** 10465–10480
[7] Riley P, Linker J A and Mikić Z 2001 J. Geophys. Res. **106** 15889–15902
[8] Arge C N, Odstrcil D, Pizzo V J and Mayer L R 2003 *Solar Wind Ten* (*American Institute of Physics Conference Series* vol 679) ed Velli M, Bruno R, Malara F and Bucci B pp 190–193
[9] Altschuler M D and Newkirk G 1969 Sol. Phys. **9** 131–149
[10] Schatten K H 1971 *Cosmic Electrodynamics* **2** 232–245
[11] Pizzo V, Millward G, Parsons A, Biesecker D, Hill S and Odstrcil D 2011 *Space Weather* **9** 03004

[12] Wang Y M, Sheeley Jr N R and Nash A G 1991 ApJ **383** 431–442
[13] Worden J and Harvey J 2000 *Sol. Phys.* **195** 247–268
[14] Schrijver C J and DeRosa M L 2003 *Sol. Phys.* **212** 165–200
[15] Upton L and Hathaway D H 2014 ApJ **780** 5 (*Preprint* 1311.0844)
[16] Arge C N, Henney C J, Koller J, Compeau C R, Young S, MacKenzie D, Fay A and Harvey J W 2010 *Twelfth International Solar Wind Conference* **1216** 343–346
[17] Hickmann K S, Godinez H C, Henney C J and Arge C N 2015 Sol. Phys. **290** 1105–1118 (*Preprint* 1410.6185)
[18] Pogorelov N V, Borovikov S N, Heerikhuisen J, Kryukov I A and Zank G P 2010 *Numerical Modeling of Space Plasma Flows, Astronum-2009* (*Astronomical Society of the Pacific Conference Series* vol 429) ed Pogorelov N V, Audit E and Zank G P p 266
[19] Linker J A, Lionello R, Downs C, Caplan R, Mikic Z, Riley P, Henney C J and Arge C 2014 *American Astronomical Society Meeting Abstracts #224* (*American Astronomical Society Meeting Abstracts* vol 224) p 323.36
[20] Smith E J and Balogh A 2003 *Solar Wind Ten* (*American Institute of Physics Conference Series* vol 679) ed Velli M, Bruno R, Malara F and Bucci B pp 67–70
[21] Riley P, Linker J A and Arge C N 2015 *Space Weather* **13** 154–169
[22] McComas D J, Barraclough B L, Funsten H O, Gosling J T, Santiago-Muñoz E, Skoug R M, Goldstein B E, Neugebauer M, Riley P and Balogh A 2000 J. Geophys. Res. **105** 10419–10434
[23] Mikić Z, Linker J A, Schnack D D, Lionello R and Tarditi A 1999 *Phys. Plasmas* **6** 2217
[24] Linker J A, Mikić Z, Bisecker D A, Forsyth R J, Gibson S E, Lazarus A J, Lecinski A, Riley P, Szabo A and Thompson B J 1999 J. Geophys. Res. **104** 9809
[25] Downs C, Linker J A, Mikić Z, Riley P, Schrijver C J and Saint-Hilaire P 2013 *Science* **340** 1196–1199
[26] Lionello R, Riley P, Linker J A and Mikić Z 2005 ApJ **625** 463–473
[27] Lionello R, Linker J A, Mikić Z and Riley P 2006 ApJ **642** L69–L72
[28] Linker J A, Lionello R, Mikić Z, Titov V S and Antiochos S K 2011 ApJ **731** 110
[29] Linker J A, Mikić Z, Lionello R, Riley P, Amari T and Odstrcil D 2003 *Phys. Plasmas* **10** 1971
[30] Lionello R, Downs C, Linker J A, Török T, Riley P and Mikić Z 2013 ApJ **777** 76
[31] Lionello R, Mikić Z and Linker J A 1999 *Journal of Computational Physics* **152** 346–358
[32] Riley P, Linker J A, Lionello R and Mikic Z 2012 *Journal of Atmospheric and Solar-Terrestrial Physics* **83** 1–10
[33] Benevolenskaya E E, Kosovichev A G and Scherrer P H 2001 ApJ **554** L107–L110
[34] Stevens M L, Linker J A, Riley P and Hughes W J 2012 *Journal of Atmospheric and Solar-Terrestrial Physics* **83** 22–31
[35] Smith E J and Balogh A 2008 Geophys. Res. Lett. **35** L22103
[36] Crooker N U, Gosling J T and Kahler S W 2002 J. Geophys. Res. **107** 3–1

The Heliotail: Theory and Modeling

N. V. Pogorelov

Department of Space Science, University of Alabama in Huntsville, Huntsville, AL 35805, U.S.A.

E-mail: nikolai.pogorelov@uah.edu

Abstract. Physical processes are discussed related to the heliotail which is formed when the solar wind interacts with the local interstellar medium. Although astrotails are commonly observed, the heliotail observations are only indirect. As a consequence, the direct comparison of the observed astrophysical objects and the Sun is impossible. This requires proper theoretical understanding of the heliotail formation and evolution, and numerical simulations in sufficiently large computational boxes. In this paper, we review some previous results related to the heliotail flow and show new simulations which demonstrate that the solar wind collimation inside the Parker spiral field lines diverted by the heliopause toward the heliotail is unrealistic. On the contrary, solar cycle effects ensure that the solar wind density reaches its largest values near the solar equatorial plane. We also argue that a realistic heliotail should be very long to account for the observed anisotropy of 1–10 TeV cosmic rays.

1. Introduction

The heliopause is a tangential discontinuity formed due to the solar wind (SW) interaction with the local interstellar medium (LISM). It is an essential ingredient of the SW–LISM interaction pattern. Mathematically, it can be considered as a component of the solution to the three-dimensional (3D), MHD Riemann problem that is used to generate an initial distribution of quantities in the computational region surrounding the Sun. Physically, this is a surface that is expected to separate the SW and LISM plasma if the effects of finite resistivity and viscosity are ignored. Numerical simulations of the SW–LISM interaction have long history. They started from simplified statements of the problem, e.g., a hypersonic, thin shock layer approximation [1, 2], and further evolved into the full gas dynamics modeling of the upwind part of the interaction pattern [3], where the effect of charge exchange of the SW plasma and LISM neutral atoms was also included. The latter effect is of great importance [4, 5, 6] because the LISM is only partially ionized with the neutral hydrogen density, $n_{\mathrm{H}\infty}$, being about three times greater than the proton density, n_∞. In principle, the SW–LISM interaction can be interpreted as a combination of blunt-body and supersonic jet flows. Both may be turbulent and experience MHD and kinetic instabilities. As shown in [7], there are similarities in the SW–LISM and comet-SW interactions, but charge exchange is a physical phenomenon that cannot be neglected in the distant heliosphere (beyond 10 AU from the Sun). A simple numerical model aimed to include the charge exchange into consideration was proposed in [8]. However, because of a very large mean free path, the neutral atom transport is essentially a kinetic process and ideally should be described by solving the Boltzmann equation. This approach was used to model the H-atom transport throughout the heliosphere initially neglecting the action of atoms on the plasma [9] and ultimately in a self-consistent fashion by solving the Euler gas dynamics equations for plasma and kinetic Boltzmann equation for neutrals iteratively [10]. The latter paper followed the approach of [11] to perform numerical simulation in a closed region surrounding the Sun. These papers manifest the first simulations of the heliotail. The axially symmetric Euler–Boltzmann simulations of the SW–LISM interaction were extended to 3D and time-dependence much later [12]. Here multi-fluid models of the heliospheric interface proved to be useful [13, 14, 15, 16] because of their simplicity. In multi-fluid models, different Euler gas dynamics equations are solved for neutral atoms born in thermodynamically different regions of the heliosphere, e.g., in the pristine LISM, in the region where the LISM plasma is modified by the presence of the heliopause, in the inner heliosheath between the heliopause and the heliospheric termination shock, and in the supersonic SW. The multi-fluid approach made it possible to perform serial time-dependent and 3D simulations.

It is now known that the heliospheric and interstellar magnetic fields (HMF and ISMF) affect the heliopause and the heliotail. MHD models of the SW–LISM interaction were used in [17, 18, 19, 20]. The paper of [20], in particular, took into account charge exchange using a simplified approach similar to [8]. Three-dimensional effect of the ISMF orientation on the shape and rotation of the heliopause were addressed in [21] in purely MHD and [22] in multifluid MHD approaches. MHD-kinetic models have been developed and used in [23]–[40], in particular, to analyze the energetic neutral atom ENA fluxes being measured by the *Interstellar Boundary Explorer* (*IBEX*) and the effect of the choice of the the BV-plane (the plane formed by the velocity and magnetic field vectors in the unperturbed LISM) on the neutral H deflection discovered and discussed in [41, 42]. In [43, 44, 45], the results of comparison between multi-fluid and MHD-kinetic models have been presented. In principle, a good qualitative agreement has been observed, although it should be understood that hydrodynamic effects are likely to produce artifacts on the neutral atom transport, especially in large computational boxes. Additionally, only MHD-kinetic models can ensure accurate energy distribution of neutral atoms. We are focused on the heliotail flow in this paper and by no means try to provide a comprehensive list of references.

2. Heliotail Theory

In [46], two extreme cases of the SW–LISM interaction have been considered. One of them assumed a subsonic, incompressible flow of the SW beyond the termination shock and a subsonic LISM. Another one, assumed the SW propagation into a magnetized vacuum. The former model is possible only if the ISMF is taken into account [47, 36]. Otherwise, the LISM is now known to be supersonic and a bow shock should, in principle, be expected [1]. The model of the SW expansion into the magnetized vacuum, on the other hand, is an idealization which does not take into account the presence of charge exchange even in a very slow LISM. However, it correctly reproduces the formation of the heliotail. Summarizing, charge exchange is known now to have a profound effect on the SW–LISM interaction and on the heliotail, in particular.

The theory of the heliotail flow was considered in detail more than 30 years ago in [48]. The knowledge of the SW–LISM interaction was very basic then as compared with nowadays, both because of the abundance of new observational data and availability of more detailed models. In [48], the SW flow beyond the termination shock was generated analytically in a way close to that used in [46]. The analysis of the solution showed the collimation of the SW flow inside the Parker spiral field deflected to the heliotail and creating tornado-like structures. The reason of such collimation was attributed to the HMF tension exerted on the SW plasma. It was suggested for this reason that a possibility of the heliotail splitting into two lobes should not be excluded. Additionally, it was acknowledged that according to [49] the parcels of plasma confined within the spiral field line should be subject to the kink instability. One should notice that such instability, as will be shown in the next section (see also [50]), eliminates the reason for the SW to be collimated. The solution similar to [48] have been proposed recently in [51] with essentially the same conclusions. Neither of those theoretical analyses took into account charge exchange processes.

Another interesting theoretical model has been proposed in [52] where the focus is on the tailward flattening of the heliopause in the direction perpendicular to both the incoming flow and the large-scale interstellar magnetic field (see, e.g., [22], where the effect of the magnetic field strength and the angle between the magnetic field and velocity vectors in the unperturbed LISM has been considered). This analytical model is an improvement over [53, 54] where the heliotail was assumed to have a circular cross-section.

It is understood, however, that the problem is too complicated to be solved analytically, especially taking into account the necessity to accurately describe charge exchange that clearly determines the length of the heliotail.

3. Heliotail Modeling

Simulations of the SW–LISM interaction require an accurate treatment of the boundary conditions at subsonic exit boundaries. This issue is especially complicated for subsonic-supersonic transitions, where solutions to the Riemann problem based on the LISM and SW parameters at the exit produce a shock slowly propagating upwind towards the Sun [55, 17, 56, 57]. It has been shown in [58] that if the LISM is supersonic (and it indeed is, according to our current knowledge, see [59]) the SW in the heliotail will ultimately become supersonic as well at distances of about 4,000 AU. The reason for this is charge exchange which is continuously fed by an inexhaustible reservoir of interstellar neutral atoms. As a result of this charge exchange, hotter SW ions are substituted with cooler secondary ions with properties of the parent neutral atoms. Newly created H atoms are energetic enough to leave the heliotail region.

In [58], a gasdynamic model of the SW–LISM interaction was used. Magnetic fields (both the HMF and the ISMF) are important once we are interested in the shape of the heliopause and its orientation in space. This is done through the magnetic pressure which occurs in the boundary conditions at the heliopause. Magnetic fields also shape the SW and LISM flows. Additionally, the HMF–ISMF coupling may result in magnetic reconnection at the heliopause. In [60, 50], we have performed simulations in the MHD-kinetic approximation assuming the unipolar HMF. Such approximation is sometimes useful if it is impossible to resolve the heliospheric current sheet (HCS), which serves as the magnetic equator for the HMF [61, 62, 63]. Although our results are in general agreement with axially-symmetric simulations without magnetic field [58], there are some differences. Primarily, we found that instability of the heliopause results in substantial mixing of the the SW and LISM plasmas (see Fig. 1). The SW plasma becomes fast magnetosonic at 4,200 AU from the Sun. The computational region was chosen to be 6,000 AU cubed. The number of particles in the Monte Carlo simulation of neutral atoms was 1.5×10^{10}. The SW quantities at 1 AU were spherically symmetric: the plasma density $n_p = 7.4$ cm^{-3}, temperature $T = 51100$ K, velocity $V_R = 450$ km/s, and the radial component of magnetic field $B_R = 37.5$ μG. The SW is of course neither steady nor spherically symmetric. The choice of the LISM properties were the same as in [64]. Notice that the LISM properties have been updated recently using *IBEX* observations [59] and modeling [40]. The top panel of this figure shows the plasma density distribution in the solar equatorial plane. The solid line is the isoline of the fast magnetosonic Mach number equal to 1. Once the flow at the exit boundary becomes superfast magnetosonic, no boundary conditions are required. This is a substantial positive byproduct of long-tail simulations. The middle panel shows the heliopause, the position of which is strictly determined with a level-set method. Yellow and blue colors correspond to the simulation with B_∞ equal to 3 μG and 4 μG, respectively. The direction of the field is at 30° to the LISM velocity vector and the BV-plane coincides with the hydrogen deflection plane defined by the neutral H velocity vectors in the unperturbed LISM and in the inner heliosphere. We note here that our

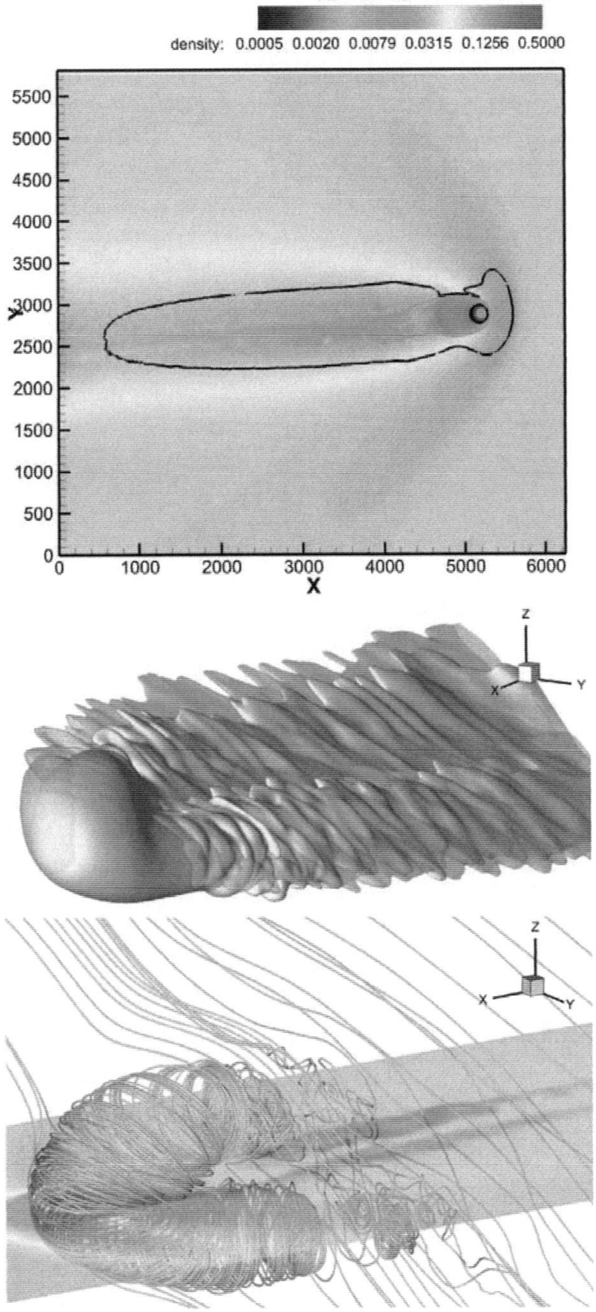

Figure 1. MHD-plasma/kinetic-neutrals simulation of the SW–LISM interaction. (Top panel) Plasma density distribution in the solar equatorial plane. The black lines outline the fast magnetosonic transition, i.e., the plasma flow is subfast magnetosonic between these lines. (Middle panel) The shape of the heliopause for two different ISMF strengths is shown (yellow and blue for $B_\infty = 3\ \mu$G and $4\ \mu$G, respectively). (Bottom panel) HMF line behavior initially exhibits a Parker spiral, but further tailward it becomes unstable. Also shown are ISMF lines draping around the heliopause. The distribution of the plasma density is shown in the semitransparent equatorial plane. (From [50] with permission of the AAS.)

coordinate system, which we call the heliospheric system, has the z-axis parallel to the Sun's rotation axis, while the x-axis belongs to the plane formed by the z-axis and $\mathbf{V}_\infty$, and is oriented upstream the LISM flow. The y-axis completes the right coordinate system. We clearly see that the heliopause is Kelvin–Helmholtz unstable. The Rayleigh–Taylor instability of its nose is not revealing itself in this simulation because this requires much higher resolution in the upwind region. The heliopause extends to the heliocentric distance of 5,000 AU showing no "croissant" structure reported in multi-fluid simulations [63]. We will return to this subject later. It is worth mentioning at this point is that our own multi-fluid solution of the same problem does not ensure the SW acceleration to superfast magnetosonic speeds. This is in contrast with the MHD-kinetic solution. The reason for this is entirely due to the hydrodynamic artifacts intrinsic to multi-neutral-fluid simulations. In particular, we have found a region inside the heliotail (and close to the equatorial plane) which is not crossed by the LISM H streamlines. The absence of (or considerable depression in) charge exchange prevents the natural evolution of the SW flow in the tail. In kinetic simulations, interstellar H atoms easily cross this region due to the chaotic component in their velocity.

The bottom panel of Fig. 1, shows the kink instability in the SW plasma described in [48, 49] and recently rediscovered in [51, 63]. According to [48], this instability is theoretically unconditional. However, numerical simulations are performed on a grid and therefore some wave lengths are inevitably suppressed. We see that the HMF lines initially belonging to the Parker spiral field are suddenly destroyed and become chaotic when the SW temperature decreases. We addressed this phenomenon in [50] and will show some additional details below.

To understand the origin of the very short, croissant-shaped heliotail in [63], we preformed our own simulation using the same set of the LISM and SW parameters. However, for simplicity and to avoid problems with the subfast magnetosonic exit boundary conditions, we assumed $B_\infty = 0$. As stated in [63], this assumption cannot eliminate the effect of the heliotail separation into two lobes as proposed in [48]. For completeness, we reproduce our figure from [50] (see Fig. 2). The heliopause is very well seen in panel e, which shows $|\nabla \times \mathbf{B}|$, which is proportional to the magnitude of the electric current (we used arbitrary units here), and is especially large near the heliopause. Panels a–c show the line $B_y = 0$. It starts at the point where the inner boundary is crossed by the Sun's rotation axis. Magnetic field vanishes on this line while it belongs to the supersonic SW region. Because we assumed zero tilt between the Sun's magnetic and rotation axes, this line indicates a singularity where the Parker spiral magnetic field degenerates into a non-magnetic line. On crossing the termination shock, the Parker spiral and the line B_y in the figure turn tailward. The HMF strength is not zero any more on this line, but it is still small because the spiral component should be dominant. One can see that the plasma density is enhanced along this line. The nature of this enhancement is described in [48]. In addition to the magnetic field strength becoming larger on the spiral field lines that start at the inner boundary with increasing spherical angle θ, the spiral field exerts tension collimating the SW flow towards the center of the spiral field (the line $B_y = 0$). Because of the kink instability [49], the spiral field starts to disappear at about 1,000 AU from the Sun along the x-axis tailward. We also notice that the start of this instability coincides with the decrease in plasma temperature, as was also discussed in [48]. Once the spiral field becomes chaotic, there is no more reason for the solar wind to remain collimated. However, as seen from panels e–f, only a narrow isthmus connects the heliotail lobes. However, the separation of the lobes is much less than it is shown in [63], and the heliotail does not resemble "a croissant." We also remember that this is exactly the region where the SW flow remains subfast magnetosonic in our multi-fluid simulations mentioned earlier. The shape of the heliotail shown in Fig. 1 is totally different: the flow collimation observed initially inside the spiral field completely disappears if neutral atoms are treated kinetically. It is worth emphasizing that no heliotail lobes are observed in the MHD-kinetic simulation of [65], where the statement of the problem is very close to our kinetic simulation.

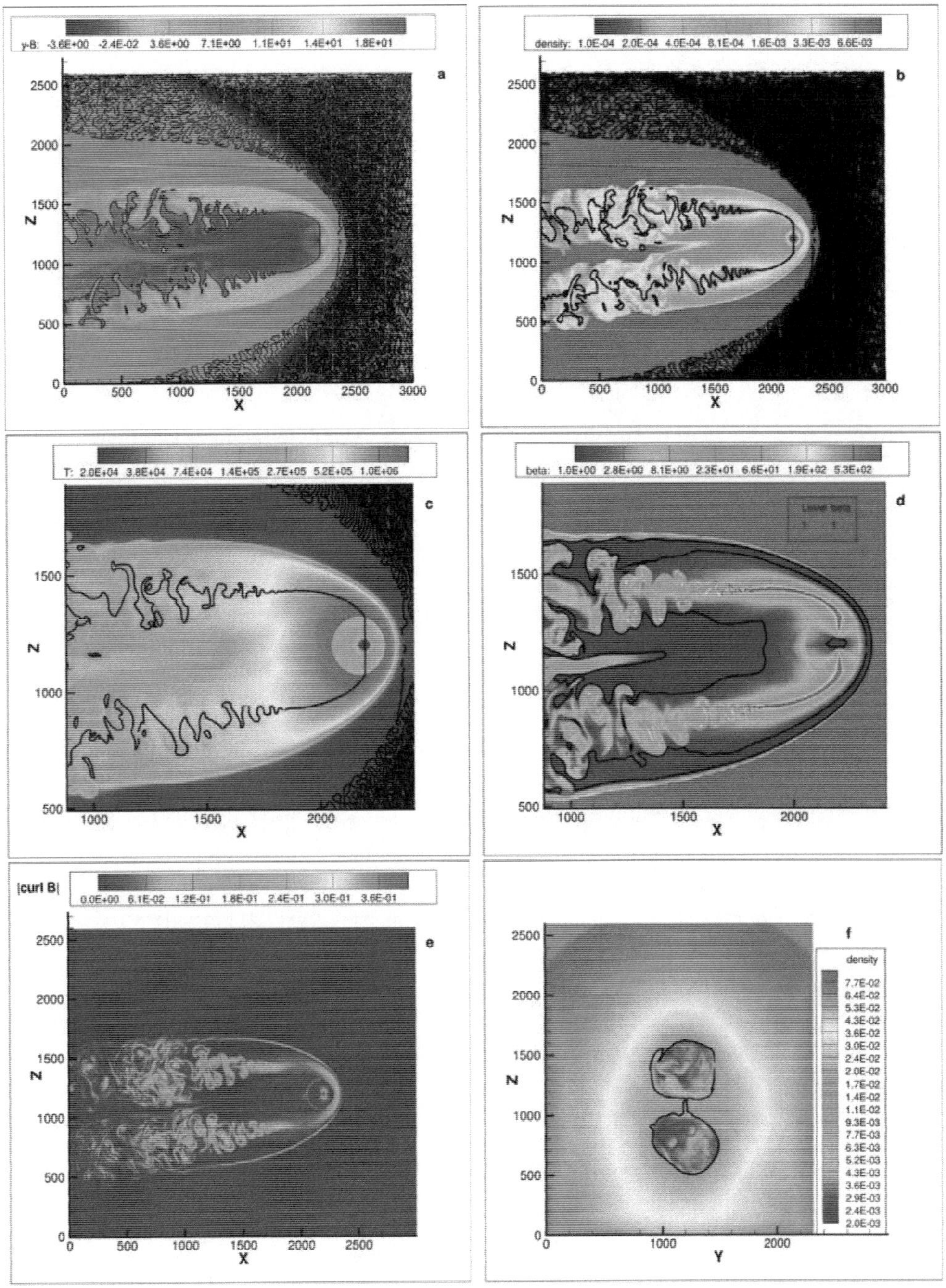

Figure 2. Meridional cuts of a multi-fluid simulation of the SW–LISM interaction with the boundary conditions from [63] without ISMF. (a) The distribution of B_y. The black lines show the level $B_y = 0$. The HMF is assumed to be unipolar. (b) The distribution of plasma density shows its increase toward the center of the Parker spiral in the tail and eventual destruction of the regular magnetic field due to kink instability. (c) The distribution of the plasma temperature. (d) The distribution of the plasma β (log scale) shows that the flow is weakly affected by the HMF, except the regions identified by the isoline $\beta = 1$. (e) The distribution of $|\nabla \times \mathbf{B}|$ shows substantial currents in the lobes. (f) The distribution of the plasma density across the tail ($x = 200$ AU) shows the northern and southern lobes. The solid line outlines the heliopause. (From [50] with permission of the AAS.)

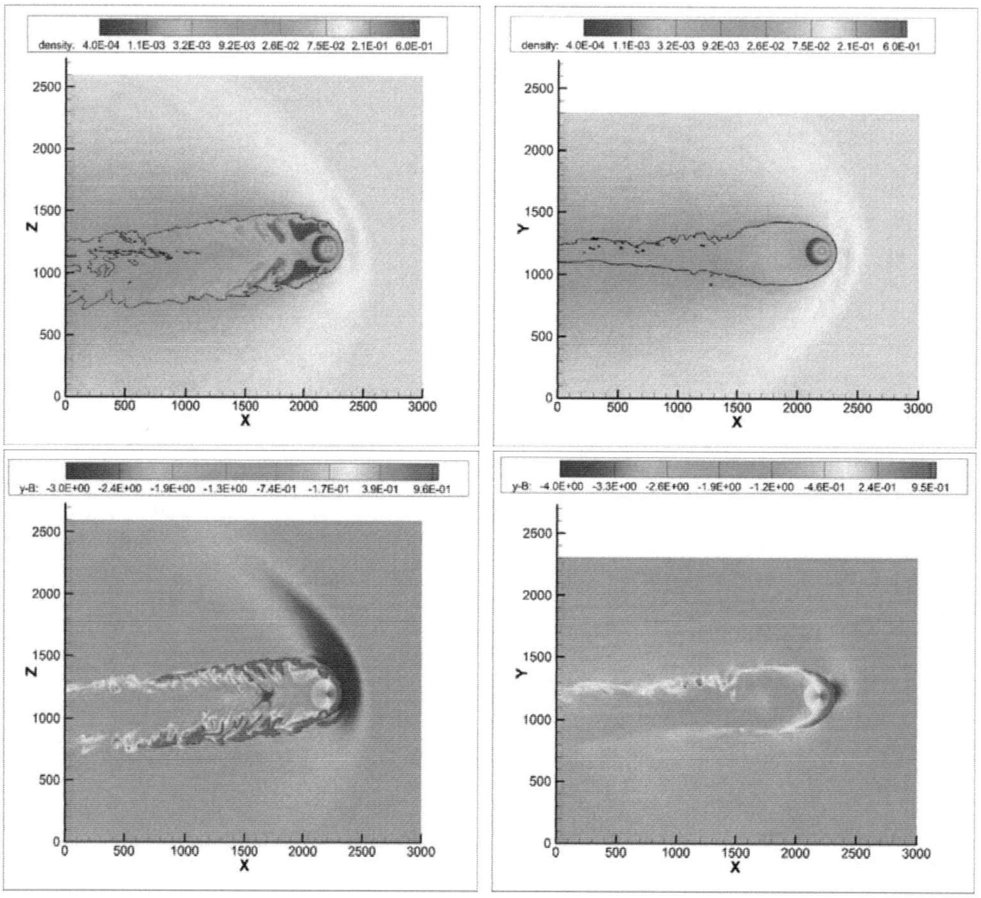

Figure 3. Meridional (left panels) and equatorial (right panels) cross-cuts of the heliosphere showing the plasma density (top panels) and the y-component of the magnetic field vector (bottom panels) in the SW–LISM simulation where the HMF is assumed to be unipolar during the solar cycle.

There is another aspect related to the multi-fluid simulation shown in Fig. 2. As shown in [50], the lobes disappear also if the HMF is dipolar while the tilt of the Sun's magnetic axis with respect to its rotation axis is zero (or small, by continuity). Such statement of the problem is more realistic than the unipolar statement. Besides, there is no difficulty in resolving the HCS in this case. This means, that the two-lobe structure predicted in [48] and shown in [63] is not only an artifact of the multi-fluid treatment of neutral atoms, but also is heavily based on the assumption of the unipolar HMF.

It is argued in [50] that the solar cycle effects related to the latitudinal variations in the boundary between slow and fast SW and the tilt between the Sun's magnetic and rotation axes (and changes in the HMF polarity every 11 years) substantially modify solutions that assume a spherically-symmetric SW. While this is surely true, it is interesting to see the solar cycle effects remaining in the framework of the unipolar HMF. This solution is shown in Fig. 3. It is immediately seen here that the collimation effects are completely swept away by the solar cycle. As one would have expected, the denser plasma is concentrated around the equatorial plane, because the slow SW is always denser than the fast wind. By choice, B_y is negative for $x - x_\odot > 0$ and positive for $x - x_\odot < 0$ in the SW region ahead of the termination shock. The flow carrying the positive polarity hits the heliopause and diverts tailward through the north-

and south-pole axes. Conversely, the SW carrying negative polarity just continues to move to the left. The boundary between these two regions is clearly seen in the figures. It is unstable as in the spherically symmetric SW simulations, but behaves quite differently. The black lines are separating the SW from the LISM. It is seen that the heliopause instability results in the clumps of LISM plasma penetrating into the SW, and vice versa. The "jets" that show up in the simulations ignoring the solar cycle disappear. One of the reasons is that the changes in the latitudinal extent of the slow wind cause changes in the HMF strength and tightening of the spiral field lines. As this frequently happens, a stronger effect made a weaker effect invisible. Another example of this can be found in [66], where the variations of the SW velocity due to the presence of the HCS in the ideal MHD simualtion without neutral atoms were completely wiped out by the flow deceleration with the heliocentric distance inside that termination shock because of charge exchange between interstellar atoms and SW protons.

While the solution that takes solar cycle effects into account is a step forward compared with the solutions assuming the unipolar HMF in a spherically symmetric SW, the changes in the angle between the Sun's rotation and magnetic axes, as well as the dipolar field flipping at solar maxima, are known to have a profound effect on the SW–LISM interaction. Figure 4 shows the same quantities as in Fig. 3 when the tilt is also the function of solar cycle with the HMF polarity changing to the opposite each solar maximum. Although the solutions look quite similar in the upwind LISM, there are noticeable differences in the heliopause shape. In essence, the heliopause is narrower in the latter simulation. First of all, this is due to the change of the HMF polarity every 11 years. The stability and numerical reconnection across the heliopause are therefore different in the dipolar-field case. Besides, clearly there exist certain regions in the heliotail where the HMF is poorly resolved due to its frequent changes of polarity caused by solar rotation. Although the energy and momentum are preserved in those regions, the numerical dissipation is elevated. Additional investigations should be performed to estimate the importance of this. Such a study is complicated by the necessity to perform simulations in computational boxes extending into the tail direction possibly to 18–20 thousand AU. Such distances are necessary if we want to explain flux anisotropy of 1–30 TeV cosmic rays observed in multiple air shower observations [67].

4. Discussion and conclusions

We have discussed very briefly the theoretical results related to the heliotail topology and flow inside it. Because of the complexity of the flow, especially due to highly nonlinear effects related to charge exchange, the applicability of such theoretical studies is limited, although the results themselves are inspiring. In particular, we have shown that the theoretical analyses of [48, 51], however sophisticated and accurate they are, inevitably miss certain details of the heliotail flow. Although some features of the above-mentioned theoretical results are indeed seen in the simplified problem statements similar to those accepted by theorists (see [50, 63]), the reality appears to be much more complicated. We have presented the results of a series of simulations which show that neutral atoms and their charge exchange with ions, as well as the solar cycle effects, turn out to be damaging to analytical solutions. On the other hand, this is not surprising because the physical processes excluded from the theoretical models described above play the major role in the SW–LISM interaction. Our comparison of the numerical models of different level of sophistication is also rather instructive. In particular, the models that try to reproduce the theoretical predictions of [48] by a radical simplification of the statement of the problem, turn out to be sensitive to the choice of the neutral-atom-transport model. We have demonstrated that multi-fluid models of the SW–LISM interaction can produce hydrodynamic artifacts that especially affect the heliotail because of its length. Again, this is not surprising because the differences between kinetic and hydrodynamic models are expected to increase at longer times and distances. It should be remembered that we are talking about collisionless plasma here,

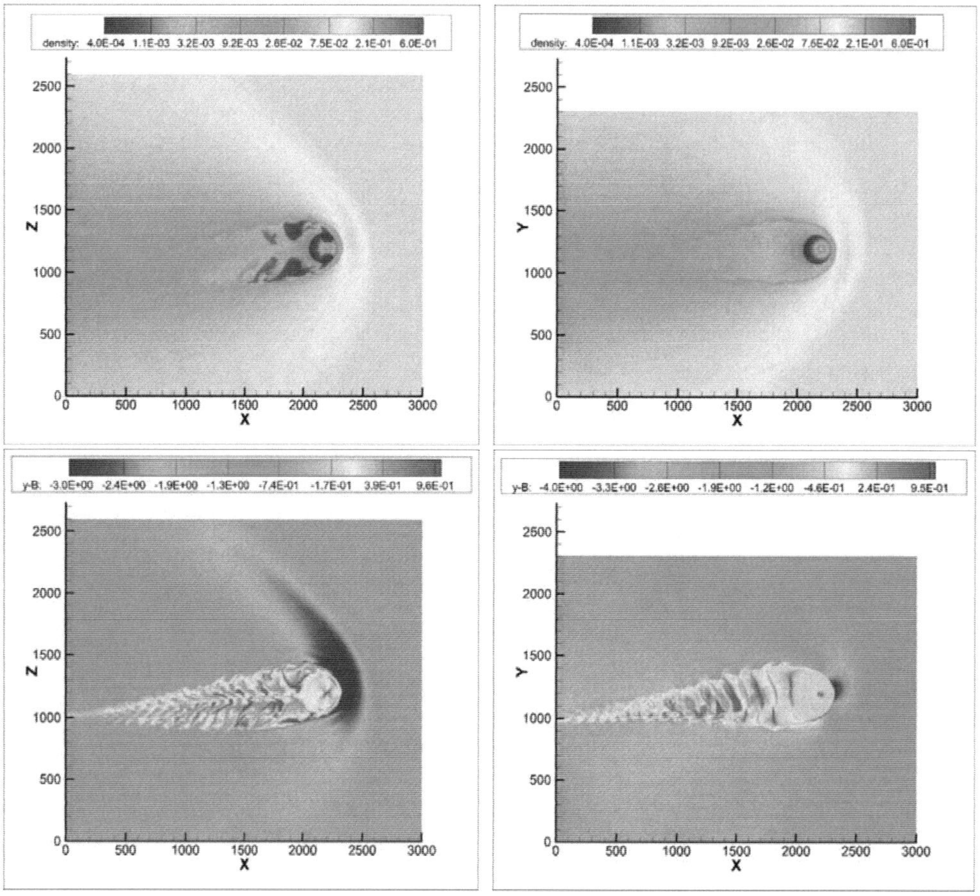

Figure 4. Meridional (left panels) and equatorial (right panels) cross-cuts of the heliosphere showing the plasma density (top panels) and the y-component of the magnetic field vector (bottom panels) in the SW–LISM simulation where the HMF is assumed to be dipolar with the angle between the Sun's magnetic and rotation axes being the function of time.

and any multi-fluid model is a simplification of real kinetic processes determining the transport of neutral atoms and their charge exchange with ions. Although good qualitative agreement between the MHD-kinetic and multi-fluid models, as described in [44, 45, 68], is possible, certain adjustments in the boundary conditions are usually necessary to have a quantitative agreement. While such adjustments are plausible if both types of simulations are performed and compared, the difference in the solutions can become large in new simulations, where no "experience" can substitute the accurate treatment of the problem. We have also demonstrated that the solar cycle destroys the flow-collimation effect described in [48, 51] even if the HMF is assumed unipolar. The unipolar assumption, however convenient it is to avoid difficulties in the HCS resolution, may be very dangerous. The MHD equations are invariant with respect to the change of sign of the **B** vector. However, solutions do depend on the change of sign of the HMF only. This occurs through the coupling between the HMF and ISMF at the heliopause. The positive and negative HMF values are also coupled across the HCS. This can be seen, e.g., in the presented simulations that take into account the flipping of the magnetic dipole of the Sun at solar maxima and the corresponding change of the HMF polarity.

Numerical modeling of the 1–30 TeV galactic cosmic ray flux in [67] requires a long heliotail. Otherwise, very high energy cosmic rays will not be affected noticeably by the presence of the heliotail. For this reason, a short, croissant-shaped heliosphere is not suitable. We have also shown here that this is incompatible with the realistic SW model. While the heliotail cannot be observed from outside, its features are seen in *IBEX* measurements [69, 53]. No SW collimation, or two-lobe structure, is necessary to see the observed heliotail features in numerical simulation, as shown, e.g., in [40].

Acknowledgments
This work was supported, in part by NASA grants NNX14AJ53G, NNX14AF41G, NNX14AF43G, NNX15AN72G, and NNX16AG83G, and DOE Grant DE-SC0008334. This work was also partially supported by the IBEX mission as a part of NASA's Explorer program. We acknowledge NSF PRAC award OCI-1144120 and related computer resources from the Blue Waters sustained-petascale computing project. Supercomputer time allocations were also provided on SGI Pleiades by NASA High-End Computing Program award SMD-15-5860 and on Stampede by NSF XSEDE project MCA07S033.

I highly appreciate discussions at the team meeting "Heliosheath Processes and Structure of the Heliopause: Modeling Energetic Particles, Cosmic Rays, and Magnetic Fields" supported by the International Space Science Institute in Bern, Switzerland.

Bibliography
[1] Baranov V B, Krasnobaev K V and Kulikovskii A G 1971 *Soviet Physics Doklady* **15** 791
[2] Baranov V B and Krasnobaev K V 1971 *Cosmic Research* **9** 568
[3] Baranov V B, Lebedev M G and Ruderman M S 1979 *Astrophys. Space Sci.* **66** 441–451
[4] Blum P W and Fahr H J 1969 *Nature* **223** 936–937
[5] Wallis M 1971 *Nature Physical Science* **233** 23–25
[6] Wallis M K 1975 *Nature* **254** 202
[7] Wallis M K and Dryer M 1976 *Astrophys. J.* **205** 895–899
[8] Baranov V B, Ermakov M K and Lebedev M G 1983 *Fluid Dynamics* **17** 754–759
[9] Baranov V B, Lebedev M G and Malama I G 1991 *Astrophys. J.* **375** 347–351
[10] Baranov V B and Malama Y G 1993 *J. Geophys. Res. Space Phys.* **98** 15
[11] Matsuda T, Fujimoto Y, Shima E, Sawada K and Inaguchi T 1989 *Progress of Theoretical Physics* **81** 810–822
[12] Izmodenov V V and Baranov V B 2006 *ISSI Scientific Reports Series* **5** 67–136
[13] Pauls H L, Zank G P and Williams L L 1995 *J. Geophys. Res. Space Phys.* **100** 21595–21604
[14] Zank G P, Pauls H L, Williams L L and Hall D T 1996 *J. Geophys. Res. Space Phys.* **101** 21639–21656
[15] Pauls H L and Zank G P 1997 *J. Geophys. Res. Space Phys.* **102** 19779–19788
[16] Fahr H J, Kausch T and Scherer H 2000 *A&A* **357** 268–282
[17] Pogorelov N V and Semenov A Y 1997 *A&A* **321** 330–337
[18] Pogorelov N V and Matsuda T 1998 *J. Geophys. Res. Space Phys.* **103** 237
[19] Ratkiewicz R, Barnes A, Molvik G A, Spreiter J R, Stahara S S, Vinokur M and Venkateswaran S 1998 *A&A* **335** 363–369
[20] Linde T J, Gombosi T I, Roe P L, Powell K G and Dezeeuw D L 1998 *J. Geophys. Res. Space Phys.* **103** 1889
[21] Pogorelov N V, Zank G P and Ogino T 2004 *Astrophys. J.* **614** 1007–1021
[22] Pogorelov N V, Zank G P and Ogino T 2006 *Astrophys. J.* **644** 1299–1316
[23] Izmodenov V, Alexashov D and Myasnikov A 2005 *A&A* **437** L35–L38
[24] Heerikhuisen J, Florinski V and Zank G P 2006 *Journal of Geophysical Research (Space Physics)* **111** A06110
[25] Heerikhuisen J, Pogorelov N V, Zank G P and Florinski V 2007 *Astrophys. J. Lett.* **655** L53–L56
[26] Pogorelov N V, Heerikhuisen J and Zank G P 2008 *Astrophys. J. Lett.* **675** L41
[27] McComas D J, Allegrini F, Bochsler P and et al 2009 *Science* **326** 959
[28] Schwadron N A, Bzowski M, Crew G B, Gruntman M, Fahr H, Fichtner H, Frisch P C, Funsten H O, Fuselier S, Heerikhuisen J, Izmodenov V, Kucharek H, Lee M, Livadiotis G, McComas D J, Moebius E, Moore T, Mukherjee J, Pogorelov N V, Prested C, Reisenfeld D, Roelof E and Zank G P 2009 *Science* **326** 966
[29] Izmodenov V V, Malama Y G, Ruderman M S, Chalov S V, Alexashov D B, Katushkina O A and Provornikova E A 2009 *Space Sci. Rev.* **146** 329–351
[30] Pogorelov N V, Heerikhuisen J, Mitchell J J, Cairns I H and Zank G P 2009 *Astrophys. J. Lett.* **695** L31–L34
[31] Heerikhuisen J and Pogorelov N V 2009 *Astrophys. J. Lett.* **695** L58–L61

[32] Heerikhuisen J, Pogorelov N V, Zank G P, Crew G B, Frisch P C, Funsten H O, Janzen P H, McComas D J, Reisenfeld D B and Schwadron N A 2010 *Astrophys. J. Lett.* **708** L126–L130
[33] Chalov S V, Alexashov D B, McComas D, Izmodenov V V, Malama Y G and Schwadron N 2010 *Astrophys. J. Lett.* **716** L99–L102
[34] Zank G P, Heerikhuisen J, Pogorelov N V, Burrows R and McComas D 2010 *Astrophys. J.* **708** 1092–1106
[35] Heerikhuisen J and Pogorelov N V 2011 *Astrophys. J.* **738** 29
[36] Pogorelov N V, Heerikhuisen J, Zank G P, Borovikov S N, Frisch P C and McComas D J 2011 *Astrophys. J.* **742** 104
[37] Izmodenov V V, Alexashov D B and Ruderman M S 2014 *Astrophys. J. Lett.* **795** L7
[38] Katushkina O A, Izmodenov V V and Alexashov D B 2015 *MNRAS* **446** 2929–2943
[39] Zirnstein E J, Heerikhuisen J, Pogorelov N V, McComas D J and Dayeh M A 2015 *Astrophys. J.* **804** 5
[40] Zirnstein E J, Heerikhuisen J, Funsten H O, Livadiotis G, McComas D J and Pogorelov N V 2016 *Astrophys. J. Lett.* **818** L18
[41] Lallement R, Quémerais E, Bertaux J L, Ferron S, Koutroumpa D and Pellinen R 2005 *Science* **307** 1447–1449
[42] Lallement R, Quémerais E, Koutroumpa D, Bertaux J L, Ferron S, Schmidt W and Lamy P 2010 *Twelfth International Solar Wind Conference* **1216** 555–558
[43] Alexashov D and Izmodenov V 2005 *A&A* **439** 1171–1181
[44] Müller H R, Florinski V, Heerikhuisen J, Izmodenov V V, Scherer K, Alexashov D and Fahr H J 2008 *A&A* **491** 43–51
[45] Pogorelov N V, Heerikhuisen J, Zank G P and Borovikov S N 2009 *Space Sci. Rev.* **143** 31–42
[46] Parker E N 1961 *Astrophys. J.* **134** 20
[47] Florinski V, Pogorelov N V, Zank G P, Wood B E and Cox D P 2004 *Astrophys. J.* **604** 700–706
[48] Yu G 1974 *Astrophys. J.* **194** 187–202
[49] Roberts P H 1956 *Astrophys. J.* **124** 430
[50] Pogorelov N V, Borovikov S N, Heerikhuisen J and Zhang M 2015 *Astrophys. J. Lett.* **812** L6
[51] Drake J F, Swisdak M and Opher M 2015 *Astrophys. J. Lett.* **808** L44
[52] Kleimann J, Röken C, Fichtner H and Heerikhuisen J 2016 *Astrophys. J.* **816** 29
[53] Schwadron N A, Adams F C, Christian E R, Desiati P, Frisch P, Funsten H O, Jokipii J R, McComas D J, Moebius E and Zank G P 2014 *Science* **343** 988–990
[54] Isenberg P A, Forbes T G and Möbius E 2015 *Astrophys. J.* **805** 153
[55] Pogorelov N V and Semenov A Y 1996 *Computational Mathematics and Mathematical Physics* **36** 395–404
[56] Pogorelov N V 2000 *Astrophys. Space Sci.* **274** 115–122
[57] Kulikovskii A G, Pogorelov N V and Semenov A Y 2001 *Mathematical Aspects of Numerical Solution of Hyperbolic Systems* (Chapman & Hall/CRC, London, UK/Boca Raton, Florida, USA)
[58] Izmodenov V V and Alexashov D B 2003 *Astronomy Letters* **29** 58–63
[59] McComas D J, Bzowski M, Frisch P, Fuselier S A, Kubiak M A, Kucharek H, Leonard T, Möbius E, Schwadron N A, Sokół J M, Swaczyna P and Witte M 2015 *Astrophys. J.* **801** 28
[60] Pogorelov N V, Borovikov S N, Bedford M C, Heerikhuisen J, Kim T K, Kryukov I A and Zank G P 2013 Modeling Solar Wind Flow with the Multi-Scale Fluid-Kinetic Simulation Suite *Numerical Modeling of Space Plasma Flows (ASTRONUM-2012)* (*Astronomical Society of the Pacific Conference Series* vol 474) ed Pogorelov N V, Audit E and Zank G P p 165
[61] Czechowski A, Strumik M, Grygorczuk J, Grzedzielski S, Ratkiewicz R and Scherer K 2010 *A&A* **516** A17
[62] Borovikov S N, Pogorelov N V, Burlaga L F and Richardson J D 2011 *Astrophys. J. Lett.* **728** L21
[63] Opher M, Drake J F, Zieger B and Gombosi T I 2015 *Astrophys. J. Lett.* **800** L28
[64] McComas D J, Alexashov D, Bzowski M, Fahr H, Heerikhuisen J, Izmodenov V, Lee M A, Möbius E, Pogorelov N, Schwadron N A and Zank G P 2012 *Science* **336** 1291
[65] Izmodenov V V and Alexashov D B 2015 *Astrophys. J. Suppl.* **220** 32
[66] Pogorelov N V, Borovikov S N, Zank G P and Ogino T 2009 *Astrophys. J.* **696** 1478–1490
[67] Zhang M, Zuo P and Pogorelov N 2014 *Astrophys. J.* **790** 5
[68] Alouani-Bibi F, Opher M, Alexashov D, Izmodenov V and Toth G 2011 *Astrophys. J.* **734** 45
[69] McComas D J, Dayeh M A, Funsten H O, Livadiotis G and Schwadron N A 2013 *Astrophys. J.* **771** 77

The Modeling of Pickup Ion or Energetic Particle Mediated Plasmas

G P Zank[1,2], P Mostafavi[1,2] and P Hunana[1,2]

[1] Center for Space Plasma and Aeronomic Research (CSPAR), University of Alabama in Huntsville, Huntsville, AL 35805, USA

[2] Department of Space Science, University of Alabama in Huntsville, Huntsville, AL 35805, USA

E-mail: garyp.zank@gmail.com

Abstract. Suprathermal energetic particles, such as solar energetic particles (SEPs) in the inner heliosphere and pickup ions (PUIs) in the outer heliosphere and the very local interstellar medium, often form a thermodynamically dominant component in their various environments. In the supersonic solar wind beyond > 10 AU, in the inner heliosheath (IHS), and in the very local interstellar medium (VLISM), PUIs do not equilibrate collisionally with the background plasma. Similarly, SEPs do not equilibrate collisionally with the background solar wind in the inner heliosphere. In the absence of equilibration between plasma components, a separate coupled plasma description for the energetic particles is necessary. Using a collisionless Chapman-Enskog expansion, we derive a closed system of multi-component equations for a plasma comprised of thermal protons and electrons, and suprathermal particles (SEPs, PUIs). The energetic particles contribute an isotropic scalar pressure to leading order, a collisionless heat flux at the next order, and a collisionless stress tensor at the second-order. The collisionless heat conduction and viscosity in the multi-fluid description results from a non-isotropic energetic particle distribution. A simpler single-fluid MHD-like system of equations with distinct equations of state for both the background plasma and the suprathermal particles is derived. We note briefly potential pitfalls that can emerge in the numerical modeling of collisionless plasma flows that contain a dynamically important energetic particle component.

1. Introduction

Most of the thermal plasma in the heliosphere and the very local interstellar medium (VLISM) is not equilibrated with the ubiquitously present energetic particle component. In the inner heliosphere, for example, energetic particles related to impulsive and gradual solar energetic particle events are pervasive, especially during solar maximum. Recently, there have been a number of reports describing observations of interplanetary shock waves that have an energetic particle component with a pressure that exceeds considerably both the corresponding thermal plasma (comprising thermal protons and electrons) and the compressed interplanetary magnetic field pressure [1–5]. Whereas in the inner heliosphere, such mediation of the thermal plasma by energetic particles appears to be associated largely with shock events, it is well-known that the outer heliosphere beyond the ionization cavity (i.e., $\gtrsim$ 8 AU) is dominated thermally by pickup ions (PUIs) (e.g., [6–10]). As reported by Decker et al. [11; 12], the inner heliosheath pressure contributed by energetic PUIs and anomalous cosmic rays far exceeds that of the thermal background plasma and magnetic field. The VLISM is also very interesting. It was noted already by Zank et al. [13] that energetic neutral H created via charge exchange in the inner heliosheath (IHS) and fast solar wind could "splash" back into the VLISM where they would experience a secondary charge exchange. The secondary charge exchange of hot and/or fast neutral H with cold ($\sim$ 6300 K [14]) VLISM protons leads to the creation of a hot or suprathermal PUI population in the VLISM. The heating of the VLISM has been discussed in detail by Zank et al. [15], who pointed out that the heating of the VLISM plasma would result in an increase of the sound speed with a concomitant weakening or even elimination of the bow shock (yielding instead a bow wave). Zirnstein et al. [16] extended the Zank et al. [17] model by including the multiple PUI populations that contribute to the heating of the VLISM [15]. Here, we restrict out attention to neutrals created in both the IHS and supersonic solar wind, i.e., with typical speeds of $\sim$ 100 km/s or $\sim$ 400 km/s, that experience secondary charge exchange in the VLISM. The PUIs form a tenuous ($n_p \simeq 5 \times 10^{-5}$ cm^{-3} [16]) suprathermal component in the VLISM.

Coulomb collisions can equilibrate a background thermal plasma and energetic protons. Assume that the background thermal plasma proton and electron distributions are Maxwellian. If we restrict our attention to suprathermal particles with energies greater than 1 - several keV (PUIs for example, and certainly the dominant component of solar energetic particles (SEPs)), then energetic particles satisfy the ordering

$$v_{ts} \ll v_p < v_{te},$$

where $v_{ts/e}$ denotes the background proton/electron thermal speed respectively and v_p the energetic particle speed. For energetic particles scattering collisionally off a Maxwellian distribution of background protons, the collision frequency is given by

$$\nu_{scat}^{ps} = \frac{n_s e^4 \ln \Lambda}{2\pi \varepsilon_0^2 m_p^2 v^3} s^{-1}, \qquad (1)$$

illustrating the well-known v^{-3} dependence with fast particle speed. By contrast, energetic particle collisional scattering off a Maxwellian electron background yields a larger collision frequency which is given by

$$\nu_{scat}^{pe} = \frac{n_e e^4 \ln \Lambda m_e^{1/2}}{2(2\pi)^{3/2} \varepsilon_0^2 (kT_e)^{3/2} m_p}. \qquad (2)$$

If the collisional time scale exceeds the characteristic flow time of the plasma region of interest, $\tau_f \simeq L/U$, where L is the size of the region and U the characteristic velocity, then the energetic particle distribution will not equilibrate with the background plasma. Expressions (1) and (2) should be used to determine whether one needs to introduce a plasma model that distinguishes energetic protons from background plasma protons.

Zank et al. [9] present detailed estimates for the equilibration times for PUIs in the supersonic solar wind of the outer heliosphere, the subsonic solar wind (the inner heliosheath), and the VLSIM using appropriate plasma parameters. In all three regions, the plasma does not equilibrate and cannot therefore be described as a magnetized single-component plasma and at least some elements of a multi-component description are necessary.

For SEPs in the inner heliosphere with energies of e.g., 5 keV and greater, the collisional frequency of thermal and energetic protons at 1 AU (using a solar wind number density $\sim$ 10 cm^{-3}) is $\nu_{scat}^{ps} \simeq 7 \times 10^{-11}$ s^{-1}. Assuming a characteristic scale length $L \sim 1$ AU for the inner heliosphere and a characteristic speed $\sim$ 400 km/s, the dynamical time scale is $\tau_f \sim 3.75 \times 10^5$ s, implying that $\tau_{scat} = (\nu_{scat}^{ps})^{-1} \gg \tau_f$. Similar results hold for fast protons scattering off electrons. Hence SEPs cannot equilibrate in the inner heliosphere and need to be treated as a distinct and separate component if the energy density becomes dynamically important. As discussed above, this is certainly the case for a number of shocks observed in the inner heliosphere [1–5].

Energetic particles such as SEPs and PUIs both drive streaming instabilities in one form or another, and experience pitch-angle scattering from both self-excited and pre-existing Alfvénic fluctuations. In the case of SEPs, a streaming instability [18–24] is responsible in part for their scattering towards isotropy (i.e., the leading order SEP velocity distribution function can be represented as isotropic and independent of particle pitch-angle with smaller pitch-angle dependent corrections, but the distribution is not equilibrated with the background thermal plasma distribution) and for PUIs, the initial ring-beam distribution is unstable to the generation of Alfvénic fluctuations that scatter PUIs towards isotropy [7; 25–27]. As we show below, pitch-angle scattering serves to introduce both a collisionless heat flux and a non-isotropic pressure tensor into the transport equations describing the energetic particles. The pressure tensor modification is expressed as a collisionless viscosity tensor.

Below, we construct an appropriate multi-component plasma description for a thermal background plasma comprising electrons and protons and a non-equilibrated energetic particle component that is subject to pitch-angle scattering by turbulence and Alfvénic fluctuations. By making various approximations, we derive successively simpler models. In so doing, we place on a more formal footing the derivation of the well-known two-fluid model of cosmic ray magnetohydrodynamics [28; 29], showing, somewhat unexpectedly and contrary to perceived wisdom, that the cosmic ray number density is in fact included implicitly in the total number density.

One conclusion that emerges from this work is that there are potential pitfalls in the numerical modeling of collisionless plasma flows that contain a dynamically important energetic particle component. Specifically, as we show below, the energetic particle transport equations introduce dissipative terms though their collisionless heat flux and viscosity and these terms can dominate those associated with the thermal background plasma. However, numerical schemes that deal with e.g., cosmic rays typically solve the gas dynamic or MHD equations and then introduce the cosmic rays through source terms. The origin of numerical dissipation is typically via the thermal equations and so discontinuities, for example, are resolved on the basis of gas dynamic dissipation rather than perhaps the more physically correct dissipation associated with energetic particles.

2. The multi-component model

In deriving a multi-component plasma model that includes energetic particles, we shall assume that the distribution functions for the background protons and electrons are each Maxwellian, which ensures the absence of heat flux or stress tensor terms for the background plasma. The exact continuity, momentum, and energy equations governing the thermal electrons and protons are therefore given by

$$\frac{\partial n_e}{\partial t} + \nabla \cdot (n_e \mathbf{u}_e) = 0; \tag{3}$$

$$m_e n_e \left(\frac{\partial \mathbf{u}_e}{\partial t} + \mathbf{u}_e \cdot \nabla \mathbf{u}_e \right) = -\nabla P_e - e n_e (\mathbf{E} + \mathbf{u}_e \times \mathbf{B}); \tag{4}$$

$$\frac{\partial P_e}{\partial t} + \mathbf{u}_e \cdot \nabla P_e + \gamma_e P_e \nabla \cdot \mathbf{u}_e = 0, \tag{5}$$

for the electrons, and

$$\frac{\partial n_s}{\partial t} + \nabla \cdot (n_s \mathbf{u}_s) = 0; \tag{6}$$

$$m_p n_s \left(\frac{\partial \mathbf{u}_s}{\partial t} + \mathbf{u}_s \cdot \nabla \mathbf{u}_s \right) = -\nabla P_s + e n_s (\mathbf{E} + \mathbf{u}_s \times \mathbf{B}); \tag{7}$$

$$\frac{\partial P_s}{\partial t} + \mathbf{u}_s \cdot \nabla P_s + \gamma_s P_s \nabla \cdot \mathbf{u}_s = 0, \tag{8}$$

for the protons. Here $n_{e/s}$, $\mathbf{u}_{e/s}$, and $P_{e/s}$ are the macroscopic fluid variables for the electron/proton number density, velocity, and pressure respectively, $\gamma_{e/s}$ the electron/proton adiabatic index, $\mathbf{E}$ the electric field, $\mathbf{B}$ the magnetic field, and e the charge of an electron.

The streaming instability for SEPs and the unstable ring-beam distribution for PUIs excite Alfvénic fluctuations. The self-generated fluctuations and *in situ* turbulence serve to scatter energetic particles in pitch-angle. The Alfvén waves and magnetic field fluctuations both propagate and convect with the bulk velocity of the system $\mathbf{U} = \mathbf{U}(\mathbf{u}_e, \mathbf{u}_s, \mathbf{u}_p, n_e, n_s, n_p, m_e, m_p)$, where n_p and $\mathbf{u}_p$ refer to energetic particle variables. The energetic particles are governed by the Fokker-Planck transport equation with a collisional term $\delta f/\delta t|_{ws}$,

$$\frac{\partial f}{\partial t} + \mathbf{v}\cdot\nabla f + \frac{e}{m_p}\left(\mathbf{E} + \mathbf{v}\times\mathbf{B}\right)\cdot\nabla_v f = \left.\frac{\delta f}{\delta t}\right|_{ws}, \tag{9}$$

for average electric and magnetic fields $\mathbf{E}$ and $\mathbf{B}$. We assume that the velocity $\mathbf{v}$ of energetic particles is always non-relativistic; the generalization to relativistic energies is straightforward [30]. On transforming the transport equation (9) into a frame that ensures there is no change in energetic particle momentum and energy due to scattering, assuming that the cross-helicity is zero, and introducing the random velocity $\mathbf{c} = \mathbf{v} - \mathbf{U}$, we obtain

$$\frac{\partial f}{\partial t} + (U_i + c_i)\frac{\partial f}{\partial x_i} + \left[\frac{e}{m_p}\left(\mathbf{E} + \mathbf{U}\times\mathbf{B}\right)_i + \frac{e}{m_p}\left(\mathbf{c}\times\mathbf{B}\right)_i - \frac{\partial U_i}{\partial t} - (U_j + c_j)\frac{\partial U_i}{\partial x_j}\right]\frac{\partial f}{\partial c_i} = \left.\frac{\delta f}{\delta t}\right|_{ws}. \tag{10}$$

The velocity $\mathbf{U}$ is still unspecified so we choose $\mathbf{U}$ such that $\mathbf{E}' \equiv \mathbf{E} + \mathbf{U}\times\mathbf{B} = 0$. This assumption corresponds to choosing

$$\mathbf{U}_\perp = \mathbf{U} - \mathbf{U}_\parallel = \frac{\mathbf{E}\times\mathbf{B}}{B^2} \equiv \mathbf{U}, \tag{11}$$

since we choose $\mathbf{U}_\parallel = 0$ ($\mathbf{U}_\parallel$ is parallel to $\mathbf{B}$ and therefore arbitrary). The use of the velocity $\mathbf{U}$ then yields

$$\frac{\partial f}{\partial t} + (U_i + c_i)\frac{\partial f}{\partial x_i} + \left[\frac{e}{m_p}\left(\mathbf{c}\times\mathbf{B}\right)_i - \frac{\partial U_i}{\partial t} - (U_j + c_j)\frac{\partial U_i}{\partial x_j}\right]\frac{\partial f}{\partial c_i} = \left.\frac{\delta f}{\delta t}\right|_{ws}. \tag{12}$$

By taking moments of (12), we can derive the evolution equations for the macroscopic energetic particle variables, such as the number density $n_p = \int f d^3c$, momentum density $n_p u_{p_i} = \int c_i f d^3c$, and energy density. Although unspecified for now, we shall assume that moments of the collisional term $\delta f/\delta t|_c$ are zero. The zeroth moment of (12) yields the continuity equation for energetic particles,

$$\frac{\partial n_p}{\partial t} + \frac{\partial}{\partial x_i}\left(n_p\left(U_i + u_{p_i}\right)\right) = 0, \tag{13}$$

where $\mathbf{u}_p$ is the energetic particle bulk velocity in the guiding center frame. For the first moment, we multiply (12) by c_j and integrate over velocity space. This yields, after a little algebra, the momentum equation for energetic particles,

$$\frac{\partial}{\partial t}\left(n_p\left(U_j + u_{p_j}\right)\right) + \nabla\cdot\left[n_p\mathbf{U}\left(U_j + u_{p_j}\right) + n_p\mathbf{u}_p U_j\right] + \frac{\partial}{\partial x_i}\int c_i c_j f d^3c = \frac{e}{m_p} n_p \varepsilon_{jkl} u_{p_k} B_l, \tag{14}$$

where ε_{ijk} is the Levi-Civeta tensor. Note the presence of the term $\int c_i c_j f d^3c$, which is the momentum flux or pressure tensor.

To close equation (14), we need to evaluate the momentum flux, which requires that we solve (12) for the energetic particle distribution function f. In solving (12), we assume 1) that the energetic particle distribution is gyrotropic, and 2) that scattering of energetic particles is sufficiently rapid to ensure that the energetic particle distribution is nearly isotropic. We can therefore average (12) over gyrophase, obtaining the "focused transport equation" for non-relativistic particles [31]. Details of the derivation can be found in Ch. 5 of Zank [30]. To solve the gyrophase-averaged transport equation requires that we specify the scattering or collisional operator. We make the simplest possible choice, which is the isotropic pitch-angle diffusion operator,

$$\frac{\partial}{\partial \mu}\left(\nu_{ws}(1-\mu^2)\frac{\partial f}{\partial \mu}\right), \tag{15}$$

where $\mu = \cos\theta$ is the cosine of the particle pitch-angle θ, and $\nu_{ws} = \tau_{ws}^{-1}$ is the scattering frequency. The form of the scattering operator (15) allows us to solve the focused transport equation using a Legendre polynomial expansion of the distribution function f. The second-order correct solution to the gyrophase-averaged Chapman-Enskog reduction of equation (12) is

$$f \simeq f_0 + \mu f_1 + \frac{1}{2}(3\mu^2 - 1)f_2; \tag{16}$$

$$f_0 = f_0(\mathbf{x}, c, t); \tag{17}$$

$$f_1 = -\frac{c\tau_{ws}}{3}b_i\frac{\partial f_0}{\partial x_i} + \frac{DU_i}{Dt}\frac{\tau_{ws}}{3}b_i\frac{\partial f_0}{\partial c}; \tag{18}$$

$$f_2 \simeq \frac{c\tau_{ws}}{15}\left(b_i b_j\frac{\partial U_j}{\partial x_i} - \frac{1}{3}\frac{\partial U_i}{\partial x_i}\right)\frac{\partial f_0}{\partial c}, \tag{19}$$

where $c = |\mathbf{c}|$ is the particle random speed, $\mathbf{b} \equiv \mathbf{B}/B$ is a directional unit vector defined by the magnetic field, and $D/Dt \equiv \partial/\partial t + U_i \partial/\partial x_i$ is the convective derivative. The expansion terms f_0, f_1, and f_2 are functions of position, time, and particle random speed c i.e., independent of pitch-angle μ (and of course gyrophase ϕ). Of particular importance is the retention of the large-scale acceleration, and shear terms. These terms are often neglected in the derivation of the transport equation describing f_0 (for relativistic particles, the transport equation is the familiar cosmic ray transport equation). The velocity derivative term in equation (18) is known as the relativistic heat inertia term [32; 33]. In the context of deriving a multi-fluid model, retaining the various flow velocity terms is essential to derive the correct multi-fluid formulation for energetic particles. We need to evaluate

$$\begin{aligned}\int c_i c_j f d^3c &= \int (c_i - u_{p_i})(c_j - u_{p_j}) f d^3c + n_p u_{p_i} u_{p_j} \equiv \int c'_i c'_j f d^3c + n_p u_{p_i} u_{p_j} \\ &\simeq \int c'_i c'_j \left(f_0 + \mu f_1 + \frac{1}{2}(3\mu^2 - 1)f_2\right) d^3c' + n_p u_{p_i} u_{p_j},\end{aligned}$$

from which we find the zeroth- and first-order expressions,

$$\int c'_i c'_j f_0 d^3c = \frac{1}{m_p}\left(\delta_{ij} P_p\right), \quad \int c'_i c'_j \mu f_1 d^3c = 0, \quad P_p \equiv m_p \frac{4\pi}{3}\int c'^2 f_0 c'^2 dc. \tag{20}$$

Consequently, the first-order energetic particle stress tensor is identically zero and the pressure is isotropic, $\delta_{ij}P_p$.

The inclusion of the second-order terms yields a non-zero collisionless stress tensor. Since the energetic particle pressure is defined in the frame of the bulk PUI velocity $\mathbf{u}_p$, the distribution function over which the integral is taken needs to evaluated in this frame. Since the expression (19) for f_2 is a function of the guiding center velocity $\mathbf{U}$, we need to transform to the frame

$\mathbf{U}_p = \mathbf{U} + \mathbf{u}_p$. On using the solution (19) for f_2, we obtain

$$\int c_x'^2 \frac{1}{2}(3\mu^2 - 1) f_2 d^3c' = \int c_y'^2 \frac{1}{2}(3\mu^2 - 1) f_2 d^3c' = \frac{\eta}{15}\left(b_i b_j \frac{\partial U_{pj}}{\partial x_i} - \frac{1}{3}\frac{\partial U_{pi}}{\partial x_i}\right); \tag{21}$$

$$\int c_z'^2 \frac{1}{2}(3\mu^2 - 1) f_2 d^3c' = -\frac{2\eta}{15}\left(b_i b_j \frac{\partial U_{pj}}{\partial x_i} - \frac{1}{3}\frac{\partial U_{pi}}{\partial x_i}\right); \tag{22}$$

$$\int c_i' c_j' \frac{1}{2}(3\mu^2 - 1) f_2 d^3c' = 0, \quad (i \neq j), \tag{23}$$

where the coefficient of viscosity η is defined as

$$\eta \equiv \frac{4\pi}{15}\int \frac{\partial}{\partial c'}(c'^4 c\tau_{ws}) f_0 dc' \simeq \frac{4\pi}{3}\int c'^2 \tau_{ws} f_0 c'^2 dc' \simeq \frac{P_p \tau_{ws}}{m_p}. \tag{24}$$

The first equality in (24) is the formal definition of the coefficient of viscosity for the energetic particle gas. If we assume (probably reasonably) that $|\mathbf{c}| \gg |\mathbf{u}_p|$, then we obtain the second equality, which may be regarded as an energetic particle pressure moment weighted by the energetic particle scattering time. Finally, if we assume that τ_{ws} is independent of c, we then obtain the "classical" form (24) of the viscosity coefficient. The pressure tensor may therefore be expressed as

$$(P_{ij}) = P_p(\delta_{ij}) + \begin{pmatrix} 1 & 0 & 0 \\ 0 & 1 & 0 \\ 0 & 0 & -2 \end{pmatrix} \frac{\eta}{15}\left(b_k b_\ell \frac{\partial U_{pk}}{\partial x_\ell} - \frac{1}{3}\frac{\partial U_{pm}}{\partial x_m}\right). \tag{25}$$

If we introduce a "viscosity matrix,"

$$(M_{k\ell}) \equiv (\eta_{k\ell}) = \left(\frac{\eta}{15} b_k b_\ell\right) \simeq \left(\frac{1}{15}\frac{P_p \tau_{ws} b_k b_\ell}{m_p}\right), \tag{26}$$

and note that $\eta_{ij} = \eta_{ji}$ and $\eta/15 = \eta_{11} + \eta_{22} + \eta_{33} = \eta_{ij}\delta_{ij}$ (since $b^2 = 1$), we can rewrite (25) in the more revealing "classical" stress tensor form,

$$\begin{aligned} \frac{\eta}{15}\left(b_k b_\ell \frac{\partial U_{pk}}{\partial x_\ell} - \frac{1}{3}\frac{\partial U_{pm}}{\partial x_m}\right) &= \frac{\eta_{k\ell}}{2}\left(\frac{\partial U_{pk}}{\partial x_\ell} + \frac{\partial U_{p\ell}}{\partial x_k}\right) - \frac{1}{3}\eta_{k\ell}\delta_{k\ell}\frac{\partial U_{pm}}{\partial x_m} \\ &= \frac{\eta_{k\ell}}{2}\left(\frac{\partial U_{pk}}{\partial x_\ell} + \frac{\partial U_{p\ell}}{\partial x_k} - \frac{2}{3}\delta_{k\ell}\frac{\partial U_{pm}}{\partial x_m}\right). \end{aligned} \tag{27}$$

The pressure tensor is therefore the sum of an isotropic scalar pressure P_p and the stress tensor, i.e.,

$$(P_{ij}) = P_p(\delta_{ij}) + \begin{pmatrix} 1 & 0 & 0 \\ 0 & 1 & 0 \\ 0 & 0 & -2 \end{pmatrix} \frac{\eta_{k\ell}}{2}\left(\frac{\partial U_{pk}}{\partial x_\ell} + \frac{\partial U_{p\ell}}{\partial x_k} - \frac{2}{3}\delta_{k\ell}\frac{\partial U_{pm}}{\partial x_m}\right) \equiv P_p\mathbf{I} + \Pi_p. \tag{28}$$

The stress tensor is a generalization of the "classical" form in that several coefficients of viscosity are present, and of course the derivation here is for a collisionless charged gas of PUIs experiencing only pitch-angle scattering by turbulent magnetic fluctuations. Use of the pressure tensor (28) yields a "Navier-Stokes"-like modification of the energetic particle momentum equation,

$$\begin{aligned} &\frac{\partial}{\partial t}(\rho_p \mathbf{U}_p) + \nabla \cdot [\rho_p \mathbf{U}_p \mathbf{U}_p + \mathbf{I} P_p] = e n_p(\mathbf{E} + \mathbf{U}_p \times \mathbf{B}) \\ &- \nabla \cdot \begin{pmatrix} 1 & 0 & 0 \\ 0 & 1 & 0 \\ 0 & 0 & -2 \end{pmatrix} \frac{\eta_{k\ell}}{2}\left(\frac{\partial U_{pk}}{\partial x_\ell} + \frac{\partial U_{p\ell}}{\partial x_k} - \frac{2}{3}\delta_{k\ell}\frac{\partial U_{pm}}{\partial x_m}\right), \\ &= e n_p(\mathbf{E} + \mathbf{U}_p \times \mathbf{B}) - \nabla \cdot \Pi_p \end{aligned} \tag{29}$$

where we used the transformation $\mathbf{U}_p = \mathbf{u}_p + \mathbf{U}$ for the remaining velocity terms in (14) and $\rho_p = m_p n_p$.

If we introduce $\mathbf{c}' \equiv \mathbf{c} - \mathbf{u}_p$ as before, we can express the heat flux $\mathbf{q}(\mathbf{x}, t)$ through the definition

$$q_i(\mathbf{x}, t) \equiv m_p \int \frac{1}{2} c'^2 c'_i f d^3c' = \frac{m_p}{2} \int c^2 c_i f d^3c - \frac{5}{2} u_{p_i} P_p - \frac{1}{2} \rho_p u_p^2 u_{p_i}. \tag{30}$$

The equation for the total energy of the energetic particles can then be derived from (12), yielding

$$\frac{\partial}{\partial t}\left(\frac{1}{2}\rho_p U_p^2 + \frac{3}{2}P_p\right) + \frac{\partial}{\partial x_i}\left[\frac{1}{2}\rho_p U_p^2 U_{p_i} + \frac{5}{2}P_p U_{p_i} + \Pi_{ij}U_{p_j} + q_i\right] = e n_p U_{p_i}\left(E_i + (\mathbf{U}_p \times \mathbf{B})_i\right), \tag{31}$$

after transforming to $\mathbf{U}_p$. To evaluate the heat flux, we have

$$\frac{1}{2}\int c'^2 c'_i f_0 d^3c = \pi \int c'^3 \mu b_i f_0 c'^2 dc' = 0,$$

and

$$\frac{m_p}{2}\int c'^2 c'_i \mu f_1 d^3c' = -\frac{2\pi}{3} m_p \int c'^2 \kappa_{ij} \frac{\partial f_0}{\partial x_j} c'^2 dc' = -\frac{1}{2}\bar{\kappa}_{ij}\frac{\partial P_p}{\partial x_j} = q_i(\mathbf{x}, t). \tag{32}$$

In (32), we introduced the spatial diffusion coefficient

$$\kappa_{ij} \equiv b_i \frac{c^2 \tau_{ws}}{3} b_j, \tag{33}$$

together with energetic particle speed-averaged form $\bar{\kappa}_{ij} \equiv K_{ij}$. The collisionless heat flux for energetic particles is therefore described in terms of the energetic particle pressure gradient and consequently the averaged spatial diffusion introduces a energetic particle diffusion time and length scale into the multi-fluid system.

For continuous flows, the transport equation for the energetic particle pressure P_p can be derived from (31), yielding

$$\frac{\partial P_p}{\partial t} + U_{pi}\frac{\partial P_p}{\partial x_i} + \frac{5}{3}P_p\frac{\partial U_{pi}}{\partial x_i} = \frac{1}{3}\frac{\partial}{\partial x_i}\left(K_{ij}\frac{\partial P_p}{\partial x_j}\right) - \frac{2}{3}\Pi_{ij}\frac{\partial U_{pj}}{\partial x_i}, \tag{34}$$

illustrating that the energetic particle heat flux yields a spatial diffusion term in the PUI equation of state together with a viscous dissipation term. The energetic particle system of equations is properly closed and correct to the second-order. Note the typo in Zank et al. [9] since we mistakenly omitted the viscous term of equation (34) in the corresponding pressure equation.

The full system of energetic particle equations can be written in the form

$$\frac{\partial \rho_p}{\partial t} + \nabla \cdot (\rho_p \mathbf{U}_p) = 0; \tag{35}$$

$$\frac{\partial}{\partial t}(\rho_p \mathbf{U}_p) + \nabla \cdot [\rho_p \mathbf{U}_p \mathbf{U}_p + \mathbf{I}P_p + \Pi] = e n_p (\mathbf{E} + \mathbf{U}_p \times \mathbf{B}); \tag{36}$$

$$\frac{\partial}{\partial t}\left(\frac{1}{2}\rho_p U_p^2 + \frac{3}{2}P_p\right) + \nabla \cdot \left[\frac{1}{2}\rho_p U_p^2 \mathbf{U}_p + \frac{5}{2}P_p \mathbf{U}_p + \Pi \cdot \mathbf{U}_p - \frac{1}{2}\mathbf{K} \cdot \nabla P_p\right] = e n_p \mathbf{U}_p \cdot \mathbf{E}, \tag{37}$$

which is the form we use below.

The full thermal electron-thermal proton-PUI multi-fluid system is therefore given by equations (3) - (8) and (35) - (37) or (34), together with Maxwell's equations

$$\frac{\partial \mathbf{B}}{\partial t} = -\nabla \times \mathbf{E}; \tag{38}$$

$$\nabla \times \mathbf{B} = \mu_0 \mathbf{J}; \tag{39}$$

$$\nabla \cdot \mathbf{B} = 0; \tag{40}$$

$$\mathbf{J} = e\left(n_s \mathbf{u}_s + n_p \mathbf{U}_p - n_e \mathbf{u}_e\right), \tag{41}$$

where $\mathbf{J}$ is the current and μ_0 the permeability of free space. The diffusion tensor is assumed to be of a simple diagonal form (i.e., we do not include the off-diagonal terms associated with drift and curvature - see the discussion in [30]) and we specify

$$\mathbf{K} = \begin{pmatrix} \kappa_\perp & 0 & 0 \\ 0 & \kappa_\perp & 0 \\ 0 & 0 & \kappa_\parallel \end{pmatrix}; \tag{42}$$

$$\kappa_\perp = \eta \frac{1}{3\Omega_p} C_0^2, \qquad \kappa_\parallel = \frac{1}{3\Omega_p} C_0^2. \tag{43}$$

We parametrize the perpendicular component of the heat conduction tensor by a term $\eta < 1$. In estimating the diffusion coefficients (43) from (33), we choose a characteristic energetic particle speed for the region of interest and assume that the scattering time can be approximated by the corresponding gyrofrequency.

3. Single-fluid-like model

For many problems, the complete multi-component model derived above is far too complicated to solve. The multi-fluid system (3) - (8) and (35) - (37) or (34), together with Maxwell's equations can be considerably reduced in complexity by making the key assumption that $\mathbf{U}_p \simeq \mathbf{u}_s$. The assumption that $\mathbf{U}_p \simeq \mathbf{u}_s$ is quite reasonable since i) the bulk flow velocity of the plasma is dominated by the background protons since the energetic particle component scatters off fluctuations moving with the background plasma speed and ii) the large-scale motional electric field forces both energetic particles and newly created PUIs to essentially co-move with the background plasma flow. Accordingly, we let $\mathbf{U}_p \simeq \mathbf{u}_s = \mathbf{U}_i$ be the bulk proton (i.e., thermal background protons and PUIs or energetic particles) velocity. The thermal proton and energetic particle/PUI continuity and momentum equations are therefore trivially combined as

$$\frac{\partial n_i}{\partial t} + \nabla \cdot (n_i \mathbf{U}_i) = 0; \tag{44}$$

$$m_p n_i \left(\frac{\partial \mathbf{U}_i}{\partial t} + \mathbf{U}_i \cdot \nabla \mathbf{U}_i\right) = -\nabla(P_s + P_p) + e n_i \left(\mathbf{E} + \mathbf{U}_i \times \mathbf{B}\right) - \nabla \cdot \Pi_p, \tag{45}$$

where $n_i = n_s + n_p$. Since the energetic particles are not thermally equilibrated with the background plasma ($T_s \neq T_p$), we need to deal separately with the P_s and P_p equations. These become

$$\frac{\partial P_s}{\partial t} + \mathbf{U}_i \cdot \nabla P_s + \gamma_s P_s \nabla \cdot \mathbf{U}_i = 0; \tag{46}$$

$$\frac{\partial P_p}{\partial t} + U_i \frac{\partial P_p}{\partial x_i} + \frac{5}{3} P_p \frac{\partial U_i}{\partial x_i} = \frac{1}{3} \frac{\partial}{\partial x_i}\left(K_{ij} \frac{\partial P_p}{\partial x_j}\right) - \frac{2}{3} \Pi_{ij} \frac{\partial U_j}{\partial x_i}. \tag{47}$$

We can combine the proton equations (44) - (47) with the electron equations (3) - (5) to obtain an MHD-like system of equations. On defining the macroscopic variables,

$$\rho \equiv m_e n_e + m_p n_i; \quad q \equiv -e(n_e - n_i); \quad \rho\mathbf{U} \equiv m_e n_e \mathbf{u}_e + m_p n_i \mathbf{U}_i; \quad \mathbf{J} \equiv -e\left(n_e \mathbf{u}_e - n_i \mathbf{U}_i\right), \tag{48}$$

we can express

$$\begin{aligned} n_e &= \frac{\rho - (m_p/e)q}{m_p(1-\xi)} \simeq \rho/m_p; \quad n_i = \frac{\rho + \xi(m_p/e)q}{m_p(1+\xi)} \simeq \rho/m_p; \\ \mathbf{u}_e &= \frac{\rho\mathbf{U} - (m_p/e)\mathbf{J}}{\rho - (m_p/e)q} \simeq \mathbf{U} - \frac{m_p}{e}\frac{\mathbf{J}}{\rho}; \quad \mathbf{u}_i = \frac{\rho\mathbf{U} + \xi(m_p/e)\mathbf{J}}{\rho + \xi(m_p/e)q} \simeq \mathbf{U}, \end{aligned} \tag{49}$$

where the smallness of the mass ratio $\xi \equiv m_e/m_p \ll 1$ has been exploited. Use of the approximations (49) allows us to combine the continuity and momentum equations in the usual way and to rewrite the thermal electron and proton pressure in terms of the single-fluid macroscopic variables. Thus,

$$\frac{\partial \rho}{\partial t} + \nabla \cdot (\rho\mathbf{U}) = 0; \tag{50}$$

$$\rho\left(\frac{\partial \mathbf{U}}{\partial t} + \mathbf{U}\cdot\nabla\mathbf{U}\right) = -\nabla(P_e + P_s + P_p) + \mathbf{J}\times\mathbf{B} - \nabla\cdot\Pi; \tag{51}$$

$$\frac{\partial P_s}{\partial t} + \mathbf{U}\cdot\nabla P_s + \gamma_s P_s \nabla\cdot\mathbf{U} = 0; \tag{52}$$

$$\frac{\partial P_e}{\partial t} + \mathbf{U}\cdot\nabla P_e + \gamma_e P_e \nabla\cdot\mathbf{U} = \frac{m_p}{e\rho}\mathbf{J}\cdot\nabla P_e + \frac{\gamma_e m_p}{e} P_e \nabla\cdot\left(\frac{\mathbf{J}}{\rho}\right), \tag{53}$$

where

$$\Pi_{k\ell} = \begin{pmatrix} 1 & 0 & 0 \\ 0 & 1 & 0 \\ 0 & 0 & -2 \end{pmatrix} \frac{\eta_{k\ell}}{2}\left(\frac{\partial U_k}{\partial x_\ell} + \frac{\partial U_\ell}{\partial x_k} - \frac{2}{3}\delta_{k\ell}\frac{\partial U_m}{\partial x_m}\right).$$

Since we may assume that the current density is much less than the momentum flux, i.e., $|\mathbf{J}| \ll |\rho\mathbf{U}|$, we can simplify (53) further by neglecting the RHS. By assuming that $\gamma_e = \gamma_s = \gamma$, we can combine the thermal proton and electron equations in a single thermal plasma pressure equation with $P \equiv P_e + P_s$,

$$\frac{\partial P}{\partial t} + \mathbf{U}\cdot\nabla P + \gamma P \nabla\cdot\mathbf{U} = 0. \tag{54}$$

Note that at this point, no assumptions about either the thermal electron or proton pressures (or temperatures) have been made.

Finally, we need an equation for the electric field $\mathbf{E}$. To do so, we multiply the respective momentum equations by the electron or proton charge, sum, and use the approximations (49) to obtain

$$\xi\left(\frac{m_p}{e}\right)^2 \frac{1}{\rho}\left[\frac{\partial \mathbf{J}}{\partial t} + \nabla\cdot(\mathbf{J}\mathbf{U} + \mathbf{U}\mathbf{J})\right] = \frac{m_p}{e\rho}\left(\nabla P_e - \mathbf{J}\times\mathbf{B} - \xi\nabla(P_s + P_p) - \xi\nabla\cdot\Pi\right) + \mathbf{E} + \mathbf{U}\times\mathbf{B}.$$

The generalized Ohm's law is therefore

$$\mathbf{E} = -\mathbf{U}\times\mathbf{B} - \frac{m_p}{e\rho}\left(\nabla P_e - \mathbf{J}\times\mathbf{B} - \xi\nabla P_p\right), \tag{55}$$

where we have retained the energetic particle pressure since in principle it can be a high temperature component of the plasma system and ξP_p may be comparable to the P_e term.

For typical cases of interest, however, the P_p term can be neglected in Ohm's law (55). Neglect of the electron pressure and Hall current term then yields the usual form of Ohm's law.

The reduced single-fluid model equations may therefore be summarized as

$$\frac{\partial \rho}{\partial t} + \nabla \cdot (\rho \mathbf{U}) = 0; \tag{56}$$

$$\rho\left(\frac{\partial \mathbf{U}}{\partial t} + \mathbf{U}\cdot\nabla\mathbf{U}\right) = -\nabla(P + P_p) + \mathbf{J}\times\mathbf{B} - \nabla\cdot\Pi; \tag{57}$$

$$\frac{\partial}{\partial t}\left(\frac{1}{2}\rho U^2 + \frac{3}{2}(P+P_p) + \frac{1}{2\mu_0}B^2\right) + \nabla\cdot\left[\frac{1}{2}\rho U^2\mathbf{U} + \frac{5}{2}(P+P_p)\mathbf{U} + \frac{1}{\mu_0}B^2\mathbf{U} - \frac{1}{\mu_0}\mathbf{U}\cdot\mathbf{BB} + \Pi\cdot\mathbf{U}_p - \frac{1}{2}\mathbf{K}\cdot\nabla P_p\right] = 0; \tag{58}$$

$$\frac{\partial P}{\partial t} + \mathbf{U}\cdot\nabla P + \gamma P \nabla\cdot\mathbf{U} = 0; \tag{59}$$

$$\mathbf{E} = -\mathbf{U}\times\mathbf{B}; \quad \frac{\partial \mathbf{B}}{\partial t} = -\nabla\times\mathbf{E}; \quad \mu_0\mathbf{J} = \nabla\times\mathbf{B}; \quad \nabla\cdot\mathbf{B} = 0. \tag{60}$$

The single-fluid description (56) - (60) differs from the standard MHD model in that a separate description for the energetic particle or PUI pressure is required. Instead of the conservation of energy equation (58), one could use the energetic particle pressure equation (47) for continuous flows. Energetic particles and PUIs introduce both a collisionless heat conduction and viscosity into the system.

The model equations (56) - (60), despite being appropriate to non-relativistic PUIs, are identical to the so-called two-fluid MHD system of equations used to describe cosmic ray mediated plasmas [29]. However, the derivation of the two models is substantially different in that the cosmic ray number density is explicitly neglected in the two-fluid cosmic ray model and a Chapman-Enskog derivation is not used in deriving the cosmic ray hydrodynamic equations. Nonetheless, the sets of equations that emerge are the same indicating that the cosmic ray two-fluid equations do in fact include the cosmic ray number density explicitly.

The single-fluid-like model may be extended to include e.g., anomalous cosmic rays (ACRs) as well as PUIs. In this case, the ACRs are relativistic particles. As noted above, the same analysis carries over, and one has an obvious extension of the model equations (56) - (60) with the inclusion of the ACR pressure. Thus, the extension of (56) - (60) is

$$\frac{\partial \rho}{\partial t} + \nabla \cdot (\rho \mathbf{U}) = 0; \tag{61}$$

$$\rho\left(\frac{\partial \mathbf{U}}{\partial t} + \mathbf{U}\cdot\nabla\mathbf{U}\right) = -\nabla(P + P_p + P_A) + \mathbf{J}\times\mathbf{B} - \nabla\cdot\Pi_p - \nabla\cdot\Pi_A; \tag{62}$$

$$\frac{\partial P}{\partial t} + \mathbf{U}\cdot\nabla P + \gamma P \nabla\cdot\mathbf{U} = 0; \tag{63}$$

$$\frac{\partial P_p}{\partial t} + \mathbf{U}\cdot\nabla P_p + \gamma_p P_p\nabla\cdot\mathbf{U} = \frac{1}{3}\nabla\cdot(\mathbf{K_p}\cdot\nabla P_p) - (\gamma_p - 1)\Pi_p : (\nabla\mathbf{U}); \tag{64}$$

$$\frac{\partial P_A}{\partial t} + \mathbf{U}\cdot\nabla P_A + \gamma P_A\nabla\cdot\mathbf{U} = \frac{1}{3}\nabla\cdot(\mathbf{K_A}\cdot\nabla P_A) - (\gamma_A - 1)\Pi_A : (\nabla\mathbf{U}); \tag{65}$$

$$\mathbf{E} = -\mathbf{U}\times\mathbf{B}; \quad \frac{\partial \mathbf{B}}{\partial t} = -\nabla\times\mathbf{E}; \quad \mu_0\mathbf{J} = \nabla\times\mathbf{B}; \quad \nabla\cdot\mathbf{B} = 0, \tag{66}$$

where we have introduced the ACR pressure P_A, the corresponding stress tensor Π_A, the ACR diffusion tensor $\mathbf{K}_A$ and adiabatic index γ_A ($4/3 \leq \gamma_A \leq 5/3$). The coupled system (61) - (66) is the simplest continuum model to describe a non-equilibrated plasma comprising a thermal proton-electron plasma with suprathermal particles (e.g., PUIs or SEPs) and relativistic energy (anomalous) cosmic rays. The system includes both the collisionless heat flux and viscosity associated with the suprathermal and relativistic particle distributions.

On reverting to equations (56) - (60), we can recover the standard form of the MHD equations if we set the heat conduction spatial diffusion tensor $\mathbf{K} = 0$ and the coefficient of viscosity $(\eta_{kl}) = 0$, which corresponds to assuming $\tau_{ws} \to 0$. If the total thermodynamic pressure $P_{total} = P + P_p$ is introduced, then we recover the standard MHD equations (dropping the subscript "total") i.e.,

$$\frac{\partial \rho}{\partial t} + \nabla \cdot (\rho \mathbf{U}) = 0; \tag{67}$$

$$\rho \frac{\partial \mathbf{U}}{\partial t} + \rho \mathbf{U} \cdot \nabla \mathbf{U} + (\gamma - 1)\nabla e + (\nabla \times \mathbf{B}) \times \mathbf{B} = 0; \tag{68}$$

$$\frac{\partial}{\partial t}\left(\frac{1}{2}\rho U^2 + e + \frac{B^2}{2\mu_0}\right) + \nabla \cdot \left[\left(\frac{1}{2}\rho U^2 + \gamma e\right)\mathbf{U} + \frac{1}{\mu_0}\mathbf{B} \times (\mathbf{U} \times \mathbf{B})\right] = 0; \tag{69}$$

$$\frac{\partial \mathbf{B}}{\partial t} = \nabla \times (\mathbf{U} \times \mathbf{B}); \quad \nabla \cdot \mathbf{B} = 0, \tag{70}$$

with an equation of state $e = \alpha n k_B T/(\gamma - 1)$. The choice of $\alpha = 2$ (or greater if incorporating the contribution of cosmic rays etc.) corresponds to a plasma population comprising protons and electrons.

In setting $\mathbf{K} = 0$ and $(\eta_{kl}) = 0$, we have implicitly assumed that energetic particles and PUIs are completely coupled to the thermal plasma. With $\mathbf{K} \neq 0$, heat conduction reduces the effective coupling of energetic particles to the thermal plasma, and their contribution to the total pressure is not as large. This will have important consequences for numerical models of e.g., the large-scale heliosphere since they incorporate PUIs into the MHD equations, without distinguishing PUIs from thermal plasma and therefore neglect heat conduction. Consequently the total pressure is over-estimated.

4. Conclusions

Energetic particles (solar energetic particles, PUIs) can be dynamically important in numerous settings [1–12]. The energetic particle population is characterized by their scattering in pitch-angle off pre-existing and self-induced Alfvénic and turbulent fluctuations. Pitch-angle scattering drives energetic particle distributions towards a state of near but not complete isotropy, and infrequent intra-particle collisions between the relatively tenuous energetic particles and more populous thermal particles ensure that the energetic particles are not equilibrated with the background plasma. This therefore necessitates distinct descriptions for the different plasma species. Our results can be summarized as follows.

(i) We have presented a systematic derivation of a multi-component plasma model that includes thermal electrons and protons and an energetic particle species such as solar energetic particles or pickup ions, coupled through Maxwell's equations. The derivation is based on a collisionless Chapman-Enskog expansion [9; 30].

(ii) The Chapman-Enskog expansion yields collisionless forms of the heat conduction tensor and (viscous) stress tensor for the energetic particles. Both the heat conduction and stress tensor expressions bear a formal resemblance to the "classical" expressions found from collisional gas dynamics.

(iii) Even though the energetic particles and the thermal plasma are not equilibrated, the multi-component model can be simplified in the MHD limit of assuming a small mass ratio m_e/m_p and current density. This yields a non-equilibrated single-fluid-like model that is reminiscent of the cosmic ray two-fluid model [28; 29]. However, the derivation presented here is fundamentally different and our approach clarifies the somewhat ad hoc derivation of the two-fluid model, showing for example that the total number density includes the cosmic ray number density. The single-fluid-like model retains distinct descriptions for energetic particle and thermal plasma pressure or temperature. We show further how to include energetic relativistic particles such as anomalous cosmic rays along with suprathermal particles such as PUIs.

(iv) Finally, in the limit of extremely rapid scattering $\tau_{ws} \to 0$, the single-fluid-like model reduces to the standard set of MHD equations in which heat conduction and viscous stress tensor terms are absent.

Several implications of our results should be noted.

(i) The models developed here provide a basic description for modeling non-equilibrated plasma systems that have distinct thermal and energetic particle components, each carrying different pressures or temperatures. The most elaborate is the full multi-component model that contains the thermal electron as well as the thermal proton and energetic particle physics. As discussed by Zank et al. [9], the presence of an energetic particle component yields a linear wave description that is much richer than the standard two-fluid description. However, a "low-frequency" approximation yields a single-fluid like description that is simpler to study and implement numerically while retaining many of the characteristics of the non-equilibrated plasma system.

(ii) The momentum and energy transport equations for energetic particles provide physically motivated collisionless dissipative terms (heat conduction and viscous stress tensor).

(iii) The presence of physically motivated dissipative terms in the energetic particle equations raises questions regarding the design of numerical algorithms that couple gas dynamic or MHD equations to an energetic particle component. Is the incorporation of numerical dissipation in the gas dynamics/MHD solver appropriate given the possibly more dominant physically based dissipative terms introduced by energetic particles?

Acknowledgments
We acknowledge the partial support of NASA grants NNX08AJ33G, Subaward 37102-2, NNX14AC08G, NNX14AJ53G, A99132BT, RR185-447/4944336 and NNX12AB30G.

References

[1] Lario D, Decker R B, Roelof E C and Viñas A F 2015 *Journal of Physics Conference Series* **642** 012014

[2] Lario D, Decker R B, Roelof E C and Viñas A F 2015 *Ap. J.* **813** 85 (*Preprint* 1509.04368)

[3] Russell C T, Mewaldt R A, Luhmann J G, Mason G M, von Rosenvinge T T, Cohen C M S, Leske R A, Gomez-Herrero R, Klassen A, Galvin A B and Simunac K D C 2013 *Ap. J.* **770** 38

[4] Terasawa T 1999 *International Cosmic Ray Conference* **6** 528

[5] Terasawa T, Oka M, Nakata K, Keika K, Nosé M, McEntire R W, Saito Y and Mukai T 2006 *Advances in Space Research* **37** 1408–1412

[6] Burlaga L F, Ness N F, Belcher J W, Szabo A, Isenberg P A and Lee M A 1994 *J. Geophys. Res.* **99** 21511

[7] Zank G P 1999 *Space Sci. Rev.* **89** 413–688

[8] Richardson J D, Paularena K I, Lazarus A J and Belcher J W 1995 *Geophys. Res. Lett.* **22** 1469–1472

[9] Zank G P, Hunana P, Mostafavi P and Goldstein M L 2014 *Ap. J.* **797** 87

[10] Zank G P 2015 *Ann. Rev. Astron. Astrophys.* **53** 449–500

[11] Decker R B, Krimigis S M, Roelof E C, Hill M E, Armstrong T P, Gloeckler G, Hamilton D C and Lanzerotti L J 2008 *Nature* **454** 67–70

[12] Decker R B, Krimigis S M, Roelof E C and Hill M E 2015 *Journal of Physics Conference Series* **577** 012006

[13] Zank G P, Pauls H L, Williams L L and Hall D T 1996 *J. Geophys. Res.* **101** 21639–21656

[14] McComas D J, Alexashov D, Bzowski M, Fahr H, Heerikhuisen J, Izmodenov V, Lee M A, Möbius E, Pogorelov N, Schwadron N A and Zank G P 2012 *Science* **336** 1291–

[15] Zank G P, Heerikhuisen J, Wood B E, Pogorelov N V, Zirnstein E and McComas D J 2013 *Ap. J.* **763** 20

[16] Zirnstein E J, Heerikhuisen J, Zank G P, Pogorelov N V, McComas D J and Desai M I 2014 *Ap. J.* **783** 129

[17] Zank G P, Heerikhuisen J, Pogorelov N V, Burrows R and McComas D 2010 *Ap. J.* **708** 1092–1106

[18] Lee M A 1983 *J. Geophys. Res.* **88** 6109–6119

[19] Ng C K, Reames D V and Tylka A J 1999 *Geophys. Res. Lett.* **26** 2145–2148

[20] Gordon B E, Lee M A, Möbius E and Trattner K J 1999 *J. Geophys. Res.* **104** 28263–28278

[21] Rice W K M, Zank G P and Li G 2003 *Journal of Geophysical Research (Space Physics)* **108** 1369

[22] Li G, Zank G P and Rice W K M 2003 *Journal of Geophysical Research (Space Physics)* **108** 1082

[23] Zank G P, Li G and Verkhoglyadova O 2007 *Space Sci. Rev.* **130** 255–272

[24] Verkhoglyadova O P, Zank G P and Li G 2015 *Phys. Rep.* **557** 1–23

[25] Lee M A and Ip W H 1987 *J. Geophys. Res.* **92** 11041–11052

[26] Williams L L and Zank G P 1994 *J. Geophys. Res.* **99** 19229–+

[27] Cannon B E, Smith C W, Isenberg P A, Vasquez B J, Murphy N and Nuno R G 2014 *Ap. J.* **784** 150

[28] Axford W I, Leer E and McKenzie J F 1982 *Astron. Astrophys.* **111** 317–325

[29] Webb G M 1983 *Astron. Astrophys.* **127** 97–112

[30] Zank G P 2014 *Transport Processes in Space Physics and Astrophysics* (*Lecture Notes in Physics, Berlin Springer Verlag* vol 877) (Springer-Verlag)

[31] Isenberg P A 1997 *J. Geophys. Res.* **102** 4719–4724

[32] Webb G M 1985 *Ap. J.* **296** 319–330

[33] Webb G M 1989 *Ap. J.* **340** 1112–1123

Minimally implicit Runge-Kutta methods for Resistive Relativistic MHD

Miguel-Á. Aloy[1] and Isabel Cordero-Carrión[2]

[1] Departamento de Astronomía y Astrofísica, Universidad de Valencia, C/ Dr. Moliner 50, 46100 Burjassot, Spain

[2] Departamento de Matemática Aplicada, Universidad de Valencia, C/ Dr. Moliner 50, E-46100 Burjassot,Spain.

E-mail: miguel.a.aloy@uv.es

Abstract. The Relativistic Resistive Magnetohydrodynamic (RRMHD) equations are a hyperbolic system of partial differential equations used to describe the dynamics of relativistic magnetized fluids with a finite conductivity. Close to the ideal magnetohydrodynamic regime, the source term proportional to the conductivity becomes potentially stiff and cannot be handled with standard explicit time integration methods. We propose a new class of methods to deal with the stiffness fo the system, which we name Minimally Implicit Runge-Kutta methods. These methods avoid the development of numerical instabilities without increasing the computational costs in comparison with explicit methods, need no iterative extra loop in order to recover the primitive (physical) variables, the analytical inversion of the implicit operator is trivial and the several stages can actually be viewed as stages of explicit Runge-Kutta methods with an effective time-step. We test these methods with two different one-dimensional test beds in varied conductivity regimes, and show that our second-order schemes satisfy the theoretical expectations.

1. Introduction

The Relativistic Resistive Magnetohydrodynamic (RRMHD) equations are a hyperbolic system of partial differential equations used to describe the fluid dynamics in the presence of magnetic fields with a finite (but potentially large) conductivity, and when the velocities involved are close to speed of light, c, or the fluid specific energies become comparable or larger than c^2. In Astrophysical contexts, all these ingredients can be found in scenarios such as compact stars or relativistic jets.

The RRMHD system of equations can be written as a set of evolution equations corresponding to the conservation of rest-mass, momentum, energy, magnetic flux and electric charge. In addition to these conserved variables, Komissarov [1] has extended the system by two extra scalar potentials which control the evolution of the solenoidal constraint for the magnetic field and the divergence of the electric field following the ideas of Dedner et al. [2].

The physical or *primitive* variables in RRMHD are

$$\mathbf{W} = (B^i, E^i, q, \rho, \epsilon, v^i), \tag{1}$$

where B^i ($i = 1, 2, 3$) are the components of the magnetic field, E^i are the components of the electric field, q is the electric charge, ρ is the rest-mass density, ϵ is the specific internal energy and v^i are the components of the flow velocity field. The conserved or numerical variables of this system are

$$\mathbf{U} = (\Phi, B^i, \Psi, E^i, q, D, \tau, P^i), \tag{2}$$

where $D = \rho W$ is the mass-density, $W = 1/\sqrt{1-v^2}$ is the Lorentz factor, $\tau = \rho h W^2 - p - \rho W + (E^2 + B^2)/2$, h is the specific enthalpy, p is the pressure, $E^2 = E^i E_i$, $B^2 = B^i B_i$, $P^i = \rho h W^2 v^i + (\mathbf{E} \times \mathbf{B})^i$ and Φ and Ψ are two scalar potentials controlling the evolution of the constrains $\nabla \cdot \mathbf{B} = 0$ and $\nabla \cdot \mathbf{E} = q$, respectively. We note that the scalar potentials can be regarded also as part of the set primitive variables.

For practical purposes, the system of RRMHD equations can be split into two sets of equations, ones containing stiff source terms and the rest of them:

$$\begin{aligned} \partial_t E^j &= \tilde{S}^j_{\mathbf{E}} - \sigma W [E^j + (\mathbf{v} \times \mathbf{B})^j - E^l v_l v^j], \\ \partial_t \mathbf{Y} &= S_{\mathbf{Y}}, \end{aligned} \tag{3}$$

where $\mathbf{Y} = (\Phi, B^i, \Psi, q, D, \tau, P^i)$, $\tilde{S}^j_{\mathbf{E}}$ and $S_{\mathbf{Y}}$ contain the corresponding fluxes and non-stiff source terms of the evolution equations, and σ is the conductivity of the system. In the ideal limit, $\sigma \to \infty$ and the electric field can be expressed in terms of the rest of the primitive variables, so that the corresponding evolution equation for the electric field is therefore not included in the system of evolved equations. In the case of a finite and high-conductivity, the term proportional to σ becomes potentially a stiff source term.

The numerical evolution of the equations with non-stiff source terms $\tilde{S}^j_{\mathbf{E}}$ and $S_{\mathbf{Y}}$ can be carried out with a TVD explicit Runge-Kutta (RK) method. In the high-conductivity regime, an alternative method has to be used instead to deal with the stiffness of the source term in the evolution equation of $\mathbf{E}$. One possibility is the use of a semi-analytic integration of the stiff source terms (and/or the scalar fields), together with a Strang-splitting scheme [1], but then the order of convergence of the numerical scheme can reduce to first-order due to the fact that the kernel of the relaxation operator is non-trivial and corresponds to a singular matrix in the linear case.

A very interesting option to numerically evolve this hyperbolic system with stiff source terms is the use of implicit-explicit (IMEX) RK methods [3, 4, 5, 6], as in the case of [7]. The problems arising in this case are, on one hand, the degradation to first or second-order for a range of values of σ, and, on the other hand, the unclear update of the variables when the primitive variables cannot be explicitly coined in terms of the conserved ones. The second point steams from the fact that the Lorentz factor and the velocity components appear in the stiff source terms. The updated values of these quantities are needed to update the electric field but, in general, the whole set of conserved variables (including the electric field) are needed to recover the velocity components. In [7] this fact was overcome by the addition of an iterative loop over the electric field, which is computationally expensive, and there is no guarantee of convergence to physically acceptable values.

We present here an alternative class of methods to the previous ones, which we name Minimally Implicit RK (MIRK) methods, to avoid numerical instabilities due to the presence of stiff terms without increasing the computational costs in comparison with explicit methods. The MIRK methods reduce to the optimal TVD explicit RK methods of Shu & Osher [8] for the $\tilde{S}^j_{\mathbf{E}}$ and $S_{\mathbf{Y}}$ operators, and implicitly evolve the stiff source terms as we explain in next section; the proposed strategy needs no iterative loop over the electric field, the analytical inversion of the implicit operator is trivial and the several stages can actually be viewed as stages from explicit RK methods with an effective time-step.

2. Minimally implicit Runge-Kutta methods

In order to derive the MIRK methods, by construction we impose that they must recover the optimal TVD explicit RK methods for the $\tilde{S}^j_{\mathbf{E}}$ and $S_{\mathbf{Y}}$ operators. We also consider updated and non-updated values of the electric field in the evaluation of the stiff source term in Eq. (3), but only non-updated values of the velocity field and Lorentz factor. The later strategy is key to avoid employing extra iterative loops to recover the primitive variables from the conserved ones (see below). In order to simplify the notation, we define $\bar{\sigma} = \sigma W$ and $S^j_{\mathbf{E}} = \tilde{S}^j_{\mathbf{E}} - \sigma W[(\mathbf{v} \times \mathbf{B})^j - E^l v_l v^j]$.

The proposed method can be written as an usual explicit, $s-$stages RK method with an effective time-step in the following way:

$$\mathbf{Y}^{(1)} = \mathbf{Y}|_n, \tag{4}$$

$$(E^k)^{(1)} = E^k|_n, \tag{5}$$

$$\mathbf{Y}^{(i)} = \mathbf{Y}|_n + \Delta t \sum_{j=1}^{i-1} \tilde{a}_{ij} S^{(j)}_{\mathbf{Y}}, \tag{6}$$

$$(E^k)^{(i)} = E^k|_n + \frac{\Delta t}{1+\Omega_i} \sum_{j=1}^{i-1} \tilde{a}_{ij} (S^k_{\mathbf{E}})^{(j)} \sum_{j,l=1}^{i-1} a_{ijl} (-\bar{\sigma})^{(l)} (E^k)^{(j)}, \quad i = 2, \ldots, s+1, \tag{7}$$

$$\Omega_i = \Delta t \sum_{j=1}^{i-1} b_{ij} \bar{\sigma}^{(j)}, \tag{8}$$

$$\mathbf{Y}|_{n+1} = \mathbf{Y}^{(s+1)}, \tag{9}$$

$$E^k|_{n+1} = (E^k)^{(s+1)}, \tag{10}$$

where $|_n$ and $|_{n+1}$ denote the values of variables at time levels t_n and t_{n+1}, respectively, $\tilde{a}_{ij}$ denotes the coefficients of the Butcher matrix associated to the explicit RK method, a_{ijl}, b_{ij} are the coefficients associated to the implicit part of the method, and Δt is the time-step. Note that, differently from standard RK-IMEX methods, there is no final *explicit* substep. Indeed, Eqs. (9) and (10), simply mean that the updated values of the variables at t_{n+1} are the ones computed after the previous $s-$substeps of the MIRK algorithm. Equivalently, in the standard notation of Butcher tables (e.g., [7]), the coefficients ω and $\tilde{\omega}$ are identically zero.

We point out that the time update of the conserved variables is explicit, i.e., all the terms in the right-hand side of the equations are evaluated in time levels of the algorithm that have already been computed. Singularly, the updated values of the electric field depend upon already known values of it and of the primitive variables. This particularity of the MIRK schemes is very useful for the RRMHD system, since it avoids the need of performing a double iterative procedure to recover the primitive variables from the conserved ones. This is a clear advantage over the usage of RK-IMEX schemes to deal with the stiffness of the RRMHD system (e.g., [7]). It is also relevant to point out that, in order to write the implicit scheme in an explicit form, a much simpler matrix than in [7] has been inverted analytically, since the terms $E^l v_l v^j$ in the expression for $S^j_{\mathbf{E}}$ are treated explicitly in MIRK schemes.

2.1. *First-order method*

The first-order MIRK method corresponds to $s = 1$ in the previous expressions, with all coefficients zero except for

$$\tilde{a}_{21} = a_{211} = 1, \quad b_{21} = (1 - c_1), \tag{11}$$

$(1 - c_1) \neq 0$, and c_1 being a real coefficient to be determined below. With this set of coefficients, the new method provides bounded values to all the conserved variables, even in the limit $\bar{\sigma} \gg 1$. After linearization of the non-stiff source terms, we study the matrix which updates the values from one time-step to the next one. Linear stability is satisfied when the absolute value of all the eigenvalues of this matrix are bounded by 1. In particular, linear stability implies that the absolute value of the determinant of the matrix that updates the values of the conserved variables from one timestep to the next one, $|\mathcal{M}_{\Delta t}|$, is also bounded by 1. This necessary (not sufficient) condition restricts the range of the coefficients very efficiently and allows to find values for the coefficient(s) very close to the optimal ones. Under the hypothesis that $|\mathcal{M}_{\Delta t}|$ is bounded by 1 when the stiff source terms are neglected, it can be checked that the choice $c_1 = 0$ yields a determinant of $\mathcal{M}_{\Delta t}$ also bounded by 1 when the stiff source terms are included, for Δt sufficient small. The final expression of the method for this optimal value of the coefficient can be written as:

$$\mathbf{Y}|_{n+1} = \mathbf{Y}|_n + \Delta t\, S_{\mathbf{Y}}|_n, \tag{12}$$

$$E^k|_{n+1} = E^k|_n + \frac{\Delta t}{1+\Delta t\,\bar{\sigma}|_n}\left[S^k_{\mathbf{E}} - \bar{\sigma}E^k\right]|_n. \tag{13}$$

We note that the factor

$$\frac{\Delta t}{1+\Delta t\,\bar{\sigma}|_n},$$

can be regarded as an effective time step, that decreases in magnitude as $\bar{\sigma}$ increases. Indeed, in the ideal limit, Eq. (13) yields $E^k \simeq -(\mathbf{v}\times\mathbf{B})^k$.

2.2. Second-order method

The second-order MIRK method corresponds to $s = 2$ in the general expression of Sect. 2. Following the same prescription as in the first-order method, i.e., imposing that the method recovers the optimal TVD second-order explicit RK method when stiff source terms are neglected, all coefficients are zero except for

$$\begin{aligned}&\tilde{a}_{21} = a_{211} = 1,\ \ b_{21} = (1-c_1),\ \ \tilde{a}_{31} = \tilde{a}_{32} = 1/2,\\ &a_{311} = c_1/2,\ \ a_{321} = (1-c_1)/2,\ \ a_{312} = 1/2,\\ &b_{32} = (c_1/2 - c_2),\end{aligned} \tag{14}$$

$(1-c_1) \neq 0 \neq (c_1/2 - c_2)$. If the values of the conserved variables $(E^i, \mathbf{Y})|_{(n)}$ are bounded at the time step n, this method guarantees that the values $(E^i, \mathbf{Y})|_{(n+1)}$ will also be bounded, even when $\bar{\sigma} \gg 1$.

In order to save RAM memory and improve on computational efficiency, the second-order MIRK scheme can be rewritten as:

$$\mathbf{Y}|_{(1)} = \mathbf{Y}|_n + \Delta t\, S_{\mathbf{Y}}|_n, \tag{15}$$

$$E^k|_{(1)} = E^k|_n + \frac{\Delta t}{1+\Delta t\,(1-c_1)\bar{\sigma}|_n}\left[S^k_{\mathbf{E}} - \bar{\sigma}E^k\right]|_n, \tag{16}$$

$$\mathbf{Y}|_{n+1} = \frac{1}{2}[\mathbf{Y}|_n + \mathbf{Y}|_{(1)} + \Delta t\, S_{\mathbf{Y}}|_{(1)}], \tag{17}$$

$$\begin{aligned}E^k|_{n+1} = {}& E^k|_{(1)} + \\ & \left.\frac{[-1+\Delta t\,\bar{\sigma}(1-c_1)]}{2[1+\Delta t\,\bar{\sigma}(c_1/2-c_2)]}\right|_{(1)}(E^k|_{(1)} - E^k|_n) + \left.\frac{\Delta t}{2[1+\Delta t\,\bar{\sigma}(c_1/2-c_2)]}\right|_{(1)}[S^k_{\mathbf{E}} - \bar{\sigma}E^k]|_{(1)}.\end{aligned} \tag{18}$$

Written in this form, the second-order MIRK scheme does not need any extra memory storage as compared with an optimal explicit second-order RK TVD scheme.

After linearization of the non-stiff source terms, we focus again on the condition $|\mathcal{M}_{\Delta t}| < 1$. We find two possible sets of values for the coefficients. If we choose $c_2 = \frac{(1-c_1)^2}{2c_1}$, we have to restrict to $c_1 < 0$ or $0.67 < c_1 < 0.75$. We find by numerical experimentation (see Sect. 3.3) that only the region $c_1 < 0$ gives stable results, with the optimal value for c_1 located around -0.1. If we consider the case $c_1 = 0$, we have to restrict to $-1 < c_2 < -0.5$. After numerical experimentation we find that the optimal value for c_2 is located around -0.97. Values of both coefficients below zero are favored, and all the options have to be verified numerically.

3. Numerical results

The MIRK methods described in the previous sections have been implemented in the MRGENESIS code [9, 10, 11, 12]. In order to compute the numerical fluxes we employ a Local Lax-Friedrichs approximate solver [1], in which the states left and right to each interface are obtained by a second-order monotone central (MC) reconstruction of primitive variables. We have also tested the new methods in combination with a more diffusive minmod reconstruction, but the results are affected by a larger diffusion close to discontinuities or big gradients (as expected).

In order to test the newly proposed MIRK schemes, we have considered two different one-dimensional tests, each of which proves a different regime of conductivities and, hence, different degrees of stiffness of the RRMHD system. In all the tests considered in the next sections, we assume that the equation of state is that of an ideal gas with a fixed adiabatic index γ.

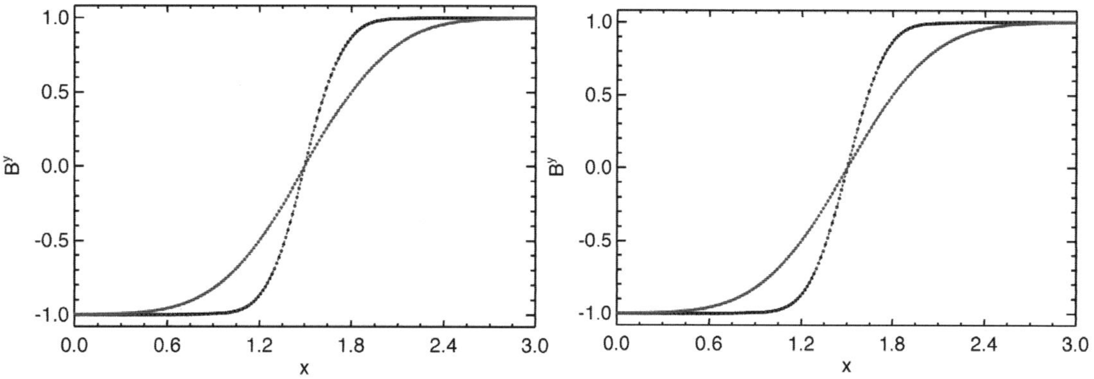

Figure 1. Comparison of the numerical and the analytical solution of the self-similar current sheet. The numerical solution (symbols) has been computed employing $N_x = 200$, CFL=0.7, the second-order MIRK scheme with $c_1 = -0.1$, $c_2 = -0.97383794$ (left panel) and $c_1 = 0$, $c_2 = -0.85$ (right panel). The analytic solution (dashed lines; difficult to see since it is below the symbols) is computed according to Eq. (19). The initial time $t = 1$ and the final time $t = 10$ are displayed with purple and red colors, respectively. There are no obvious differences in the numerical solutions among the two sets of parameters (c_1, c_2) in this resistive (i.e., low σ) test.

3.1. Self-similar current sheet

Komissarov [1] proposed a test to prove the resistive regime in which $\sigma \simeq 100$, i.e., relatively far away from the ideal limit. This is the so-called self-similar current sheet, where we have a magnetic field $\mathbf{B} = (0, B(x,t), 0)$. The magnetic pressure is much smaller than the gas pressure and $B(x, 0)$ changes sign within a thin current layer. Except for the magnetic field, the rest of the primitive variables are set to be constant in the whole domain: $p = 50$, $\rho = 1$, $\gamma = 4/3$, $\mathbf{E} = \mathbf{0}$, $\mathbf{v} = \mathbf{0}$ and $\sigma = 100$. This test has an analytic (self-similar) solution for $t > 0$ of the form

$$B(x,t) = B_0 \,\mathrm{erf}\left(\frac{(x - x_0)}{2}\sqrt{\frac{\sigma}{t}}\right), \tag{19}$$

where $B_0 = 1$, $x_0 = 1.5$, the initial time is set to $t = 1$, and the error function is defined in the usual way:

$$\mathrm{erf}(x) = \frac{2}{\sqrt{\pi}} \int_0^x e^{-t^2} \mathrm{d}t.$$

We employ 200 computational cells to cover the domain $[0, 3]$. We let the system to evolve until $t = 10$ and compare our results with the analytic solution in Fig. 1. In this case we have used the second order MIRK scheme with $c_1 = -0.1$ and $c_2 = -0.97383794$ (Fig. 1; left) and also with $c_1 = 0$ and $c_2 = -0.85$ (Fig. 1; right). As can be observed, there is basically no difference between the numerical and the analytical solutions. The results do not qualitatively change if the MIRK coefficients are slightly changed with respect to their canonical values. We note that the CFL employed in this test is 0.7.

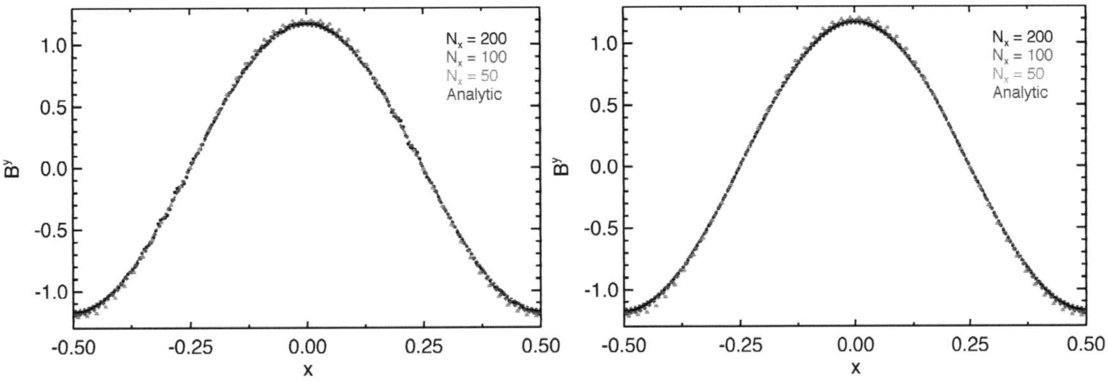

Figure 2. Comparison of numerical solutions of the circularly polarized Alfvén waves test for different resolutions, $N_x = 50, 100$ and 200 (in orange, red and black, respectively). The numerical solution has been computed employing the second-order MIRK scheme with $c_1 = 0$, $c_2 = -0.85$ and CFL=0.45 (right panel) and with $c_1 = -0.1$, $c_2 = -0.97383794$ and CFL=0.4 (left panel).

3.2. Circularly polarized Alfvén waves

A second one-dimensional test that proves the ideal limit assuming a very large value for the conductivity is considered by Palenzuela et al. [7]. This test was originally proposed by Del Zanna et al. [13] and has an exact solution in ideal Relativistic Magnetohydrodynamics. The solution describes the propagation of large amplitude Alfvén waves along a uniform background magnetic field with strength B_0, in a domain with periodic boundary conditions. In the ideal limit, the velocity along the $x-$direction is zero ($v_x = 0$), and the exact solution reads:

$$(B_y, B_z) = \eta_A B_0 \left(\cos\left[k(x - v_A t)\right], \sin\left[k(x - v_A t)\right]\right), \tag{20}$$

$$(v_y, v_z) = -\frac{v_A}{B_0}(B_y, B_z), \tag{21}$$

where $B_x = B_0$, k is the wave vector, η_A is the wave amplitude, and v_A is the relativistic Alfvén speed, that in this case is

$$v_A^2 = \frac{2B_0^2}{h + B_0^2(1 + \eta_A^2)} \left\{ 1 + \sqrt{1 - \left[\frac{2\eta_A B_0^2}{h + B_0^2(1 + \eta_A^2)}\right]^2} \right\}^{-1}. \tag{22}$$

Though there is no exact solution in the RRMHD regime for circularly polarized Alfvén waves, in the limit of very large conductivity, Eqs. (20) and (21) are approximately satisfied. Thus we compute the evolution from the initial state and until the train of waves completes one period. In the ideal limit, the solution should be identical to that in the initial state and we expect that if $\sigma \gg 1$, also the resistive solution shall be quite similar to the initial one. For this test we fix $\gamma = 2$, $k = 2\pi$, $B_0^2 = 4/3$, $p = \rho = \eta_A = 1$, which leads to a period $T = 2$, and an Alfvén speed $v_A = 1/2$.

Figure 2 shows the comparison of the analytic versus the numerical solutions with different resolutions. It is evident that our numerical solution lies on top of the analytic one, even for very moderate resolutions. As in the previous test, we have computed the solution with different MIRK coefficients. We find that, in this case, much more demanding from the numerical point of view (because of the large value of the conductivity), the fact that we do have only approximately optimal MIRK coefficients (see Sect. 2) has a clear impact on the stability of the method. While the set of coefficients $(c_1, c_2) = -(0.1, -0.97383794)$ yields unstable results for CFL values larger than approximately 0.4 (Fig. 2; left), the set of parameters $(c_1, c_2) = (0, -0.85)$ is still stable for CFL=0.45. In any case, the range of allowed values of the CFL yielding stable results depends on the magnitude of σ. From an extensive numerical experimentation we conclude that for $\sigma \lesssim 5 \times 10^6$, CFL values up to about 0.7 yield stable results for $(c_1, c_2) = (0, -0.85)$ and small variations of these parameters. For $\sigma \gtrsim 5 \times 10^6$ progressively smaller values of the CFL coefficient must be used to maintain the stability of the numerical solution.

3.3. Order of convergence

We have also performed a number of convergence tests employing as reference the one dimensional problems at hand, varying the value of the second-order MIRK coefficients, c_1 and c_2. In Fig. 3 we display the L_1-norm errors for the B_y component of the magnetic field as a function of the resolution. The precise definition of the error for a given test including N_x uniform cells reads:

$$\mathrm{Error}_{N_x} = \frac{L}{N_x} \sum_{i}^{N_x} |B_i^y - B_{(\mathrm{ref}),i}^y|, \tag{23}$$

where L is the domain size, and $B_{(\mathrm{ref}),i}^y$ is a numerical solution computed at sufficiently high resolution to be considered as a reference solution. In our case, $B_{(\mathrm{ref}),i}^y$ is computed with $N_x = 3200$.

Our second-order MIRK scheme converges to a bit more than second-order for the self-similar current sheet test for a relatively large range of coefficients c_1 and c_2 (Fig. 3). The order of convergence is slightly smaller than 2 ($p = 1.83$) for $c_1 = -0.1$ and $c_2 = (1 - c_1)^2/(2c_1) = -6.05$ and a bit larger than 2 ($p = 2.19$) for $c_1 = 0$ and $c_2 = -0.9$. Thus, in the limit of moderate conductivity the method performs as expected.

In Fig. 4 we display the L_1-norm errors for the B_y component of the magnetic field as a function of the resolution for the test of circularly polarized Alfvén waves. We observe that at low resolution and for very high conductivity values, $\sigma = 10^6$, the second-order schemes have an order of convergence closer to first-order than to second-order. However, as the resolution increases, we find that the order of the method evolves towards the nominal second-order of the algorithm. We also find by numerical experimentation that the best values of the second-order MIRK coefficients lie in the range $-0.1 \le c_1 \le 0$, and $-1 < c_2 < -0.85$. We believe that the difference between the theoretical and numerical order of convergence are triggered by the very large conductivity, which not only increases the stiffness of the system of equations, but also introduces severe looses of accuracy as a result of the very different orders of magnitude among different variables in arithmetic operations (e.g., to compute the electric currents).

Journal of Physics: Conference Series **719** (2016) 012015 doi:10.1088/1742-6596/719/1/012015

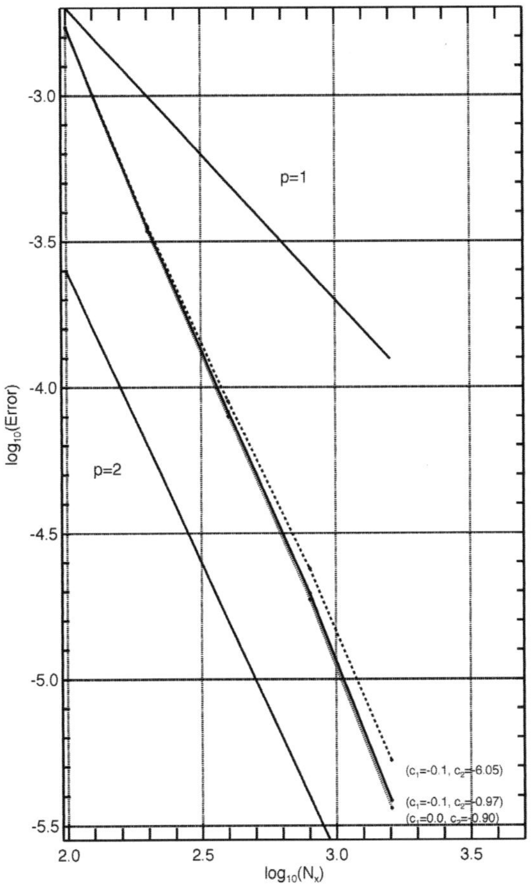

Figure 3. Comparison of L_1-norm errors (defined in Eq. 23) for the B^y component of the magnetic field in the self similar current sheet test. The different line styles correspond to different choices of the parameters c_1 and c_2 of the second-order MIRK scheme (see legends). All the numerical tests have been computed with CFL=0.7 and $\sigma = 10^2$. As a reference, we add two solid lines showing a theoretical evolution of the error as in a first-order scheme ($p = 1$) and in a second-order one ($p = 2$).

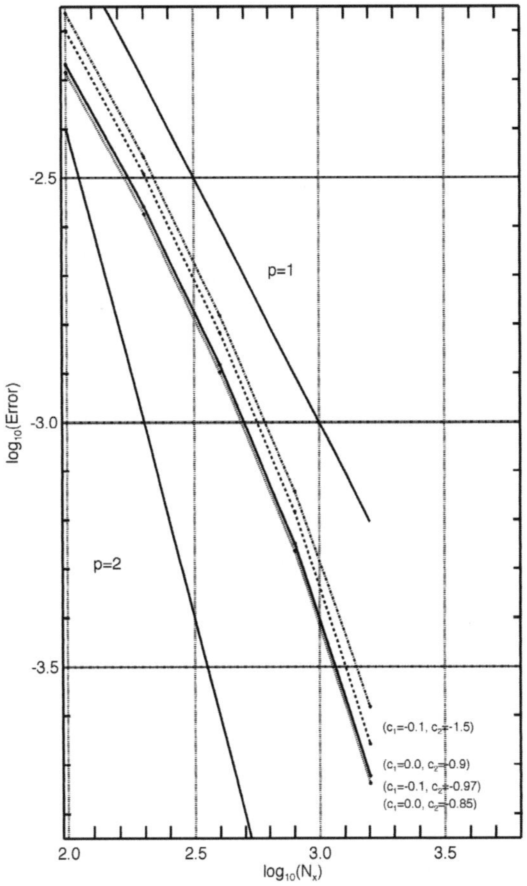

Figure 4. Same as Fig. 3 but for the test of large amplitude circularly polarized Alfvén waves. The different line styles correspond to different choices of the parameters c_1 and c_2 of the second-order MIRK scheme (see legends). All the numerical tests have been computed with CFL=0.1 and $\sigma = 10^6$. For clarity, we do not include the case $c_1 = 0$, $c_2 = -0.97383794$, since it basically lies on top of the case $c_1 = -0.1$, $c_2 = -0.97383794$.

4. Conclusions

We propose here a new kind of RK methods, named minimally implicit RK methods, to numerically evolve some systems of hyperbolic equations with stiff source terms. As a first target for the new MIRK methods, we have considered the numerical evolution of the RRMHD equations as a motivation of the development of the methods. However, we foresee applications to other systems of hyperbolic equations with stiff source terms. At the moment, we are working on applying MIRK methods to coupled system of General Relativistic Radiation Hydrodynamics. In this system, the neutrino transport equations become very stiff in the opaque regime (when neutrinos are trapped in the matter; see, e.g., [14, 15, 16]).

MIRK methods recover the optimal TVD explicit RK algorithm when the stiff source terms are neglected. They have a very simple analytical inversion of the implicit operators included to deal with the stiff source terms, and their structure can be viewed as an explicit RK method with an effective time-step.

Currently, one dimensional numerical test problems have been considered, but we are working on their application to multidimensional scenarios. Our aim is to improve the current treatment of the resistivity employed in ideal MHD simulations (where the source of such resistivity is not physical but numerical). In this line, we aim to improve the treatment of resistivity in some local numerical simulations of the Kevin-Helmholtz instability in the context of neutron star mergers [17], as well as in global models (e.g., [18]). Likewise, we aim to incorporate physical resistive dissipation to the problem of internal shocks in blazars [19, 20, 21, 22] and in the shocks associated to afterglows [23, 24, 25].

MIRK methods present some advantages in comparison with the well-known RK IMEX methods: CPU time saving, RAM memory saving and accuracy; these properties have to be explored in more detail. We also seek to compare our methods with Strang-splitting ones.

Acknowledgments

Authors acknowledge financial support from the European Research Councill (ERC) through the Starting Independent Researcher Grant CAMAP-259276, the Spanish Government (grants AYA2013-40979-P, AYA2015-66899-C2-1-P and SAF2013-49284-EXP) and the Regional Government (Generalitat Valenciana, grant Prometeo-II/2014/069).

References

[1] Komissarov S S 2007 *MNRAS* **382** 995–1004 (*Preprint* 0708.0323)
[2] Dedner A, Kemm F, Kröner D, Munz C D, Schnitzer T and Wesenberg M 2002 *JCP* **175** 645–673
[3] Ascher U M, Ruuth S J and Wetton B T 1995 *SIAM Journal on Numerical Analysis* **32** 797–823
[4] Ascher U M, Ruuth S J and Spiteri R J 1997 *Applied Numerical Mathematics* **25** 151–167
[5] Pareshi L 2001 *SIAM Journal on Numerical Analysis* **39** 1395
[6] Pareshi L and Russo G 2005 *J. Sci. Comput.* **25** 129
[7] Palenzuela C, Lehner L, Reula O and Rezzolla L 2009 *MNRAS* **394** 1727–1740 (*Preprint* 0810.1838)
[8] Shu C W and Osher S 1988 *Journal of Computational Physics* **77** 439–471
[9] Antón L, Miralles J A, Martí J M, Ibáñez J M, Aloy M A and Mimica P 2010 *ApJS* **188** 1–31 (*Preprint* 0912.4692)
[10] Leismann T, Antón L, Aloy M A, Müller E, Martí J M, Miralles J A and Ibáñez J M 2005 *A&A* **436** 503–526
[11] Mimica P, Aloy M A, Agudo I, Martí J M, Gómez J L and Miralles J A 2009 *ApJ* **696** 1142–1163 (*Preprint* 0811.1143)
[12] Mimica P, Giannios D, Metzger B D and Aloy M A 2015 *MNRAS* **450** 2824–2841 (*Preprint* 1501.00361)
[13] Del Zanna L, Zanotti O, Bucciantini N and Londrillo P 2007 *A&A* **473** 11–30 (*Preprint* 0704.3206)
[14] Obergaulinger M, Janka H T and Aloy M A 2014 *MNRAS* **445** 3169–3199 (*Preprint* 1405.7466)
[15] Obergaulinger M, Just O, Janka H T, Aloy M A and Aloy C 2014 A New Two-moment Scheme with Algebraic Closure for Energy-dependent Multi-flavor Neutrino Transport in Supernovae *8th Int. Conf. of Numerical Modeling of Space Plasma Flows (ASTRONUM 2013)* (*ASP Conf. Ser.* vol 488) ed Pogorelov N V, Audit E and Zank G P p 255
[16] Just O, Obergaulinger M and Janka H T 2015 *MNRAS* **453** 3386–3413 (*Preprint* 1501.02999)
[17] Obergaulinger M, Aloy M A and Müller E 2010 *A&A* **515** A30 (*Preprint* 1003.6031)
[18] Rezzolla L, Giacomazzo B, Baiotti L, Granot J, Kouveliotou C and Aloy M A 2011 *ApJL* **732** L6 (*Preprint* 1101.4298)
[19] Mimica P, Aloy M A and Müller E 2007 *A&A* **466** 93–106 (*Preprint* astro-ph/0611765)
[20] Mimica P and Aloy M A 2010 *MNRAS* **401** 525–532 (*Preprint* 0909.1328)
[21] Mimica P and Aloy M A 2012 *MNRAS* **421** 2635–2647 (*Preprint* 1111.6108)
[22] Rueda-Becerril J M, Mimica P and Aloy M A 2014 *MNRAS* **438** 1856–1869 (*Preprint* 1310.5441)
[23] Giannios D, Mimica P and Aloy M A 2008 *A&A* **478** 747–753 (*Preprint* 0711.1980)
[24] Mimica P, Giannios D and Aloy M A 2009 *A&A* **494** 879–890 (*Preprint* 0810.2961)
[25] Mimica P, Giannios D and Aloy M A 2010 *MNRAS* **407** 2501–2510 (*Preprint* 1004.2720)

The *ideal* tearing mode: theory and resistive MHD simulations

L. Del Zanna[1,2,3], S. Landi[1,2], E. Papini[4], F. Pucci[5], and M. Velli[6]
[1]Dipartimento di Fisica e Astronomia, Università degli Studi di Firenze, Italy.
[2]INAF - Osservatorio Astrofisico di Arcetri, Firenze, Italy.
[3]INFN - Sezione di Firenze, Italy.
[4]Max-Planck-Institut für Sonnensystemforschung, Göttingen, Germany.
[5]Università di Roma Tor Vergata, Dipartimento di Fisica, Roma, Italy.
[6]Earth Planetary and Space Sciences, University of California, Los Angeles, USA.
E-mail: luca.delzanna@unifi.it

Abstract. Classical MHD reconnection theories, both the stationary Sweet-Parker model and the tearing instability, are known to provide rates which are too slow to explain the observations. However, a recent analysis has shown that there exists a critical threshold on current sheet's thickness, namely $a/L \sim S^{-1/3}$, beyond which the tearing modes evolve on fast macroscopic Alfvénic timescales, provided the Lunquist number S is high enough, as invariably found in solar and astrophysical plasmas. Therefore, the classical Sweet-Parker scenario, for which the diffusive region scales as $a/L \sim S^{-1/2}$ and thus can be up to $\sim$ 100 times thinner than the critical value, is likely to be never realized in nature, as the current sheet itself disrupts in the elongation process. We present here two-dimensional, compressible, resistive MHD simulations, with S ranging from 10^5 to 10^7, that fully confirm the linear analysis. Moreover, we show that a secondary *plasmoid* instability always occurs when the same critical scaling is reached on the local, smaller scale, leading to a cascading explosive process, reminiscent of the flaring activity.

1. Introduction: from *slow* to *ideal* reconnection

Magnetic reconnection is thought to be the primary mechanism providing fast energy release, readily channeled into heat and particle acceleration, in astrophysical and laboratory magnetically dominated plasmas. Within the macroscopic regime of resistive magnetohydrodynamics (MHD), however, classical reconnection models predict timescales, in highly conducting plasmas, which are too slow to explain bursty phenomena such as solar flares in the corona or tokamak disruptions. In the following we summarize the main results of classical theories and the latest developments in this subject.

Given a plasma flow pattern $\vec{v}$, the magnetic field evolution is governed within resistive MHD by the induction equation

$$\frac{\partial \vec{B}}{\partial t} = \nabla \times (\vec{v} \times \vec{B}) + \eta \nabla^2 \vec{B}, \tag{1}$$

where η is the magnetic diffusivity, supposed to be a constant. A simple dimensional analysis allows one to separate the characteristic timescales of fluid advection and Ohmic diffusion

$$\tau_A = \frac{L}{c_A}, \quad \tau_D = \frac{L^2}{\eta}, \quad S = \frac{L\, c_A}{\eta} = \frac{\tau_D}{\tau_A}, \tag{2}$$

where S is the Lundquist number, that is the magnetic Reynolds number defined via the macroscopic quantities L and c_A, the Alfvén speed. Astrophysical tenuous plasmas are characterized by very large values of S, say up to $S \sim 10^{12}$, so that diffusion occurs on very long timescales and only where currents are strong, that is in current sheets.

The Sweet-Parker model (hereafter SP) of two-dimensional, steady, incompressible reconnection [1, 2] predicts, for an aspect ratio L/a (L is the current sheet length or breadth, identified with the macroscopic scale, and a its width), a rate of reconnected flux

$$R = \frac{v_{\rm in}}{c_A} = \frac{a}{L} \sim S^{-1/2} \tag{3}$$

leading to time scales $\tau/\tau_A \sim S^{1/2}$, far too slow to explain solar flares with $\tau \sim 0.1 - 1\tau_A \sim 10^3$s.

The linear stability of a current sheet of infinite length was investigated by [3], who discovered that the equilibrium is unstable to the *tearing mode*, leading to the formation of X-points and plasmoids during the reconnection process. If measured on top of the only available scale, the sheet half width a (from now on starred quantities refer to a rather than L, so they indicate local properties rather than global), the instability growth rate $\gamma = 1/\tau$ is, again, far too slow

$$\gamma\,\tau_A^* \sim {S^*}^{-1/2}, \quad k_{\rm max} a \sim {S^*}^{-1/4} \quad (\tau_A^* = a/c_A, \quad S^* = a\,c_A/\eta), \tag{4}$$

where $k_{\rm max}$ is the wave number at which the instability peaks. The scaling for γ is exactly the same found for the stationary rate R in the SP model, so once more we would need extremely small scales to explain the observations.

In order to make a step forward, one could also investigate the behaviour of the extremely thin current sheets predicted by the SP stationary theory. As first demonstrated by means of 2D MHD simulations [4], reconnecting SP-like sites become unstable once the Lundquist number exceeds a critical value of order $S \sim 10^4$, and are subject to fast tearing modes and plasmoid formation when their aspect ratio L/a becomes large enough, also increasing the local reconnection rate. Recent detailed linear analyses and simulations have confirmed these findings [5, 6, 7, 8, 9, 10, 11, 12, 13]. In particular, the SP current sheet, of inverse aspect ratio $a/L \sim S^{-1/2}$, in the presence of the typical inflow/outflow pattern characterizing steady reconnection, was shown to be tearing unstable with growth rate and wave number of maximal instability

$$\gamma\,\tau_A \sim S^{1/4} \gg 1, \quad k_{\rm max} L \sim S^{3/8} \gg 1. \tag{5}$$

However, the existence of instabilities with growth rates scaling as a positive power of S poses severe conceptual problems, since the ideal limit, corresponding to $S \to \infty$ would lead to infinitely fast instabilities, while it is well known that in ideal MHD reconnection is impossible. Moreover, the positive power of S also for $k_{\rm max}L$ indicates an explosive behaviour of the tearing mode, with the formation of a large number of plasmoids. From these paradoxical findings it should be clear that, since SP current sheets are violently unstable in the ideal MHD limit $S \to \infty$, this could only imply that such extremely elongated equilibrium structures with $a/L \sim S^{-1/2}$ (10^{-6} in corona) simply cannot form in nature.

This issue was resolved by [14] (PV hereafter), who studied the stability of current sheets with generic inverse aspect ratios $a/L \sim S^{-\alpha}$. The authors showed that a critical exponent separates current sheets subject to slow instabilities, with growth rates scaling as a negative power of S, from the unphysical fast instabilities scaling as a positive power of S. For this generic dependence the growth rate is

$$a/L \sim S^{-\alpha} \Rightarrow \gamma \sim {\tau_A^*}^{-1} {S^*}^{-1/2} = \tau_A^{-1} S^{-1/2} S^{3/2\alpha} \tag{6}$$

thus, there is a critical value for what they called the *ideal* tearing mode, given by

$$\alpha = 1/3 \Rightarrow \gamma \sim \tau_A^{-1}, \tag{7}$$

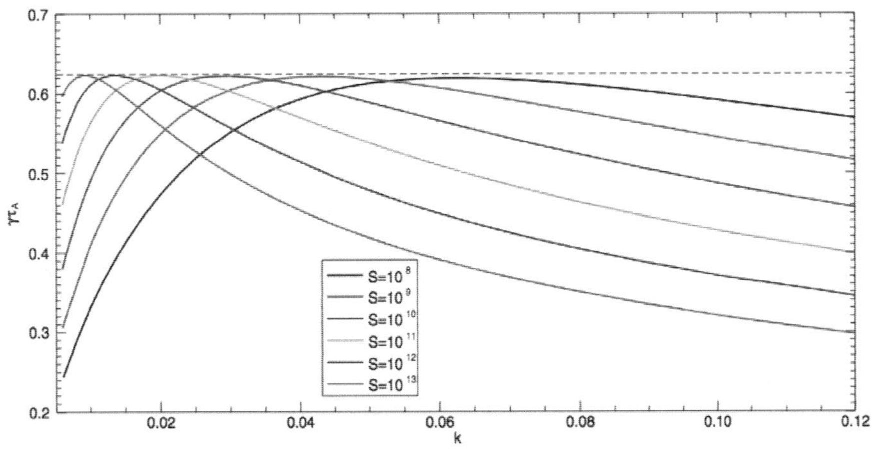

Figure 1. The instability dispersion relation $\gamma\tau_A$ as function of ka for various (large) values of the Lundquist number S. Notice the decrease of $k_{\max}a$, where the instability peaks, and the asymptotic value of $\gamma_{\max}\tau_A \simeq 0.62$.

for which the growth rate of the fastest reconnecting mode becomes independent on the Lundquist number. Notice that for $S = 10^{12}$ the threshold $a/L \sim S^{-1/3} = 10^4$ is 100 times larger than the SP one. Thus, in this limit reconnection starts to occur on *ideal* timescales, and the SP configuration is never established in the thinning process. This scenario should also survive in the presence of moderate viscosity [15].

Going in deeper details, consider a current sheet defined by the equilibrium magnetic field $B_{0y}(x)$ (for example the Harris sheet), the usual tearing mode linear analysis for fluctuations $\sim f(x)\exp(\gamma t + iky)$ in an incompressible medium leads to the system [3]

$$\gamma\,(v_x'' - k^2 v_x) = ik[B_{0y}(b_x'' - k^2 b_x) - B_{0y}'' b_x], \tag{8}$$

$$\gamma\, b_x = ikB_{0y}v_x + S^{-1}({b_x}'' - k^2 b_x), \tag{9}$$

where v_x and b_x are the linear fluctuations across the sheet (the other components are retrieved by $\nabla\cdot\vec{v} = \nabla\cdot\vec{b} = 0$). When letting $x \to x/a$ and $ka \to k$ with $a \sim S^{-1/3}$, the dispersion relation $\gamma(k)$ for varying S clearly shows curves with an *ideal* asymptotic limit $\gamma_{\max} \simeq 0.62\tau_A^{-1}$ for $S \gg 1$, as can be seen in figure 1, taken from PV.

The wave number $k_{\max}$ corresponding to the maximum growth rates is seen to decrease with S as

$$a/L \sim S^{-1/6} \Rightarrow k_{\max}a \sim S^{-1/6}, \tag{10}$$

as expected. However, once the *macroscopic* normalization against the current sheet length L is recovered, we do expect a number of islands actually increasing with S

$$k_{\max}L \sim S^{1/6}, \tag{11}$$

thus even this ideal tearing mode leads to the disruption of the current sheet in a large number of islands, well before the SP sheet configuration (thinner than this one) can be reached.

2. Numerical simulations: setup

In order to prove the analytical theory by PV, [16] performed 2D resistive MHD simulations of the same scenario, that is they studied numerically the stability of a PV current sheet with

$a/L \sim S^{-1/3}$, subject to initially small velocity perturbations. Here we summarize and report the main results.

The compressible, resistive MHD equations in the form

$$\frac{\partial \rho}{\partial t} + \nabla \cdot (\rho \mathbf{v}) = 0, \tag{12}$$

$$\frac{\partial \mathbf{v}}{\partial t} + (\mathbf{v} \cdot \nabla)\mathbf{v} = \frac{1}{\rho}\left[-\nabla p + (\nabla \times \mathbf{B}) \times \mathbf{B}\right], \tag{13}$$

$$\frac{\partial T}{\partial t} + (\mathbf{v} \cdot \nabla)T = (\Gamma - 1)\left[-(\nabla \cdot \mathbf{v})T + \frac{1}{S}\frac{|\nabla \times \mathbf{B}|^2}{\rho}\right], \tag{14}$$

$$\frac{\partial \mathbf{B}}{\partial t} = \nabla \times (\mathbf{v} \times \mathbf{B}) + \frac{1}{S}\nabla^2 \mathbf{B}, \tag{15}$$

are integrated numerically, where S is the Lundquist number defined above, $\Gamma = 5/3$ is the adiabatic index, and other quantities retain their obvious meaning. Physical quantities are normalized using Alfvénic units, namely a characteristic length scale L, a characteristic density ρ_0, and a characteristic magnetic field strength $B_0/\sqrt{4\pi}$ (the background values measured far from the current sheet). Velocities are then expressed in terms of the Alfvén speed $c_A = B_0/\sqrt{4\pi\rho_0}$, time in terms of $\tau_A = L/c_A$, the fluid pressure in terms of $B_0^2/4\pi$. Note that we are using the energy equation (14) written for the normalized temperature $T = p/\rho$, where we use as a reference value $T_0 = (m/k_\mathrm{B})c_A^2$. With the given normalizations the Lundquist number is basically the inverse of the magnetic diffusivity, namely $S = c_A L/\eta \rightarrow \eta^{-1}$.

The initial condition at $t = 0$ for our two-dimensional simulations of the tearing instability is a main magnetic field along the direction y, with $\rho = 1$ and $\vec{v} = 0$ everywhere, asymptotic pressure (and temperature) $p = T = \beta/2$ and field magnitude $B = 1$ far from $x = 0$, the current sheet center where $B_y = 0$. The generic equilibrium can be conveniently described as

$$\mathbf{B} = \tanh\left(\frac{x}{a}\right)\hat{\mathbf{y}} + \zeta\,\mathrm{sech}\left(\frac{x}{a}\right)\hat{\mathbf{z}}, \quad p = T = \frac{\beta}{2} + \frac{1-\zeta^2}{2}\mathrm{sech}^2\left(\frac{x}{a}\right), \tag{16}$$

where $\zeta = 0$ for the Harris sheet case with $B_z = 0$ and pressure equilibrium (PE hereafter), with p varying to preserve $p + B^2/2$ constant, and $\zeta = 1$ for the force-free equilibrium (FFE hereafter) with magnetic field rotating across the sheet. Intermediate cases of mixed fluid/magnetic pressure equilibrium with $0 < \zeta < 1$ are also possible.

The compressible, resistive MHD equations (12-15) are solved in a rectangular numerical box $[-L_x, L_x] \times [0, L_y]$ with resolution N_x and N_y respectively. In the x-direction, in order to resolve the steep gradients inside the current sheet using a reasonable number of grid points, we limit our domain to a few times the current sheet thickness, i.e. we set $L_x = 20a$, with $a = S^{-1/3}$: this is a good compromise between the high resolution required inside the current sheet and the need to have boundaries sufficiently far from the reconnecting region. Along the y-direction the length is chosen in order to resolve for the fastest growing modes of the instability (see [16] for details). Finally, incompressible velocity perturbations of amplitude $\varepsilon \sim 10^{-3}$are applied at $t = 0$ in order to trigger the instability.

The numerical simulations are performed by integrating equations (12-15) with an MHD code developed by our group. Along the current sheet, where periodicity is assumed, spatial integration is performed by using pseudo-spectral methods, while in the x-direction integration is performed by the use of a fourth-order scheme based on compact finite-differences [17]. The boundary conditions in the non-periodic direction are treated with the method of projected characteristics, here assuming non-reflecting boundary conditions. Time integration is performed using a third-order Runge-Kutta method. Details of the code are described in [18]. The

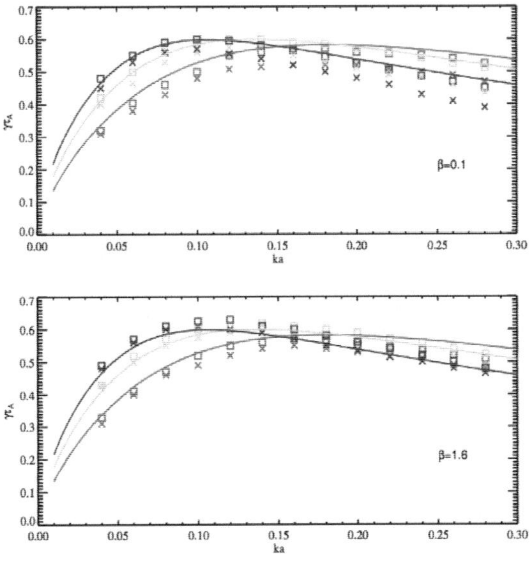

Figure 2. The instability dispersion relation (growth rate as a function of k, normalized against $a^{-1} = S^{1/3}$) for different values of the asymptotic beta (top panel: $\beta = 0.1$; bottom panel: $\beta = 1.6$), and Lundquist numbers (red color: $S = 10^5$; green color $S = 10^6$; blue color: $S = 10^7$). Solid lines are the theoretical expectations, symbols are for numerical results (crosses: PE; squares: FFE).

resolution is adapted to the Lundquist number we use: for $S = 10^5$ and $S = 10^6$ we choose $N_x = 1024$ and $N_y = 128$, while for $S = 10^7$ the number of cells in the x direction is increased up to $N_x = 2048$. In the periodic direction we use $N_y = 128$ for the single-mode runs (in order to reduce the computational costs as many simulations are required to reproduce the instability dispersion relation curves), while we take $N_y = 256$ in the nonlinear reference simulation. We have verified that this relatively low resolution along the periodic direction is adequate, due the extreme accuracy of Fourier methods and the rather smooth gradients observed in the y direction. In spite of the relatively high values of S, in addition to the instability evolution, the initial equilibrium diffuses on timescales which, although long compared to the instability one, are still sufficient to underestimate the growth rates of linear modes. To avoid this problem, only for the single mode linear analysis, the diffusion term of the initial equilibrium (not that related to the perturbations, obviously needed for the tearing instability evolution) is removed from the induction equation at all times, as explained in [19].

3. Numerical simulations: single mode analysis

A first set of simulations of the tearing instability in current sheets with $a = S^{-1/3}$ is performed to confirm the expected linear behavior presented in Section 1. We choose to test the two limits of our initial equilibria for the current sheet, namely PE ($\zeta = 0$) and FFE ($\zeta = 1$). Moreover, we investigate both cases with $\beta < 1$ ($\beta = 0.1$) and $\beta > 1$ ($\beta = 1.6$), where β is the asymptotic plasma beta in equation (16), even if no differences are expected at the linear level from the incompressible, analytical analysis. Finally, three different values of the Lundquist number are tested here, namely $S = 10^5$, 10^6, and 10^7, for a total of 12 sets of simulations of the linear phase of the tearing instability, with the aim of reproducing the expected dispersion relations numericall, as shown in figure 2. As anticipated in the previous section, for each value of ka

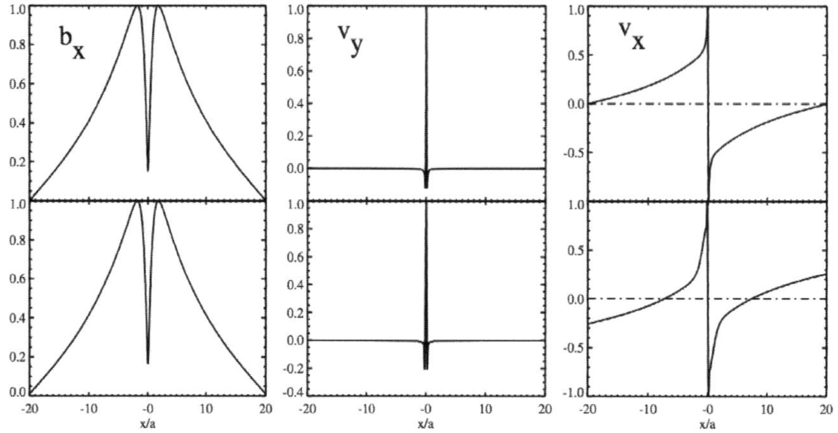

Figure 3. Profiles across the current sheet of the tearing instability eigenmodes b_x (b_y is automatically determined by the solenoidal constraint), v_y, and v_x. Top panel: analytical linear theory; bottom panel: numerical results. All profiles are normalized to their maximum value.

we vary L_y while always selecting a single mode $m = 1$. The growth rate of the instability is computed by measuring the x-averaged amplitude of the component b_x of the perturbed magnetic field ($b_x = 0$ at the initial time).

The first thing to notice by inspecting the computed dispersion relations is that, as predicted by the classical linear theory, for each value of S the curves have a maximum at a given k, the peak location decreasing in k as S increases. The growth rate of the instability (normalized to the inverse of the large-scale Alfvén time τ_A) has peaks ranging from $\gamma \approx 0.5$ for $S = 10^5$ to $\gamma > 0.6$ for $S = 10^7$. In general we find that the simulations with $\beta = 1.6$ (bottom panel) are more precise in matching the analytical results than those with $\beta = 0.1$ (top panel), since a large beta is a condition closer to incompressibility (formally corresponding to an infinite value for the sound speed). Moreover, we find that simulations of the FFE scenario (squares) yield higher values of the growth rates, closer to the results from the analytic theory, as compared to those employing the PE settings (crosses): this is probably due to the fact that the purely force-free equilibrium leads to intrinsically less compressible fluctuations. Finally, rather large discrepancies are observed for small scales (large values of ka), especially in the PE case.

In figure 3 we plot the profiles of the perturbations b_x (b_y is determined by $\nabla \cdot \mathbf{B} = 0$), v_y and v_x, all normalized to their respective maximum, across the current sheet in the x direction. In the top panels we show the analytical results, that is the eigenmodes of the linear analysis (here the PV calculations have been recomputed by imposing $v_x = 0$ for $x = \pm 20a$), and in the lower panels we report the numerical solutions for a simulation in the FFE scenario with $S = 10^7$ and $ka = 0.10$, at a given time of the linear evolution of the instability. In order to recover the theoretical eigenmodes, velocity and magnetic field perturbations are shown with a $\pi/2$ shift in ky, as expected. Notice the steep gradients arising within the current sheet ($|x| \leq a$), where a high resolution is needed to resolve the small scales developed during the instability evolution. The eigenmodes are very well reproduced: the magnetic field perturbations are identical to the analytical expected ones, while in the velocity perturbations the only major difference is due to the non-reflecting free-outflow boundary conditions imposed, that do not force $v_x = 0$ at $x = \pm a$ as in the analytical solutions and result in a slightly different profile even in the vicinity of the reconnecting region.

4. Numerical simulations: fully nonlinear case

Now we describe the nonlinear stages of the evolution of the tearing instability. Since we are interested in its late development, where interaction and merging of plasmoids is expected, we trigger the instability by selecting an initial spectrum of modes, rather than a single one as in the previous set of simulations, and we choose a maximum mode number $m_{\max} = 10$. We also choose $L_y = 1$ and $S = 10^7$, so the modes with $ka \simeq 0.029\,m$; $m = 1, 10$ are all excited. From the theoretical curves in the previous section we expect mainly a competition between modes $m = 3$ and $m = 4$ as the fast growing ones. As a reference run, we analyze the instability of a force-free equilibrium with constant temperature (FFE, $\zeta = 1$), and we select the case with $\beta = 1.6$. This combination was shown to provide a linear phase which is the closest to the analytical expectations (see figure 2). The resolution employed for this run is 2048×256, which is very high if one consider that the code employs high-order methods (compact finite-differences along x and Fourier transforms along y, where periodical boundary conditions apply).

The complete evolution is shown in figure 4, where 2D snapshots for selected times are provided, of the region $|x/a| < 10$. Note that the figure has different scales in x and y, the x axis is normalized against a, the y axis against L, with $L/a = S^{1/3} \simeq 215$). Panels on the left show the J_z current component, whereas panels on the right show the distorted field lines superimposed on a map of temperature T. It is easy to recognize the end of the linear phase of the tearing instability (top row), with the dominant $m = 3$ mode still clearly apparent. When the tearing instability growth is over, the nonlinear phase sets in leading to further reconnection events and eventually to island coalescence, departing from the dominant phase with $m = 3$. First, due to the attraction of current concentrations of the same sign, the X-points elongate and stretch along the y direction. This process leads to a strong increase of the electric current concentrations and thus to the formation of new, elongated current sheets (see the second row, details of this phase will be discussed further on). Beyond $t \simeq 9$ (third row) the evolution has become fully nonlinear and we clearly observe the process leading to the creation of a single, large magnetic island as arising from coalescence. The situation is very dynamic, especially near the major X-point where explosive expulsion of smaller and smaller islands is observed, typical of the plasmoid instability. These islands then move towards the largest one, which is continually fed and thus further increases its size in a sort of inverse cascade eventually leading to the largest $m = 1$ mode. Notice also the temperature enhancement at the reconnection sites. At time $t = 9.5$ (bottom row) both the current concentration and the plasma temperature around the X-point are so high that we need to saturate their values in order to retain an appropriate dynamical range in the color bar. The initial macroscopic current sheet is basically disrupted in a series of highly dynamical features. The current and temperature enhancements are stronger at the X-point and at the boundaries of the magnetic islands. Moreover, from the major reconnecting site we clearly see the production of magnetosonic waves, which propagate and soon steepen into shocks.

Let us now investigate in more detail the situation right after the end of the linear phase of the tearing instability, when the islands coalescence is about to set in and the local current sheets have just formed and start to further evolve. In figure 5 (left panel) we show the zoom around the X-point (which is just about to develop) near $y = 0.2$ at $t = 8.25$, the first row of figure 4. Here we display the electric current by using the same scaling in both x and y directions, normalized against the macroscopic current sheet's width $a = S^{-1/3} \simeq 1/200 \simeq 5 \times 10^{-3}$, thus the position of the reconnection zone is now expressed as $y \simeq 40a$. The local current sheet has just formed as the result of a stretching process in the y direction (and shrinking across the other direction), as a typical output of the nonlinear phase of the tearing instability. We are now in the phase in which, in the center of the current sheet, reconnection is about to take place, leading to a topology change in the magnetic structure and to the disruption of the current sheet itself, initially into two smaller strips. It is very interesting to measure the aspect ratio

Journal of Physics: Conference Series **719** (2016) 012016 doi:10.1088/1742-6596/719/1/012016

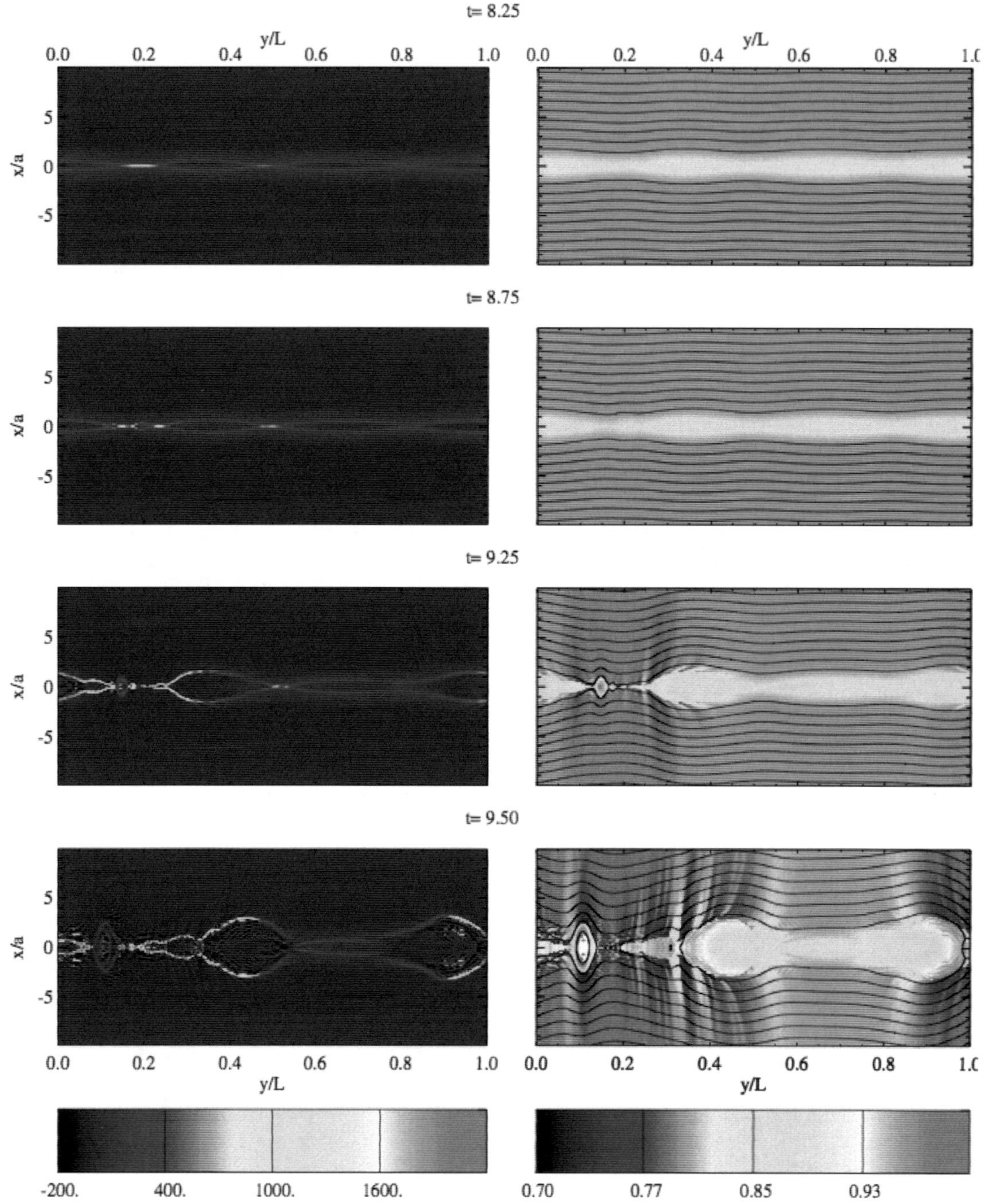

Figure 4. Nonlinear evolution of the tearing instability of a current sheet (FFE, $\zeta = 1$, $\beta = 1.6$) with $a = S^{-1/3}$. On the left panels we show the intensity of the J_z electric current component, while on the right panels the magnetic fieldlines and the plasma temperature are displayed. Notice that the x and y scales are different and that we are showing the inner region $|x| \leq 10a$, while the computation extends out to $|x| = 20a$.

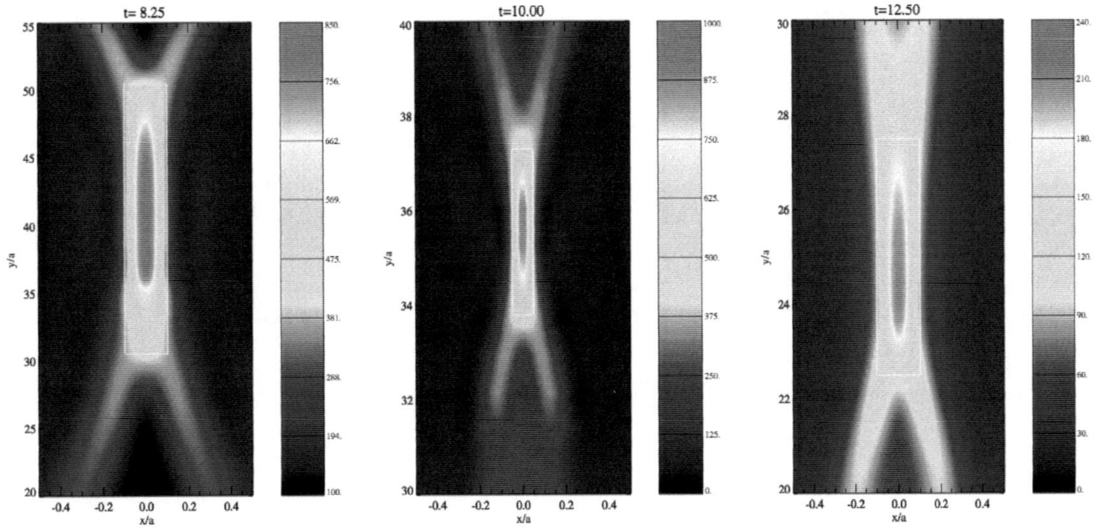

Figure 5. Zoom of the most prominent X-point reconnecting region, during the early nonlinear stage of the tearing instability. Colors refer to the strength of the J_z component of the electric current (the scaling is different in each plot). From left to right we show the situation for decreasing values of the Lundquist numbers: $S = 10^7$, $S = 10^6$, $S = 10^5$.

of this reconnection site. From the white box in the figure, defined by the rectangular region where the J_z component has values which are roughly half those of the central peak, we estimate $L^*/a^* \simeq 200$, where we have used an asterisk to indicate the *local values* and to differentiate with respect to the macroscopic ones (we also recall that in the whole paper we identify a as the *half* width of a current sheet). Additional simulations for $S = 10^6$ and $S = 10^5$, reported in the second and third panels, lead to local aspect ratios of $L^*/a^* \simeq 80$ and $L^*/a^* \simeq 50$, respectively.

If we now compare these numbers with $S^{*\alpha}$, trying to find a value of α that fits the data best, it is easy to see that the value $\alpha = 1/3$ is a very good guess. Here $S^* = (L^*/L)S$ is the local Lundquist number, which is obviously smaller than the macroscopic one, due to the much smaller length of the local current sheet (to be measured in each case). Therefore, based on our very limited data set, we derive the scaling

$$L^*/a^* = k\, S^{*1/3}, \tag{17}$$

where $k \simeq 2.1 - 2.3$, that is of order unity, as expected. In figure 6 we report the numerical results, assuming a 10% error on both L^*/a^* and S^*, for the sequence of increasing (macroscopic) Lundquist numbers $S = 10^5$, $S = 10^6$, and $S = 10^7$. The scaling in equation (17) is over-plotted, as a solid line, for the average value of $k = 2.2$.

5. Conclusions

In the present paper we have summarized the main results of the analytical analysis by [14] and of the numerical simulations by [16], who studied, by means of compressible, resistive MHD simulations, the linear and nonlinear stages of the tearing instability of a current sheet with inverse aspect ratio $a/L \sim S^{-1/3}$, where $S \gg 1$ is the Lundquist number measured on the macroscopic scale (the current sheet length) and the asymptotic Alfvén speed. Our results confirm on the one hand the linear analysis of PV of the *ideal* tearing mode, leading to extremely fast growth rates with $\gamma \sim \tau_A^{-1}$ when S is sufficiently large, and on the other hand the nonlinear

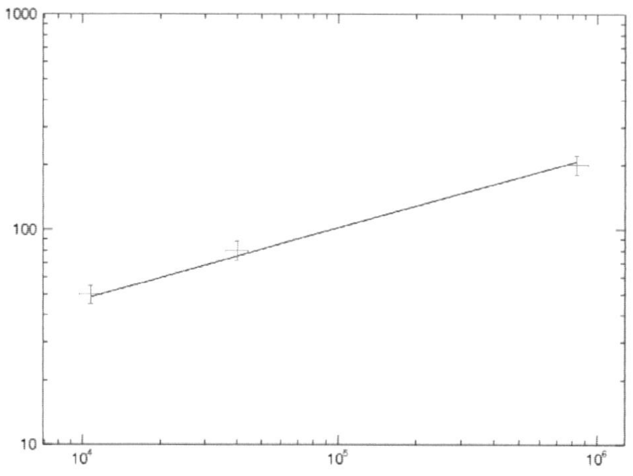

Figure 6. Comparison of numerical results (crosses) for the local aspect ratio L^*/a^* of current sheets undergoing secondary reconnection events, together with the theoretical expectation of equation (17) plotted for $k = 2.2$ (the average value derived from data). The corresponding values of the macroscopic Lundquist number are $S = 10^5$, $S = 10^6$, and $S = 10^7$.

simulations show that the evolution follows what appears to be a quasi-self-similar path, with subsequent collapse, current sheet thinning, elongation, and destabilization, starting from the X-points formed in the original sheet. As scales become smaller, and the local Lundquist numbers decrease, the dynamical time-scales decrease too, leading to explosive behavior.

These findings are very important, in our opinion. For the first time we clearly see in simulations that, even in the nonlinear stages of the tearing instability, the new current sheets that form locally become unstable when the inverse aspect ratio of these structures reaches the critical threshold of $a^*/L^* \sim S^{*-1/3}$ (where the star indicates the local, smaller scale), precisely the same limit found by PV for the fast reconnection of the initial, macroscopic current sheet. After that, a new *ideal* tearing instability starts, with time occurring on a *faster* timescale $\tau_A^* = L^*/c_A$, since typically $L^* \ll L$. Furthermore, when smaller and smaller scales are produced nonlinearly, as observed at the time proceeds, each time the newly formed local current sheets elongate and reach their own critical value, that corresponding to equation (17), faster and faster reconnection will arise producing a cascading, accelerating process: this, we believe, is the real nature of the plasmoid (or super-tearing) instability.

There are many space and astrophysical applications of the *ideal tearing*, including geomagnetic storms in our planet, coronal heating and coronal mass ejections from the Sun. As future work we plan to move to 3D simulations [20, 21] and to investigate the tearing instability of thin current sheets within relativistic MHD [22, 23, 24]. Fast reconnection in magnetically dominated plasmas is of paramount importance in high-energy astrophysics too [25], invariably invoked in models for magnetar flares, acceleration of Poynting-dominated jets, and dissipation in pulsar winds [26, 27, 28].

M.V. was supported by the NASA Solar Probe Plus Observatory Scientist grant. The research leading to these results has received funding from the European Commissions Seventh Framework Programme (FP7/2007-2013) under the grant agreement SHOCK (project number 284515).

References

[1] Sweet P A 1958 *Electromagnetic Phenomena in Cosmical Physics* (*IAU Symposium* vol 6) ed Lehnert B p 123
[2] Parker E N 1957 *J. Geophys. Res.* **62** 509–520
[3] Furth H P, Killeen J and Rosenbluth M N 1963 *Physics of Fluids* **6** 459–484
[4] Biskamp D 1986 *Physics of Fluids* **29** 1520–1531
[5] Loureiro N F, Schekochihin A A and Cowley S C 2007 *Physics of Plasmas* **14** 100703 (*Preprint* astro-ph/0703631)
[6] Lapenta G 2008 *Physical Review Letters* **100** 235001 (*Preprint* 0805.0426)
[7] Samtaney R, Loureiro N F, Uzdensky D A, Schekochihin A A and Cowley S C 2009 *Physical Review Letters* **103** 105004 (*Preprint* 0903.0542)
[8] Bhattacharjee A, Huang Y M, Yang H and Rogers B 2009 *Physics of Plasmas* **16** 112102 (*Preprint* 0906.5599)
[9] Cassak P A, Shay M A and Drake J F 2009 *Physics of Plasmas* **16** 120702
[10] Huang Y M and Bhattacharjee A 2010 *Physics of Plasmas* **17** 062104 (*Preprint* 1003.5951)
[11] Uzdensky D A, Loureiro N F and Schekochihin A A 2010 *Physical Review Letters* **105** 235002 (*Preprint* 1008.3330)
[12] Cassak P A and Shay M A 2012 *Space Sci. Rev.* **172** 283–302
[13] Loureiro N F and Uzdensky D A 2015 *ArXiv e-prints* (*Preprint* 1507.07756)
[14] Pucci F and Velli M 2014 *ApJLett* **780** L19
[15] Tenerani A, Rappazzo A F, Velli M and Pucci F 2015 *ApJ* **801** 145 (*Preprint* 1412.0047)
[16] Landi S, Del Zanna L, Papini E, Pucci F and Velli M 2015 *ApJ* **806** 131 (*Preprint* 1504.07036)
[17] Lele S K 1992 *Journal of Computational Physics* **103** 16–42
[18] Landi S, Velli M and Einaudi G 2005 *ApJ* **624** 392–401
[19] Landi S, Londrillo P, Velli M and Bettarini L 2008 *Physics of Plasmas* **15** 012302
[20] Bettarini L, Landi S, Velli M and Londrillo P 2009 *Physics of Plasmas* **16** 062302 (*Preprint* 0906.5383)
[21] Landi S and Bettarini L 2012 *Space Sci. Rev.* **172** 253–269
[22] Del Zanna L, Zanotti O, Bucciantini N and Londrillo P 2007 *A&A* **473** 11–30 (*Preprint* 0704.3206)
[23] Bucciantini N and Del Zanna L 2013 *MNRAS* **428** 71–85 (*Preprint* 1205.2951)
[24] Del Zanna L, Bugli M and Bucciantini N 2014 *Astronomical Society of the Pacific Conference Series* (*Astronomical Society of the Pacific Conference Series* vol 488) ed Pogorelov N V, Audit E and Zank G P p 217 (*Preprint* 1401.3223)
[25] Kagan D, Sironi L, Cerutti B and Giannios D 2015 *Space Sci. Rev.* (*Preprint* 1412.2451)
[26] Tavani M 2013 *Nuclear Physics B Proceedings Supplements* **243** 131–140
[27] Porth O, Komissarov S S and Keppens R 2013 *MNRAS* **431** L48–L52 (*Preprint* 1212.1382)
[28] Olmi B, Del Zanna L, Amato E and Bucciantini N 2015 *MNRAS* **449** 3149–3159 (*Preprint* 1502.06394)

Journal of Physics: Conference Series **719** (2016) 012017 doi:10.1088/1742-6596/719/1/012017

Characteristics of a current sheet shear mode in collisionless magnetic reconnection

Keizo Fujimoto

Division of Theoretical Astronomy, National Astronomical Observatory of japan
2-21-1 Ohsawa, Mitaka, Tokyo 181-8588, Japan

E-mail: keizo.fujimoto@nao.ac.jp

Abstract. The current study shows the characteristics of the kink-type electromagnetic mode excited in the thin current layer formed around the x-line during the quasi-steady phase of magnetic reconnection. The linear wave analyses are carried out for the realistic current sheet profile which differs significantly from the Harris current sheet. It is found that the peak growth rate is very sensitive to the current sheet width even though the relative drift velocity at the center of the current sheet is fixed. This indicates that the mode is excited by the velocity shear rather than the relative drift velocity. Thus, the mode is termed here a current sheet shear mode. It is also shown that the wavenumber k_y has a clear mass ratio dependency as $k_y\lambda_i \propto (m_i/m_e)^{1/4}$, implying the coupling of the ion and electron dynamics, where λ_i is the ion inertia length.

1. Introduction

Magnetic reconnection is a fundamental process in magnetized plasmas, generating fast plasma jets at the expense of the magnetic field energy. Although the process takes place in a microscopic region called the diffusion region formed around the magnetic x-line, the impact of the process extends over large distances beyond the kinetic scales from the x-line. Because of this multi-scale nature, it is believed that magnetic reconnection plays a significant role in large-scale dynamics of global phenomena such as geo-magnetospheric substorms and solar flares.

One of the main issues of magnetic reconnection is the mechanism breaking the frozen-in condition around the x-line and providing the large electric resistivity sufficient for a fast reconnection in collisionless plasmas. In two-dimensional (2D) reconnection, it has been demonstrated that the momentum transport due to the Speiser-type motion of the electrons gives rise to the electron viscosity around the x-line which leads to an effective resistivity [1, 2, 3, 4]. Although the electron viscosity gives sufficient dissipation for supporting the reconnection electric field under the thin current layer on the order of the electron inertia length, such a thin current sheet has not been observed neither in the laboratory experiments [5] nor in the geo-magnetosphere [6]. The observations have also shown intense activities of electromagnetic (EM) waves around the x-line [7]. The frequency range is usually from ω_{ci} to ω_{LH}, where ω_{ci} and ω_{LH} are the ion cyclotron frequency and the lower hybrid frequency, respectively. These observational facts imply the existence of the anomalous momentum transport due to the wave activities that are not incorporated in the 2D systems. In fact, kinetic simulations in three-dimensional (3D) system have revealed that kink-type EM waves are excited at the center of the current layer not only in the Harris sheet [8, 9, 10] but also in the thin current layer formed

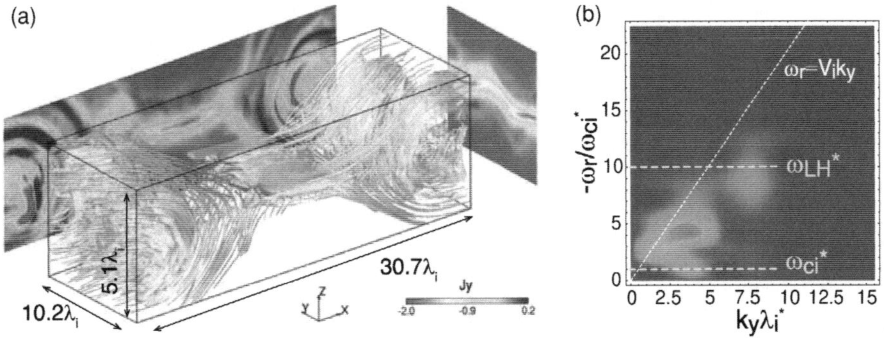

Figure 1. Simulation results. (a) 3D snapshot of an isosurface for $J_{xz} = \sqrt{J_x^2 + J_z^2} = 0.80\, en_0 V_A$ with the magnetic field lines in yellow curves and 2D profiles of J_y at $x = 41.0\lambda_i$ (the center of the x system) and $y = 0.0$. (b) Wave spectrum of B_x at the dominant x-line averaged over the y axis during a fast reconnection. The phase velocity of V_{i0} is indicated by dotted curve in (b).

during a quasi-steady reconnection with anti-parallel configuration [11, 12]. These EM modes are important because they give the anomalous momentum transport and effective resistivity at the x-line. However, the generation mechanisms of the waves are poorly understood.

The linear properties of the waves responsible for the anomalous momentum transport have been investigated, so far, for the Harris-type current sheet [13, 10, 14]. In the Harris sheet, the pressure and current density gradients are supported by the density gradient with the uniform temperature and bulk velocity profiles [15]. Meanwhile, during the quasi-steady reconnection, the current sheet profile is far from the Harris sheet. For the case of symmetric reconnection, the density profile is almost uniform, so that the pressure and current density profiles are maintained through the temperature and bulk velocity gradients, respectively. Furthermore, the current sheet develops a two-scale structure consisting of the electron and ion current sheets. Therefore, it is very questionable that the previous results for the Harris sheet are still applicable in the quasi-steady phase of reconnection.

The purpose of this paper is to investigate the properties of the EM mode excited in the quasi-steady reconnection initiated with anti-parallel configuration. The linear wave analyses are employed for the realistic current sheet equilibrium based on the two-fluid equations. It is shown that the mode is excited due to the velocity shears rather than the relative drift velocity between the ion and electron. Thus, the mode is termed here a current sheet shear mode.

2. Simulation

The evidence of the kink-type EM mode has been recently reported in a large-scale particle-in-cell (PIC) simulation in 3D system [12]. Here we show an example of the simulation results. The simulation model employs adaptive mesh refinement (AMR) to achieve efficient multi-scale computation of magnetic reconnection [16]. The simulation is carried out for the Harris-type current sheet under the anti-parallel magnetic field configuration, which is common for magnetotail reconnection of the Earth. The domain size along the x-line is $L_y = 10.2\lambda_i$, where λ_i is the ion inertia length based on n_0 the number density at the center of the initial current sheet. The ion-to-electron mass ratio is $m_i/m_e = 100$. Magnetic reconnection is initiated with a small perturbation to the magnetic field B_x and B_z, which produces the x-line at the center of the xz domain. More detailed setup is described in Fujimoto and Sydora [12].

Figure 1a shows a snapshot of the 3D current layer during a quasi-steady fast reconnection. It is evident that a kink mode is excited in the electron current layer formed around the x-line. The

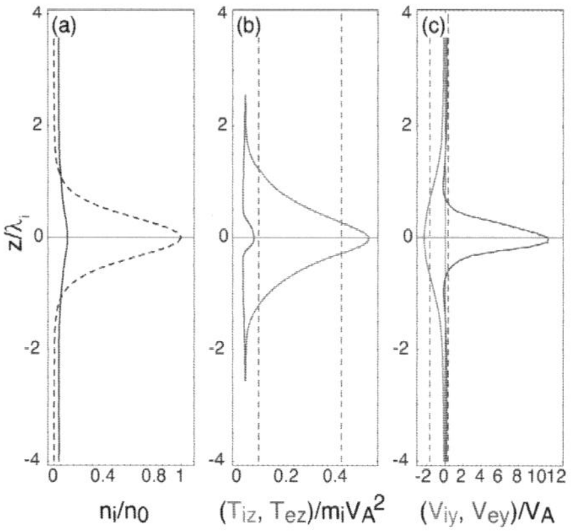

Figure 2. 1D profiles averaged over the y axis at the dominant x-line in the simulation for (a) ion density, (b) ion (red curves) and electron (blue curves) temperatures, and (c) ion (red curves) and electron (blue curves) out-of-plane bulk velocities. The solid and dashed curves indicate the profiles during a fast reconnection and of the Harris current sheet, respectively.

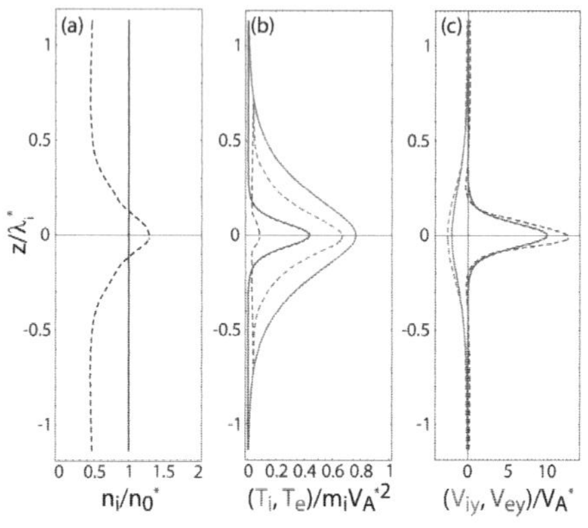

Figure 3. Example of the model current sheet (solid curves) used for the linear analyses of (a) number density, (b) ion (red curves) and electron (blue curves) temperatures, and (c) ion (red curves) and electron (blue curves) out-of-plane bulk velocities. The dashed curves indicate the profiles during a fast reconnection.

mode is an EM mode and propagates in the y direction (i.e., current density direction) almost with the ion flow velocity V_{iy}. The wave spectrum in the ω-k_y space (Fig. 1b) reveals that the mode has a frequency between ω_{ci}^* and ω_{LH}^*, consistent with the observations [7], where the asterisk indicates the local values. As discussed in the previous paper [12], this mode provides the anomalous resistivity at the x-line, so that the understanding of the wave characteristics is very important in reconnection physics.

3. Linear analysis

Most of the previous linear analyses of the kink-type EM modes have been carried out for the Harris current sheet [13, 10, 14]. As shown in Fig. 2 (dashed curves), the Harris sheet is characterized by the non-uniform density profile with the uniform temperature and bulk velocity profiles to make the pressure and current density gradients. However, these profiles are drastically changed after the onset of the fast reconnection. During the quasi-steady phase of reconnection, the density is almost uniform across the current sheet, and the pressure and current density gradients are generated due to the non-uniform temperature and bulk velocity profiles (see solid curves in Fig. 2). Therefore, it is very questionable that the previous results for the Harris sheet are still applicable in the quasi-steady reconnection. In fact, since there is little density gradient, the lower hybrid drift instability [14] is hardly excited in this current layer.

In the present study, a more realistic equilibrium is introduced to mimic the current sheet profiles during the quasi-steady reconnection. The new equilibrium is based on the two-fluid equations, and is provided by the following relations:

$$n_i(z) = n_e(z) = const., \quad (1)$$

$$B_x(z) = B_{0i}\tanh(z/\delta_i) + B_{0e}\tanh(z/\delta_e), \quad (2)$$

$$V_s(z) = -V_{s0}/\cosh^2(z/\delta_s), \text{ and} \quad (3)$$

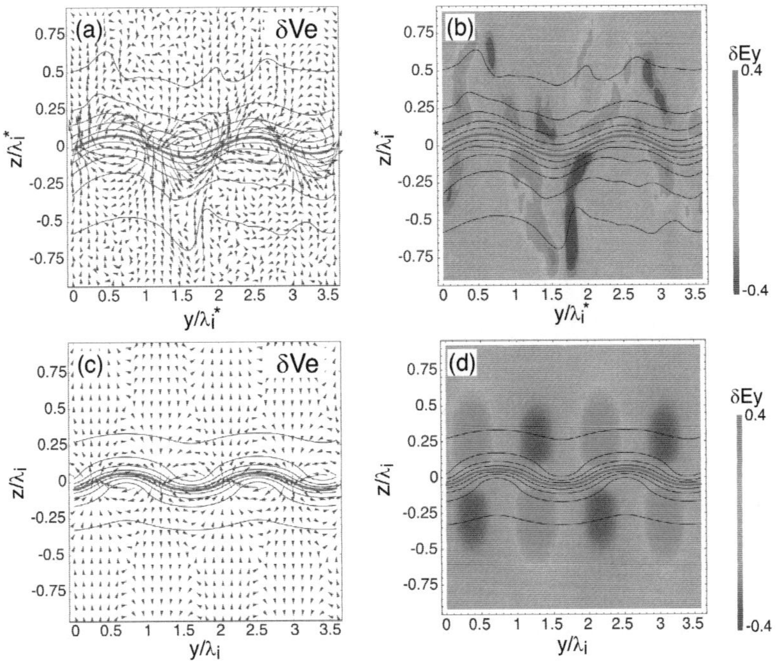

Figure 4. (a, b) Simulation results in the yz plane through the dominant x-line of (a) $\delta\vec{V}_e$ (arrows) the deviation of the electron bulk velocity from the values averaged over the y axis at each z location and (b) δE_y (color contours) the deviation of E_y. (c, d) Eigenfunctions solved through the theoretical linear analyses of (c) $\delta\vec{V}_e$ (vectors) and (d) δE_y (color contours). The black curves represent the contours of B_x, and the red curves show the contours at $B_x = 0$.

$$T_s(z) = -\int^z q_s V_s(\xi) B_x(\xi)\, d\xi, \tag{4}$$

where s indicates the species and there must be the relation $V_{s0} = (1/\mu_0 n q_s)(B_{0s}/\delta_s)$ to satisfy the Ampère's law. In Fig. 3, an example of the equilibrium profiles is shown for the case of $B_{0i}/B_0^* = 2.0$, $B_{0e}/B_0^* = 1.0$, $\delta_i/\lambda_i^* = 0.35$, $\delta_e/\lambda_e^* = 1.0$, and $m_i/m_e = 100$, where B_0^* is assumed to be the magnetic field at the inflow edge of the electron diffusion region. One can see that the profiles in the simulation (dashed curves) are reproduced in the model current sheet, providing the two-scale structure with the temperature and bulk velocity gradients. The wave dispersion is obtained by solving the linearized two-fluid equations for $\omega = \omega_r + i\gamma$ numerically using an initial value approach starting with small random noise [13].

In Fig. 4, the simulation results and the eigenfunctions obtained by the linear analyses are compared for $\delta\vec{V}_e$ and δE_y, where δA in the simulation is calculated by $\delta A \equiv A - \langle A\rangle_y$ with $\langle\cdot\rangle_y = \int_0^{L_y} \cdot\, dy/L_y$. The current sheet structure predicted by the linear analyses is well consistent with the simulation results, regarding the kinked structure, electric field structure, and the associated electron flow structure. We have also confirmed that the dispersion relation in the ω-k_y space matches the simulation result in Fig. 1b. These consistencies demonstrate that the two-fluid approximation is valid for this EM mode.

To investigate further the characteristics of the wave, we calculate the δ_i dependency of the peak growth rate. It is important to note that the peak bulk velocities V_{i0} and V_{e0} are fixed in this analysis. Therefore, the relative drift velocity between the species is not changed at the center of the current layer. As shown in Fig. 5a, the peak growth rate is drastically

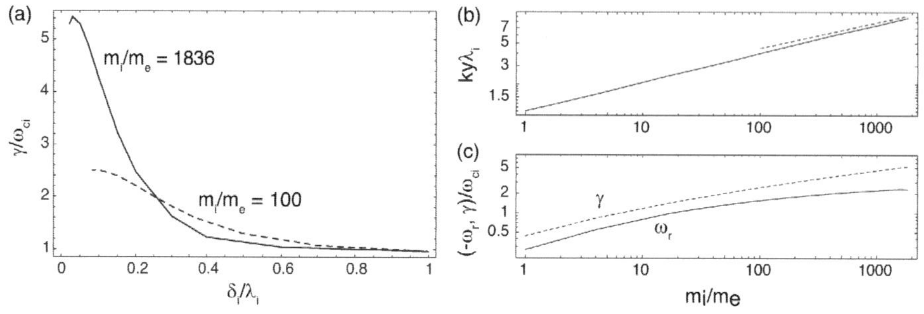

Figure 5. Theoretical linear properties of the electromagnetic wave. (a) Dependency of the growth rate γ on the width of the ion current sheet δ_i for the cases of $m_i/m_e = 100$ (dashed curves) and $m_i/m_e = 1836$ (solid curves). (b, c) Dependencies on m_i/m_e of (b) the wavenumber k_y and (c) the real frequency ω_r (solid curve) and growth rate γ (dashed curve). The dashed curve in (b) indicates the relation $k_y\lambda_i = \sqrt{2}(m_i/m_e)^{1/4}$.

decreased as δ_i the width of the ion current layer increases for both the cases of $m_i/m_e = 100$ and $m_i/m_e = 1836$, where $\delta_e = \lambda_e^*$ is assumed in each mass ratio. This indicates that the mode is very sensitive to the velocity shear rather than the relative drift velocity, implying that the mode is driven by the velocity shear. The wave characteristics differs from that of the drift kink mode, where the peak growth rate depends on the relative drift velocity as $\gamma \propto |V_{i0} - V_{e0}|$ [13].

We next consider the dependency on m_i/m_e the ion-to-electron mass ratio, assuming $\delta_i = \delta_e = \lambda_e^*$. As displayed in Figs. 5b and 5c, the EM mode has a clear m_i/m_e dependency in spatio-temporal scales. In particular, it is found that the wavenumber k_y has a relation of $k_y\lambda_i^* \simeq \sqrt{2}(m_i/m_e)^{1/4}$, so that the wavelength is scaled with $\lambda \propto \sqrt{\lambda_i^*\lambda_e^*}$ an intermediate scale between the ion and electron scales. The peak growth rate increases as m_i/m_e increases. This is because the current sheet width (shear scale) decreases and the peak velocity increases with the mass ratio. Even though the width of the ion current layer δ_i is larger than λ_e^*, the asymptotic value of γ is comparable between the cases of $m_i/m_e = 100$ and 1836 (Fig. 5a). Thus, the mode survives even in the realistic mass ratio. The clear m_i/m_e dependency suggests that the mode differs from the ion-ion kink mode, where the spatio-temporal scales are determined only by the ion physics [10].

4. Summary

The current study has investigated the characteristics of the kink-type EM mode excited in the electron current layer formed around the x-line during the quasi-steady phase of magnetic reconnection. The mode is important in reconnection physics because it is known to provide the anomalous resistivity at the x-line. The linear wave analyses are carried out for the realistic current sheet which differs significantly from the Harris current sheet. The new equilibrium current sheet consists of two-scale structure of the ion and electron current sheets. The pressure and current density gradients are supported by the temperature and bulk velocity gradients, respectively, with uniform number density. The wave dispersion was gained by solving the linearized two-fluid equations numerically.

We first confirmed that the eigenfunctions obtained in the linear analyses are consistent with the simulation results, which warrants the two-fluid approximation for this mode. Further analyses revealed that the peak growth rate decreases drastically as the current sheet width increases, even though the relative drift velocity is fixed at the center of the current layer. Thus, it is reasonable to conclude that the mode is driven by the velocity shears rather than the relative drift velocity, so that the mode is termed here a current sheet shear mode. We also found that the mode has a clear mass ratio dependency of $k_y\lambda_i \propto (m_i/m_e)^{1/4}$, implying the coupling of the ion and electron dynamics.

Since the current linear analyses employ the numerical approach, the theoretical background behind the mass ratio dependency is not completely understood. The analytical analysis is needed in near future to clarify more the generation mechanism of the current sheet shear mode.

Acknowledgments

The simulations were carried out by Fujitsu FX1 installed at the ITC, Nagoya University, with support from the HPC joint research project at Nagoya University and a computational joint research program at the STE Laboratory, Nagoya University.

References

[1] Speiser T W 1970 *Planet. Space Sci.* **18** 613–22
[2] Cai H J and Lee L C 1997 *Phys. Plasmas* **4** 509–20
[3] Hesse M, Schindler K, Birn J and Kuznetsova M 1999 *Phys. Plasmas* **6** 1781–95
[4] Fujimoto K and Sydora R D 2009 *Phys. Plasmas* **16** 112309
[5] Ji H, Ren Y, Yamada M, Dorfman S, Daughton W and Gerhardt S P 2008 *Geophys. Res. Lett.* **35** L13106
[6] Wygant J R *et al* 2005 *J. Geophys. Res.* **110** A09206
[7] Zhou M *et al* 2009 *J. Geophys. Res.* **114** A02216
[8] Zhu Z and Winglee R M 1996 *J. Geophys. Res.* **101** 4885–97
[9] Horiuchi R and Sato T 1999 *Phys. Plasmas* **6** 4565–74
[10] Karimabadi H Daughton W Pritchett P L and Krauss-Varban D 2003 *J. Geophys. Res.* **108** 1400
[11] Fujimoto K 2009 *Phys. Plasmas* **16** 042103
[12] Fujimoto K and Sydora R D 2012 *Phys. Rev. Lett.* **209** 265004
[13] Pritchett P L, Coroniti F V and Decyk V K 1996 *J. Geophys. Res.* **101** 27413–29
[14] Daughton W 2003 *Phys. Plasmas* **10** 3103–19
[15] Harris E G 1962 *Nuovo Cimento* **23** 115–21
[16] Fujimoto K 2011 *J. Comput. Phys.* **230** 8508–26

The Biermann catastrophe of numerical MHD

C Graziani[1], P Tzeferacos, D Lee, D Q Lamb, K Weide, M Fatenejad and J Miller

Flash Center for Computational Science, Department of Astronomy & Astrophysics, University of Chicago, Chicago, IL, 60637

E-mail: [1]carlo@oddjob.uchicago.edu

Abstract. The Biermann Battery effect is frequently invoked in cosmic magnetogenesis and studied in High-Energy Density laboratory physics experiments. Unfortunately, direct implementation of the Biermann effect in MHD codes is known to produce unphysical magnetic fields at shocks whose value does not converge with resolution. We show that this convergence breakdown is due to naive discretization, which fails to account for the fact that discretized irrotational vector fields have spurious solenoidal components that grow without bound near a discontinuity. We show that careful consideration of the kinetics of ion viscous shocks leads to a formulation of the Biermann effect that gives rise to a convergent algorithm. We note a novel physical effect a *resistive magnetic precursor* in which Biermann-generated field in the shock "leaks" resistively upstream. The effect appears to be potentially observable in experiments at laser facilities.

1. Introduction

Dynamo theories of proto- and extra-galactic primordial magnetic fields, which endeavor to explain how those fields achieved their current strength and structure, work by amplifying small initial seed fields by means of turbulent plasma motions [1, 2]. However, the induction equation of resistive magnetohydrodynamics (MHD),

$$\frac{\partial \mathbf{B}}{\partial t} = \nabla \times \left\{ \mathbf{u} \times \mathbf{B} - \frac{\eta c^2}{4\pi} \nabla \times \mathbf{B} \right\}, \tag{1}$$

always admits the solution $\mathbf{B}(\mathbf{x}, t) = 0$. This poses a problem for the generation of the required seed fields, as they cannot be created in ideal MHD starting from a field-free state.

There have been several proposals for generating the required seed fields from mechanisms such as primordial phase transitions, or from processes occurring during inflation [3, 4]. The Biermann battery effect [5] provides another popular resolution of this problem [2, 3, 6]. The effect, which arises in consequence of the large difference in the electron and ion mass, is attributable to small-scale charge separation in the plasma. Pressure forces produce much larger accelerations of electrons than of ions, and the relative acceleration of the two components results in charge separation that must be balanced by an electric field

$$\mathbf{E}_B \equiv -(en_e)^{-1} \nabla P_e, \tag{2}$$

where n_e and P_e are the electron number density and pressure, respectively. Since this field is not, in general, irrotational, it can act as a source of magnetic field in the induction equation,

$$\frac{\partial \mathbf{B}}{\partial t} = \nabla \times \left\{ \mathbf{u} \times \mathbf{B} - \frac{\eta c^2}{4\pi} \nabla \times \mathbf{B} + \frac{c}{en_e} \nabla P_e \right\}, \tag{3}$$

generating a non-zero **B** from an initially unmagnetized state.

The Biermann battery effect has been successfully invoked in numerical simulations exploring the generation of seed fields in cosmological ionization fronts [7, 8], protogalaxies [9], and Pop-III star formation [10]. Magnetic field generation by the Biermann mechanism is also of significant interest in direct-drive and indirect-drive inertial confinement fusion, where strong gradients behind the converging shock can lead to dynamically important field strengths [11], and more generally in the field of High-Energy Density Physics, where the effects of field generation [12] and amplification [13] can be examined in a laboratory setting at laser facilities, in experiments where large gradients are produced in strong plasma shocks [14, 15, 16].

Unfortunately, a straightforward implementation of the Biermann effect in finite-volume Eulerian, purely Langrangian, and ALE codes, whether as a split source term or as a flux term, leads to non-convergent results [17]. In symmetric situations such as planar or spherical shocks, where no field should arise, such implementations produce anomalous field generation that grows without bound with resolution [17]. This behavior is observed across a range of different MHD codes (see the discussion of codes in [17]). This is the Biermann catastrophe of numerical MHD. We show below that this failure is intimately related to the failure of such codes to correctly model the structure of the plasma shock.

In a gasdynamic/MHD formulation, where the shocks are modeled as zero-width discontinuities of the flow, the trouble arises from the behavior of the Biermann flux, which is to say from the electric field, Equation (2), in the vicinity of a shock. The gradient ∇P_e, which analytically speaking acquires a Dirac δ component at the shock, is ascribed a numerical magnitude that grows without bound at the shock with increasing resolution. It is this divergence that is connected with failure of MHD codes to correctly predict shock-driven magnetic field generation in supposedly simple test cases, where the correct value of the generated field is zero. The investigation of this failure is one of the central concerns of this article.

In what follows, we appeal to the kinetic theory of the structure of plasma shocks to guide us to an accurate and convergent treatment of the effect in MHD codes. We clarify the origin of the Biermann catastrophe as a numerical effect attributable to the difficulty of discretizing the source term of Equation (2) in the vicinity of a shock. We show that the numerical anomaly can be eliminated by leveraging the continuity of the electron temperature T_e across shocks. Reformulation of the Biermann source term in terms of T_e allows the singularity to be isolated, and the flux of magnetic field due to the Biermann effect to be rewritten in a manifestly finite form suitable for translation into a convergent numerical algorithm.

The Biermann effect is due to electron-ion charge separation, and is sensitive to departure from thermal equilibrium between electrons and ions. Such a departure is precisely what occurs at shocks, so that a correct treatment of the effect at shocks necessarily requires that the disequilibrium be modeled. For this, a 2-temperature plasma model is mandatory. There is no consistent treatment of the Biermann effect at shock in full electron-ion thermal equilibrium.

We also point out a novel and interesting effect associated with the Biermann effect near a shock: A *resistive magnetic precursor* is generated in resistive MHD, wherein magnetic field generated by the Biermann effect in the shock "leaks" resistively into the upstream fluid whose physical extent is proportional to the resistivity. The effect is potentially observable in laboratory conditions at high-intensity laser facilities such as Vulcan, Omega, and NIF. An appropriately-designed experiment at such facilities could currently observe the precursor, providing a clean experimental validation test of the Biermann effect in plasma shocks.

2. The Biermann battery at shocks

We begin by reviewing some essential results from the kinetic theory of shocks in plasmas. The basic theory of the fluid structure of planar shocks in plasmas was set out in [18], while the electromagnetic structure of such shocks was discussed in [19], and more recently in [20]. An extremely lucid presentation of these results may be found in Chapter VII of [21].

The essential ingredients to be imported from the kinetic theory of plasma shocks in order to fashion a working MHD model of the Biermann effect are (1) the loss of thermal equilibrium between electrons and ions at shocks; and (2) the adiabatic behavior of electrons.

As discussed on p.36 of [22], a strong shock disturbs the thermal equilibrium between electrons and ions in a plasma. That equilibrium is maintained by electron-ion collisions, and operates over timescales τ_{ei} that are long compared to the shock-crossing time of a parcel of fluid entering the ion viscous shock in consequence of the large ratio m_i/m_e [23]. As a result, it is essential to describe the fluid in terms of an additional degree of freedom – the electron temperature, T_e – with respect to the usual equilibrium MHD model.

Fortunately, it is unnecessary to model the fluid using new inertial degrees of freedom to describe the electron fluid. A single inertial component for the fluid as a whole gives an adequate description of the fluid structure near the shock, and a completely satisfactory description of the fluid in smooth flow regions, where electron-ion collisions restore local thermal equilibrium [22, 21]. This makes it easier to adapt existing MHD codes to treat the Biermann effect correctly, since it is much easier to add a scalar degree of freedom for T_e than it would be to deal with two velocity fields, one each for the electron fluid and for the ion fluid.

The large ion-to-electron mass ratio, which we recall from the discussion preceding Equation (2) is responsible for the charge separation that produces the Biermann effect, has the further consequence that thermal conductivity due to electron-electron collisions dominates the heat transport in the fluid. At the same time, the forces between electrons and ions as the fluid crosses the ion viscous shock front are effectively dissipation-free, as they are simply electrostatic fields generated by charge separation, and collisional dissipation processes are too slow to act during the shock-crossing timescale.

These observations lead to the conclusion that while the ions undergoing shock compression experience the usual entropy-generating processes, the electrons are compressed *adiabatically* by the electrostatic forces exerted upon them by the ions, and consequently do not suffer entropy increments due to shock compression. The electron entropy is thus a passively-advected scalar, except for the dissipative effect of electron thermal conduction, and for the slow (relative to timescales relevant to the shock) effect of electron-ion collisional equilibration. Electron thermal conduction effectively rules out any sudden change in T_e at the shock, since such a change would produce an enormous restoring heat flux to heal the discontinuity.

The fluid structure that follows from these considerations is of a sudden discontinuous compression of the ions at the shock, accompanied by a smooth increase in T_e, which is continuous throughout the shock (in the Eulerian limit where the shock width tends to zero, T_e acquires a discontinuous derivative in the direction of the shock normal). The electron temperature exhibits a *thermal precursor* that leads the shock. The size λ_T of the precursor region may be calculated by balancing advection against heat diffusion in the shock frame:

$$\lambda_T \approx \frac{\kappa_e}{\rho c_{v,e} D}, \tag{4}$$

where κ_e is the electron thermal conductivity, $c_{v,e}$ is the electron specific heat at constant volume per unit mass, and D is the shock speed [18, 21]. The effect of the precursor region on accretion shocks in galaxy clusters has been recently studied in [24].

The additional degree of freedom represented by T_e must be modeled numerically by adding an equation to the usual equations of compressible MHD. That is, in addition to the usual conservation laws for mass, momentum, and energy, and to the Equation (3) for $\mathbf{B}$, we need to choose a variable whose evolution equation allows us to track T_e. While there are many such variables available (T_e itself, or the ion thermal energy, for example), and while all such evolutions would agree in smooth flow, not all such equations give the correct behavior at shocks. In order to guarantee the nearly-adiabatic behavior of electrons, it is necessary to choose the *specific electron entropy* s_e as our variable, and use the evolution equation

$$\frac{\partial \rho s_e}{\partial t} + \nabla \cdot (\rho \mathbf{u} s_e) = -{T_e}^{-1} \nabla \cdot (-\kappa_e \nabla T_e) + \frac{\rho c_{v,e}}{T_e \tau_{ei}}(T_i - T_e), \tag{5}$$

which expresses the conservation of electron entropy s_e (up to heat conduction and electron-ion heat exchange). As we pointed out above, this is required by the same approximation $m_e/m_i \to 0$ that gives rise to the Biermann effect in the first place.

3. The Biermann Effect At Shocks

By inspection of the Biermann source term in Equation (3) we see that it is proportional to $\nabla n_e \times \nabla P_e$. It follows that field generation by the Biermann effect can only occur if the gradients of n_e and P_e are not aligned. This means that in shocks with planar, cylindrical, or spherical symmetry, the field generation rate is zero. We must therefore treat non-symmetric shock surfaces in order to analyze non-trivial cases of field generation. Accordingly, in this subsection, we set up the basic kinematics of the Biermann effect at general shock surfaces.

To describe the shock surface, we introduce a level function $\Psi(\mathbf{x}, t)$, and use it to define the shock surface $\Psi(\mathbf{x}, t) = 0$. Note that the function Ψ is of no dynamical significance, but is rather simply a mathematical convenience for describing the shock surface. We will assume that the shock is moving in the direction of the normal vector $\mathbf{n} \equiv \nabla\Psi/|\nabla\Psi|$, so that the region $\Psi(\mathbf{x}, t) > 0$ is upstream, whereas the region $\Psi(\mathbf{x}, t) < 0$ is downstream.

We denote the local shock speed along $\mathbf{n}$ by D. By considering the motion of the level surface $\Psi(\mathbf{x}, t) = 0$ it is not difficult to show that

$$D = -\frac{\partial \Psi}{\partial t} / \left| \nabla \Psi \right|. \tag{6}$$

The treatment of shock kinematics is complicated by the fact that, as we have noted, the standard flux term for magnetic field is suspect, because it is not well-defined at a discontinuity. The usual route to the Rankine-Hugoniot shock conditions, in which one balances the fluxes on either side of the shock in the rest frame of the shock, is therefore precluded. We adopt an alternative route that circumvents this difficulty.

To describe a moving MHD discontinuity that coincides with the surface $\Psi(\mathbf{x}, t) = 0$, we will express a field quantity X by the decomposition $X = X_u(\mathbf{x}, t)\Theta(\Psi) + X_d(\mathbf{x}, t)\Theta(-\Psi)$, where X_u and X_d are continuous functions, and where $\Theta(\Psi)$ is a Heaviside step function. After substituting into field equations, we will find some terms proportional to the Dirac distribution $\delta(\Psi)$, resulting from differentiation of the Heaviside functions. We will refer to the collected terms of this form as the "shock part" of the evolution equations. Such terms embody Rankine-Hugoniot jumps, which can thus be efficiently extracted for these non-symmetric shocks. This trick is also useful for deducing hydrodynamic flux terms, as will also be evident below.

It is convenient to reformulate the Biermann source term using the ideal equation of state to replace n_e with T_e, the electron temperature. Assuming an ideal gas equation of state, the reformulated electric field is

$$\mathbf{E}_B \equiv -(k_B/e) T_e \nabla \ln P_e, \tag{7}$$

so that the source term due to the Biermann effect in the induction equation is

$$\left. \frac{\partial \mathbf{B}}{\partial t} \right|_B = (c k_B/e) \nabla T_e \times \nabla \ln P_e. \tag{8}$$

An important reason for this reformulation is that as we saw in §2, in the presence of electron thermal conduction, T_e *is continuous at the shock*, whereas n_e is not [18, 21]). As we saw, the continuity of T_e is a consequence of the high mobility of electrons relative to ions, and is therefore part and parcel of the same approximation that led to the Biermann source term (Equation 2) in the first place. It is central to the developments that follow.

Using this reformulation, we may write for the induction equation

$$\frac{\partial \mathbf{B}}{\partial t} = \nabla \cdot \{\mathbf{Bu} - \mathbf{uB}\} + \frac{ck_B}{e} \nabla T_e \times \nabla \ln P_e. \tag{9}$$

Plugging in the discontinuous form for the various field variables, and respecting the continuity of T_e and of the normal magnetic field, we obtain

$$\nabla \ln P_e = \Theta(\Psi)\nabla \ln P_{e,u} + \Theta(-\Psi)\nabla \ln P_{e,d} + [\ln P_e]_d^u \delta(\Psi)\nabla\Psi \tag{10}$$

$$\nabla T_e = \Theta(\Psi)\nabla T_{e,u} + \Theta(-\Psi)\nabla T_{e,d} \tag{11}$$

$$\mathbf{B} = B_n \mathbf{n} + \mathbf{B}_{Tu}\Theta(\Psi) + \mathbf{B}_{Td}\Theta(-\Psi) \tag{12}$$

$$\mathbf{u} = \mathbf{u}_u\Theta(\Psi) + \mathbf{u}_d\Theta(-\Psi) \tag{13}$$

After plugging Equations (10-13) and Equation (6) into Equation (9) and isolating the Dirac δ-function term in the resulting equation, we obtain the corrected Rankine-Hugoniot equation

$$[(D - u_n)\mathbf{B}_T]_d^u + B_n [\mathbf{u}_T]_d^u + \frac{ck_B}{e} \nabla T_e \times \mathbf{n} [\ln P_e]_d^u = 0. \tag{14}$$

In this equation, $\mathbf{u}_T$ and $\mathbf{B}_T$ are respectively the fluid velocity and magnetic field component transverse to the shock, while u_n and B_n are the longitudinal velocity and field components. The notation $[\ldots]_d^u$ means the difference of the upstream and downstream values at the shock location. The first two terms in Equation (14) comprise the usual jump condition for the induction equation (see Chapter 7 of [25] for example). The final term is the contribution to the jump condition from the Biermann effect, which, in consequence of the continuity of T_e is seen to produce a finite, well-defined discontinuity at the shock.

We conclude from the above development that the Biermann source term is mathematically well-defined even at weak solution discontinuities, and yields definite finite predictions to which a properly designed numerical algorithm should be expected to converge.

4. Correcting the algorithm

The origin of the Biermann catastrophe may be diagnosed by means of a calculation described in [26]. It can be shown that even in simulations in which one expects that $\nabla P_e \propto \nabla n_e$, the discrete representation of the curl of the "irrotational" Biermann electric field, $\nabla \times \left({n_e}^{-1} \nabla P_e \right)$, diverges as $\mathcal{O}(\Delta^{-1})$ at a discontinuity. Consequently, at discontinuities, the "naively discretized" Biermann electric field grows *spurious solenoidal behavior* that diverges with increasing resolution.

To design a better algorithm we need a flux of $\mathbf{B}$ along $\mathbf{n}$ that is well-behaved at shocks, unlike the flux $c\mathbf{n} \times \mathbf{E}_B = -(ck_B/e)T_e\mathbf{n} \times \nabla \ln P_e$ that would be inferred from the naive Biermann term. To obtain such a flux, we can reverse the usual logic of the Rankine-Hugoniot jump conditions: instead of deriving the jump from the flux, we can infer the flux from the jump!

In terms of the flux $\mathbf{F}_\Phi$ of a generic field variable Φ, the Rankine-Hugoniot condition at a shock traveling with speed D along the unit normal $\mathbf{n}$ is given by

$$D \times [\Phi]_d^u = [\mathbf{n} \cdot \mathbf{F}_\Phi]_d^u . \tag{15}$$

Comparing this expression to the corrected Rankine-Hugoniot jump condition for $\mathbf{B}$ given in Equation (14) we may immediately infer the corrected flux of $\mathbf{B}$:

$$\mathbf{n} \cdot \mathbf{F}_\mathbf{B} = (\mathbf{n} \cdot \mathbf{u})\mathbf{B}_T - (\mathbf{n} \cdot \mathbf{B})\mathbf{u}_T + \frac{ck_B}{e} \ln P_e \mathbf{n} \times \nabla T_e. \tag{16}$$

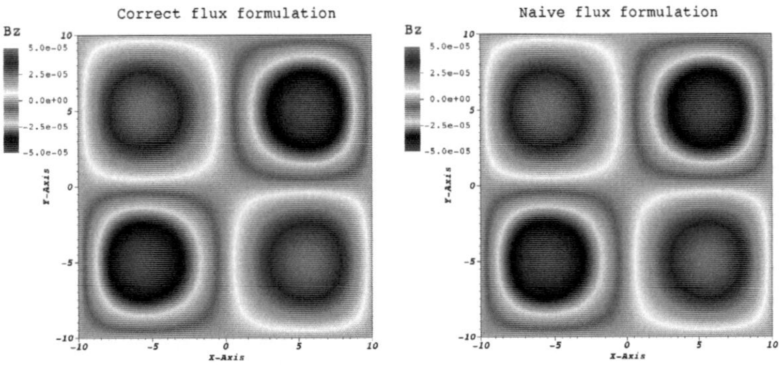

Figure 1. Smooth test in a cartesian box. In both panels we show the magnetic field generated after one computational step for the correct and naive flux formulations (left and right respectively). Both methods recover the analytical solution of Equation (17).

The third term in this expression is the correction due to the Biermann term, which is manifestly well-defined at discontinuities. From the form of Equation (16), only the tangential components of **B** are subject to change according to the Rankine-Hugoniot condition, as expected.

Discretization of the Biermann effect based on this flux expression is convergent *so long as the mesh resolves the scales on which T_e is continuous.* The scale at which T_e is continuous is the scale length λ_T of the electron thermal conduction precursor region [18, 19, 21], discussed in §2. This length scale is characteristic of the variation of temperature perpendicular to the shock; however, it is also the length scale over which heat diffuses transversely during the time required for the shock to travel a distance λ_T (and hence traverse its own thermal precursor). It is therefore the scale to be resolved in order for the discretization of Equation (16) to converge. At coarser resolutions, T_e appears as discontinuous at shocks as all the other fluid variables, and the discretization discussed here will not yield converged results for **B**. Only as the thermal precursor zone is resolved can convergence be expected.

5. Numerical Verification

5.1. Implementation Using the FLASH Code

We have implemented the corrected Biermann Effect algorithm within the publically-available FLASH hydrodynamic simulation framework [16, 27, 28]. FLASH is a modular and extensible multiphysics scientific simulation software package that has been widely used for reactive compressible flows typical of astrophysical situations, High-Energy Density Physics (HEDP) applications, cosmology, computational fluid dynamics, and fluid–structure interactions. In particular, FLASH makes available both a 2-temperature single-fluid model and resistive MHD, which makes it an ideal platform for testing the proposed algorithm.

We incorporated the new flux term represented by Equation (16) into the FLASH MHD solver. FLASH also supports electron entropy advection, Equation (5), with the standard Sackur-Tetrode equation for entropy (see, for example, p. 77 of [29]) implemented in the EOS.

5.2. Smooth flow

The first test problem considers a smooth electron density and pressure configuration for which the analytic calculation of the magnetic field generation due to the Biermann Effect is trivial. We employ this test to verify the implementations of the naive and correct flux formulations, which should both be able to recover the analytic solution. This non-dimensional 2D problem

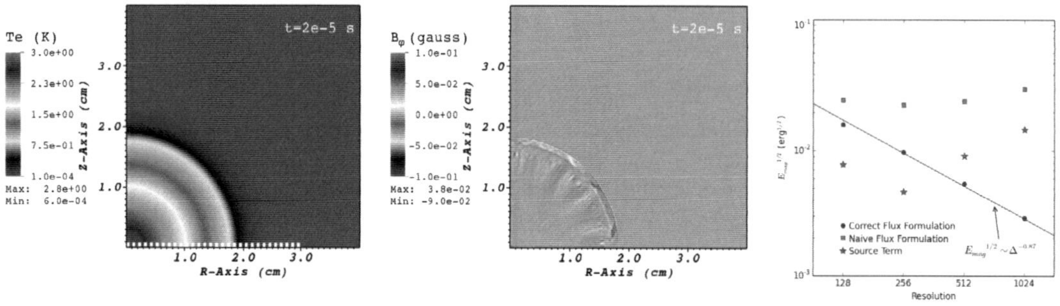

Figure 2. Null test using a spherical shock. Left panel: electron temperature distribution for a near-Sedov explosion. Middle panel: B_ϕ due to numerical noise. Right panel: Total magnetic energy as a function of simulation resolution, showing convergence of the new algorithm and convergence failure of the old source term and flux term algorithms.

is drawn from [30] and assumes an initial condition of $\mathbf{B} = 0$, $\mathbf{u} = 0$, $n_e = n_0 + n_1 \, cos(k_x \, x)$, and $P_e = P_0 + P_1 \, cos(k_y \, y)$, where $k_x = k_y = \pi/10$, $n_0 = P_0 = 1$, and $n_1 = P_1 = 0.1$. The cartesian domain spans from -10 to 10 in x and y, with periodic boundary conditions on both directions. The resulting magnetic field generation rate in the z direction will then be

$$\frac{\partial B_z}{\partial t} = -\frac{k_x \, k_y \, n_1 \, P_1 \, sin(k_x \, x) \, sin(k_y \, y)}{[n_0 + n_1 \, cos(k_x \, x)]^2} \tag{17}$$

As in [30], we use a resolution of [160 × 160] points and take one step, $\Delta t = 0.05$, to recover the analytic solution for both implementations. Both results agree with the analytic solution (Figure 1) within an error of $\sim 2 \times 10^{-8}$ (see also figure 6 in [30]).

5.3. Shocked flow

In the simulations that follow, we assume fully-ionized Hydrogen – $A = 1$, $Z = 1$, adiabatic index $\gamma = 5/3$, $c_{v,e} = \frac{3}{2} k_B N_A$, where N_A is Avogadro's number and k_B is Boltzmann's constant. Simulations are conducted in cylindrical coordinates, assuming azimuthal symmetry, and are therefore 2-dimensional. In addition, we impose a reflection boundary on the R-axis (where R is perpendicular distance from axis of rotational symmetry, which is to say, R is the cylindrical radius) so that the domain represents a hemisphere of a solution with reflection symmetry about that axis. In every case, the domain is 4 cm radius cylinder that extends 4 cm in the z direction from the R axis. The boundary conditions at $R = 4$ cm and at $z = 4$ cm are outflow. Magnetic fields are always azimuthal in these verification tests – since the gradients of T_e and P_e are always in the $R - z$ plane the Biermann effect only generates non-zero field along the azimuthal direction. These simulations are described in greater detail in [26].

It is not easy to construct a non-trivial analytic verification solution of a plasma shock magnetized by the Biermann effect. However, a trivial solution may be constructed by imposing symmetry requirements that align the gradients ∇P_e and ∇T_e, thus guaranteeing zero field generation. We use a spherical Sedov-like explosion as an example of such a test. This is in fact the test that revealed the Biermann catastrophe in the first place [17].

The results of this study are shown in Figure 2. The left panel shows the electron temperature T_e and the middle panel shows the magnetic field strength generated by the corrected treatment using the flux in Equation (16). We see spurious non-zero field due to numerical noise. However,

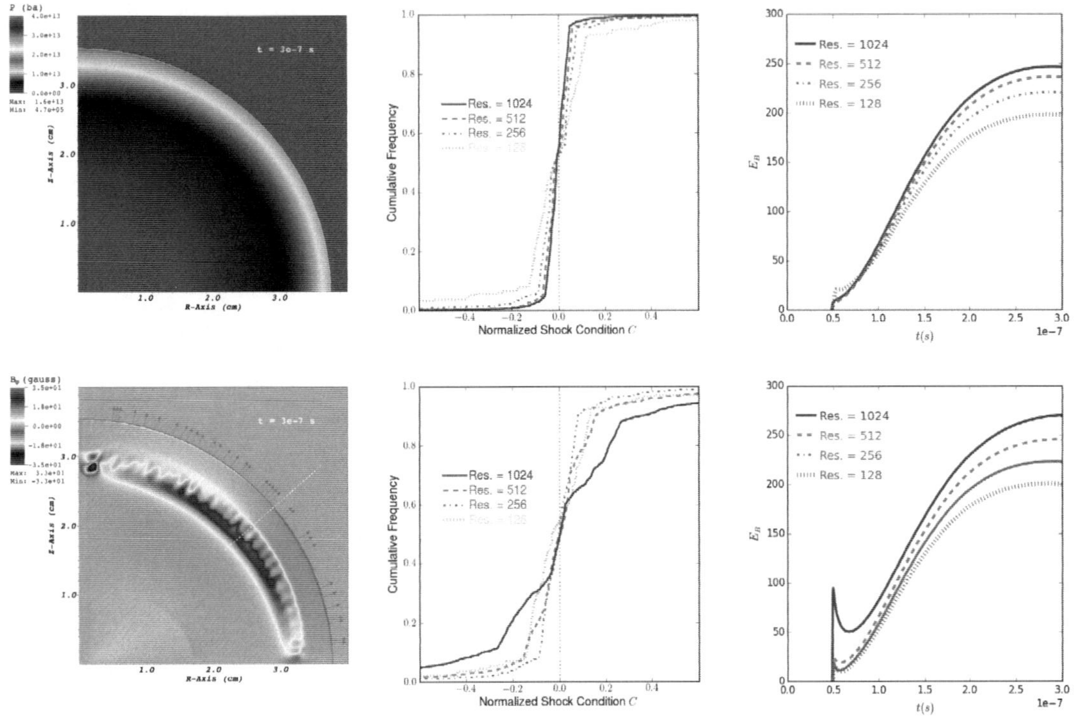

Figure 3. Results for an ellipsoidal shock. Top left: pressure distribution; Bottom Left: magnetic field distribution. The solid red line displays the location of the shock, while the red arrows display the direction of the shock normal; Middle Panels: Cumulative distribution of the normalized magnetic shock condition C, defined by Equation (21), evaluated at points along the shock surface, for four different resolutions, illustrating convergence to the correct jump condition for the corrected algorithm (top panel), and convergence failure for the naive algorithm (bottom panel); Right Panels: Total magnetic energy as a function of simulation time for the four resolutions studied, for the corrected algorithm (top panel) and for the naive algorithm (bottom panel).

the right panel shows that this field is converging to zero with increasing resolution. This figure displays the square-root of the total magnetic energy in the domain. Since the analytic solution for the magnetic field strength is zero everywhere, this quantity functions as an un-normalized L^2-norm of the difference between the analytic and numerical solutions. The blue circles show the convergence with resolution of the correct formulation. This convergence behavior is in contrast to the behavior of two algorithms that use the uncorrected Biermann term (a flux-based formulation, and a source term), which fail altogether to converge.

Next we exhibit simulations designed to produce non-zero values of the magnetic field strength. We start an ellipsoidal shock surface, obtained by distorting the spherical Sedov solution. The state of the shock after a period during which magnetic field is generated is illustrated by the pressure colormap plot in the top-left panel of Figure 3. The lower-left panel of Figure 3 shows the magnetic field intensity generated according to the corrected flux, ranging into the tens of Gauss. The solid red line in the figure shows the shock location, while the red arrows display the shock unit normal vector.

We do not have an analytical solution for the magnetic field distribution to compare to the output of these simulations. We do, however, have the relation between shock quantities expressed by Equation (14), which determines the jump condition for **B**. By locating points adjoining the shock, and computing the local shock velocity at those points, we can then verify that Equation (14) in fact obtains, to some accuracy limited by the numerical approximation.

To perform this verification, we first locate cells adjoining the shock by the method described in the Appendix of [26], which yields a 1-cell wide shock surface by fitting the mass, momentum, and energy Rankine-Hugoniot conditions to the neighboring data, using a speed-of-sound weighted inner-product on the state space, and treating the shock speed D as a free fit parameter. The fitted shock speed is then used in the verification of Equation (14).

In the absence of normal component field B_n, Equation (14) becomes

$$(D-u_d)B_d-(D-u_u)B_u+\frac{ck_B}{e}\left(n_z\frac{\partial T_e}{\partial R}-n_R\frac{\partial T_e}{\partial z}\right)[\ln P_{e,d}-\ln P_{e,u}]\equiv a_d-a_u+b_d-b_u=0, \quad (18)$$

where we've defined "Advection" terms $a_{d,u}$ and "Biermann" terms $b_{d,u}$ by

$$a_{d,u} \equiv (D-u_{d,u})B_{d,u} \quad (19)$$

$$b_{d,u} \equiv \frac{ck_B}{e}\left(n_z\frac{\partial T_e}{\partial R}-n_R\frac{\partial T_e}{\partial z}\right)\ln P_{d,u} \quad (20)$$

The sum of terms in Equation (18) is required to be zero. In a discretized numerical PDE integration, this really means that the terms must cancel up to some truncation or rounding precision, which is expressed relative to the largest of the magnitudes in play. We therefore define the "shock condition parameter" C by

$$C \equiv \frac{a_d-a_u+b_d-b_u}{\max(|a_d|,|a_u|,|b_d|,|b_u|)}. \quad (21)$$

We calculate the value of C at cells along the shock front, at each of our four resolution levels, for both the correct flux and the "naive" flux implementations. At the highest resolution (1024×1024) the electron precursor length is spanned by about 50 cells, while at the lowest resolution (128×128) it is only spanned by about 6 cells. We therefore expect substantial improvement in the computation of field generation as resolution is progressively refined.

We display cumulative distributions of C for corrected and "naive" flux implementations in the middle panels of Figure 3. It is evident from these figures that the distribution of C is in fact centered near zero for both flux implementations. The width of the distributions behave very differently, however. In the case of the correct Biermann flux implementation, the distributions get narrower with each refinement of resolution – the steepening of the cumulative profile shows that RMS deviations of C from zero shrink progressively with increasing resolution. This provides some evidence of convergence with resolution to the expected result. In the case of the naive flux implementation, there is no such evidence of convergence.

The right panels of Figure 3 show the evolution of total magnetic energy in the domain as a function of time, for the different resolutions and for the two flux implementations. The correct flux implementation appears to be converging (although not in any strong sense converged) by this measure, even at late time, whereas any evidence of convergence in magnetic field energy is simply lacking for the naive flux implementation.

6. New Physics: The Resistive Magnetic Precursor

Recall the physical basis for the electron conduction shock precursor: a balance of heat diffusion out of the shock and heat advection into the shock, in the presence of impulsive heating at the shock. A similar effect occurs for **B** in the presence of finite resistivity η: the balance of magnetic

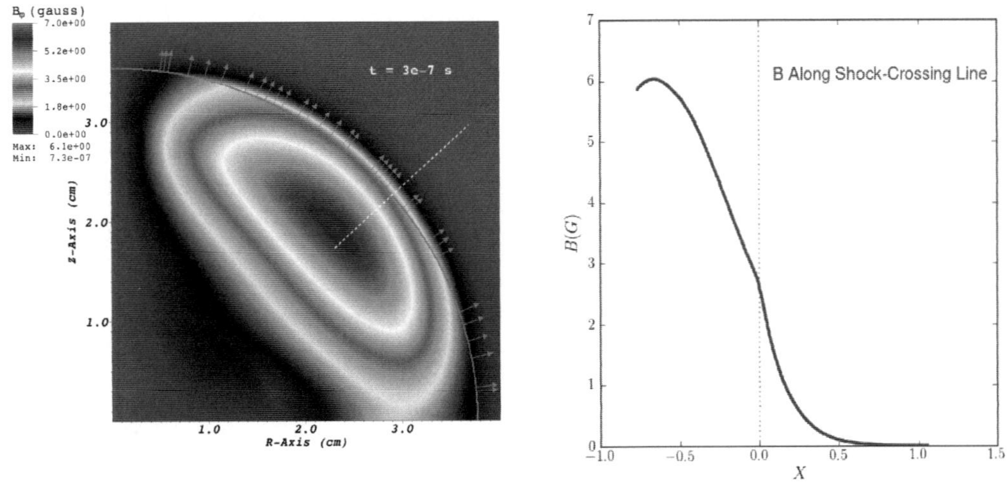

Figure 4. Left Panel: Magnetic field distribution due to passage of shock, in the presence of finite resistivity. The simulation differs from the one shown in Figure 3 only by the presence of non-zero resistivity. The solid red line shows the location of the shock, while the vectors represent the unit normal to the shock. The magnetic precursor can be clearly seen ahead of the shock surface. The white dashed shock-crossing line illustrates the location of the data displayed in the right panel. Right Panel: Magnetic field strength magnitude along the white dashed shock-crossing line shown in the left panel. The position of the shock is shown by the vertical solid red line. The predicted exponential decay of the precursor is evident.

diffusion out of the shock with magnetic advection into the shock in the presence of impulsive magnetogeneration by the BBT at the shock gives rise to a *resistive magnetic shock precursor.*

As shown in [26], the precursor is exponential, with a skin depth $\lambda_B = \frac{c^2\eta}{4\pi D}$. Assuming Spitzer-type diffusion, λ_B scales as follows:

$$\lambda_B = 16.4\,\mathrm{cm} \times Z \times \frac{\ln\Lambda}{10} \times \left(\frac{D}{10^6\,\mathrm{cm\,s^{-1}}}\right)^{-1} \times \left(\frac{k_B T_e}{1\mathrm{eV}}\right)^{-3/2}$$

The scales chosen here are not atypical of the conditions that can be created in a plasma at a large laser facility. Potentially this effect could be the basis of an HEDP validation experiment.

We confirm the presence of this precursor using simulations that maintain the simulation parameters of the ellipsoidal shock described in the previous section, but also turn on the resistivity η to a finite positive value chosen to yield an easily-discernible precursor

The left panel of Figure 4 shows the distribution of magnetic field strength across the domain. The shock location is shown by the solid red line, and the superposed vectors indicate the shock-normal direction. The magnetic field is evidently smoothed out by the resistivity, as can be seen by a direct comparison with the bottom-left panel of Figure 3. The precursor is evident in this figure, as the substantial amount of magnetic field that has "leaked" ahead of the shock.

The diagonal white dashed shock-crossing in the left panel of Figure 4 illustrates the line along which data was extracted to produce the right panel of Figure 4. This figure shows the magnetic field values plotted along that line. The vertical solid red line at $X = 0$ marks the location of the shock. The exponential decay of the field is easily visible in this figure.

References

[1] Kronberg P P 1994 *Reports on Progress in Physics* **57** 325 URL `http://stacks.iop.org/0034-4885/57/i=4/a=001`

[2] Kulsrud R M and Zweibel E G 2008 *Reports on Progress in Physics* **71** 046901 URL `http://stacks.iop.org/0034-4885/71/i=4/a=046901`

[3] Widrow L M 2002 *Reviews of Modern Physics* **74** 775–823 (*Preprint* `astro-ph/0207240`)

[4] Widrow L M, Ryu D, Schleicher D R G, Subramanian K, Tsagas C G and Treumann R A 2012 *Space Sci. Rev.* **166** 37–70 (*Preprint* `1109.4052`)

[5] Biermann L 1950 *Zeitschrift Naturforschung Teil A* **5** 65

[6] Kulsrud R M, Cen R, Ostriker J P and Ryu D 1997 *The Astrophysical Journal* **480** 481 URL `http://stacks.iop.org/0004-637X/480/i=2/a=481`

[7] Subramanian K, Narasimha D and Chitre S M 1994 *MNRAS* **271** L15

[8] Gnedin N Y, Ferrara A and Zweibel E G 2000 *The Astrophysical Journal* **539** 505 URL `http://stacks.iop.org/0004-637X/539/i=2/a=505`

[9] Davies G and Widrow L M 2000 *The Astrophysical Journal* **540** 755 URL `http://stacks.iop.org/0004-637X/540/i=2/a=755`

[10] Xu H, OShea B W, Collins D C, Norman M L, Li H and Li S 2008 *The Astrophysical Journal Letters* **688** L57 URL `http://stacks.iop.org/1538-4357/688/i=2/a=L57`

[11] Srinivasan B and Tang X Z 2013 *Physics of Plasmas (1994-present)* **20** 056307

[12] Gregori G, Ravasio A, Murphy C D, Schaar K, Baird A, Bell A R, Benuzzi-Mounaix A, Bingham R, Constantin C, Drake R P, Edwards M, Everson E T, Gregory C D, Kuramitsu Y, Lau W, Mithen J, Niemann C, Park H S, Remington B A, Reville B, Robinson A P L, Ryutov D D, Sakawa Y, Yang S, Woolsey N C, Koenig M and Miniati F 2012 *Nature* **481** 480–483

[13] Meinecke J, Doyle H W, Miniati F, Bell A R, Bingham R, Crowston R, Drake R P, Fatenejad M, Koenig M, Kuramitsu Y, C Kuranz C, Lamb D Q, Lee D, MacDonald M J, Murphy C D, Park H S, Pelka A, Ravasio A, Sakawa Y, Schekochihin A A, Scopatz A, Tzeferacos P, Wan W C, Woolsey N C, Yurchak R, Reville B and Gregori G 2014 *Nature Physics* **10** 520–524

[14] Fryxell B, Kuranz C, Drake R, Grosskopf M, Budde A, Plewa T, Hearn N, Hansen J, Miles A and Knauer J 2010 *High Energy Density Physics* **6** 162–165

[15] Tzeferacos P, Fatenejad M, Flocke N, Gregori G, Lamb D, Lee D, Meinecke J, Scopatz A and Weide K 2012 *High Energy Density Physics* **8** 322–328

[16] Tzeferacos P, Fatenejad M, Flocke N, Graziani C, Gregori G, Lamb D, Lee D, Meinecke J, Scopatz A and Weide K 2015 *High Energy Density Physics* **17, Part A** 24 – 31

[17] Fatenejad M, Bell A, Benuzzi-Mounaix A, Crowston R, Drake R, Flocke N, Gregori G, Koenig M, Krauland C, Lamb D, Lee D, Marques J, Meinecke J, Miniati F, Murphy C, Park H S, Pelka A, Ravasio A, Remington B, Reville B, Scopatz A, Tzeferacos P, Weide K, Woolsey N, Young R and Yurchak R 2013 *High Energy Density Physics* **9** 172 – 177 ISSN 1574-1818 URL `http://www.sciencedirect.com/science/article/pii/S1574181812001280`

[18] Shafranov V D 1957 *Soviet Physics JETP* **5** 1183–1188

[19] Jaffrin M and Probstein R 1964 *Physics of Fluids* **7** 1658

[20] Amendt P, Milovich J, Wilks S, Li C, Petrasso R and Séguin F 2009 *Plasma Physics and Controlled Fusion* **51** 124048

[21] Zel'dovich, YB and Raizer, YP 2002 *Physics of shock waves and high-temperature hydrodynamic phenomena* (Dover Pubns)

[22] Drake R 2006 *High-Energy-Density Physics: Fundamentals, Inertial Fusion, and Experimental Astrophysics* High pressure shock compression of condensed matter (Springer) ISBN 9783540293149 URL `http://books.google.com/books?id=P-30eLJZN1YC`

[23] Spitzer L 1962 *Physics of Fully Ionized Gases ... Second Revised Edition* [Interscience Tracts on Physics and Astronomy. no. 3.] (New York, London) URL `http://books.google.com/books?id=06okMwEACAAJ`

[24] Smith B, O'Shea B W, Voit G M, Ventimiglia D and Skillman S W 2013 *ApJ* **778** 152

[25] Gurnett D and Bhattacharjee A 2005 *Introduction to Plasma Physics: With Space and Laboratory Applications* (Cambridge University Press) ISBN 9780521364836 URL `http://books.google.com/books?id=VcueZlunrbcC`

[26] Graziani C, Tzeferacos P, Lee D, Lamb D Q, Weide K, Fatenejad M and Miller J 2015 *The Astrophysical Journal* **802** 43

[27] Fryxell B, Olson K, Ricker P, Timmes F X, Zingale M, Lamb D Q, MacNeice P, Rosner R, Truran J W and Tufo H 2000 *ApJS* **131** 273–334

[28] Dubey A, Antypas K, Ganapathy M K, Reid L B, Riley K, Sheeler D, Siegel A and Weide K 2009 *Parallel Computing* **35** 512–522

[29] Kittel C and Kroemer H 1980 *Thermal physics* (Macmillan)
[30] Tóth G, Van der Holst B, Sokolov I V, De Zeeuw D L, Gombosi T I, Fang F, Manchester W B, Meng X, Najib D, Powell K G *et al.* 2012 *Journal of Computational Physics* **231** 870–903

Grid coupling mechanism in the semi-implicit adaptive Multi-Level Multi-Domain method

M. E. Innocenti[1], **C. Tronci**[2], **S. Markidis**[3], **G. Lapenta**[1]

[1] Center for mathematical Plasma Astrophysics, Department of Mathematics, K.U. Leuven (University of Leuven), Celestijnenlaan 200B, B-3001 Leuven, Belgium

[2] Department of Mathematics, University of Surrey, Guildford GU2 7XH, United Kingdom

[3] PDC Center for high Performance Computing, KTH Royal Institute of Technology, Stockholm, Sweden

E-mail: mariaelena.innocenti@wis.kuleuven.be

Abstract. The Multi-Level Multi-Domain (MLMD) method is a semi-implicit adaptive method for Particle-In-Cell plasma simulations. It has been demonstrated in the past in simulations of Maxwellian plasmas, electrostatic and electromagnetic instabilities, plasma expansion in vacuum, magnetic reconnection [1, 2, 3]. In multiple occasions, it has been commented on the coupling between the coarse and the refined grid solutions. The coupling mechanism itself, however, has never been explored in depth. Here, we investigate the theoretical bases of grid coupling in the MLMD system. We obtain an evolution law for the electric field solution in the overlap area of the MLMD system which highlights a dependance on the densities and currents from both the coarse and the refined grid, rather than from the coarse grid alone: grid coupling is obtained via densities and currents.

1. Introduction

Fully kinetic simulations of space plasmas have greatly benefitted, over the last decades, of the almost predictable technological improvements commonly known as Moore's law: semiconductor technology doubles the number of transistors per unit area every 18 months [4]. It is however becoming clear that the technological trend that allowed computational scientists to perform larger and more resolved simulations every year is not going to last. [5] blames the prohibitive power consumption of High Performance Computing infrastructures, the need to improve the resiliency of the booming number of components and the increasing gap between the increment of computing performance and that of memory and storage. Even innovative architectures, such as GPUs and Xeon Phi's, may not be an easy solution to smoothly achieve exascale computing ($> 10^{18}$ floating point operations per second): as remarked in [6], very few applications can fully exploit concurrency. Hence, *algorithmic improvement*, as opposed to *computational improvement* alone, becomes a key factor for larger, more resolved fully kinetic plasma simulations in the coming years.

The development of more efficient algorithms for Particle In Cell (PIC) simulations of plasmas [7, 8] has moved, over the decades, in two directions: (semi-) implicit and adaptive algorithms.

Explicit PIC algorithms are subject to rather strict stability constraints which result from the explicit temporal discretisation of the equations for field and particle advancement. Fully implicit and semi-implicit methods can afford more relaxed stability constraints. They can focus on processes of interest at scales larger than the ones (Debye length, inverse electron plasma frequency) which explicit algorithms need to resolve for stability. The stability constraint of (semi)-implicit algorithms is usually of the form

$$\sigma < v_{th,e} dt/dx < 1, \tag{1}$$

with dt and dx temporal and spatial resolution, $v_{th,e}$ the average thermal electron velocity and $\sigma \sim 0.01$. This is achieved through rather advanced ways of dealing with the coupling between the field and particle system of equations. The coupling arises from the implicit time discretisation. In fully implicit methods [9, 10] this coupling is not removed and the equation for the field and particle systems are solved concurrently. In semi-implicit methods, the coupling is removed by approximating the particle sources in Maxwell's equations around the particle position at the previous (Implicit Moment Methods – IMM [11, 12, 13]) or next (direct implicit methods [14, 15, 16, 17]) time step.

Adaptive methods rely on *locally* increased resolution as a way to contain computational costs. The alternative, in fact, would be to use high resolution everywhere. Adaptive techniques can be grouped into two main families. Moving Mesh Adaptation algorithms [18, 19, 20, 21] keep the number of grid points fixed; points are attracted in the "interesting" part of the domain, identified monitoring a relevant quantity. Adaptive Mesh Refinement (AMR) algorithms [22, 23] change the number of grid points on the fly, again according to the variation of a monitor quantity. As the number of cells changes, particles, whose shape function is usually tailored on the local grid size, are split or coalesced.

We recently introduced the Multi-Level Multi-Domain (MLMD) method [1, 2, 3], a method for fully kinetic PIC simulations which combines the benefits of implicit and adaptive algorithms. Different grid levels are simulated with different spatial and temporal resolution: comparing the spatial and temporal resolution used at the different grid levels in previously published work, e.g. [2, 3], one can see that rather large "jumps" in resolution are obtained between the levels. Large resolution jumps are used because the processes one wants to resolve at the different levels operate at different scales, for example ion vs. electron scales. Large and sudden changes in resolution could prove a fatal challenge for an explicit Particle In Cell algorithm, where stability constraints require to resolve fractions of the Debye length and of the inverse electron plasma frequency. The problem is circumvented with the IMM, which is used in each grid. The stability region of the IMM is larger than that of explicit methods, see Equation 1. Hence, large jumps in temporal and spatial resolution between the grid can be achieved, while all grids stay in the stability region of the method. The different grid levels are coupled through information exchange operations: exchange of field and boundary conditions from the less resolved to the more resolved grid and projection of the more resolved field solution to the less resolved grid level. This last operation in particular has profound implication for moment conservation in the MLMD system, as explored in [24]. It also constitutes one of the key ingredients of the grid coupling mechanism. In this paper, we will explore how the projection operation results into grid coupling in the MLMD system.

The paper is structured as follows. Section 2 recaps the basics of the MLMD method and recalls the types of scenarios to which it has been applied so far. Section 3 is the core of the paper. It analyses the mechanism that couples the grids simulated at different resolution levels. This analysis has never been shown before. It is demonstrated that the projection (restriction) operation implemented in the MLMD method (Equation 32-33 in [1]) results into grid coupling because it introduces a dependancy of the mixed grid electric field solution on the refined grid density and currents. In Section 4, conclusions are drawn.

2. The Multi-Level Multi-Domain method

Figure 1 shows a sketch of a two-level MLMD system.

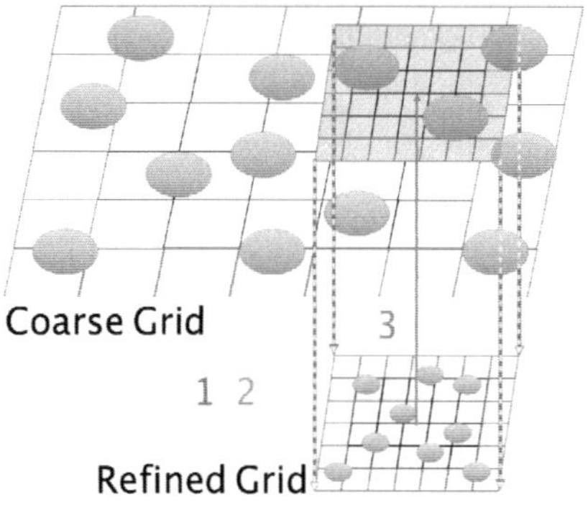

Figure 1. Sketch of a two-level MLMD system. Modified from [3]

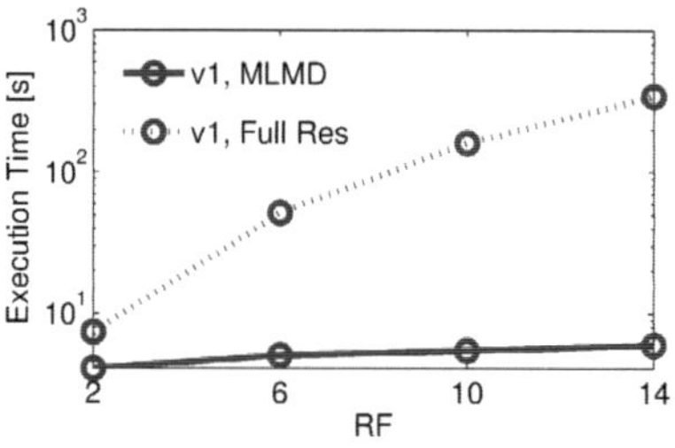

Figure 2. Execution times in seconds for MLMD (solid line) and corresponding single level (dotted line) simulations as a function of the RF. Modified from [3]

The Coarse Grid (CG) covers the entire domain and is simulated with "lower" temporal and spatial resolution. The Refined Grid (RG) covers a fraction of the entire domain and is resolved with "higher" resolution. Small scale processes (for example, electron scale processes as opposed to ion scale processes) need higher resolution to be properly reproduced in the simulation. Consequently, the Refined Grid is placed in correspondence of the areas where one knows, from previous experience, that smaller scale processes are going to develop. Small scale processes are therefore simulated twice, by the Coarse and by the Refined Grid. They are reproduced without approximation (i.e., at the appropriate resolution) by the RG. They are simulated in an approximated way by the Coarse Grid where, as typical of semi-implicit methods, selective damping and spatial compression affect the under-resolved scales [25]. Larger scale processes are simulated without approximation by both grids.

The jumps in spatial and temporal resolution between the two grids are indicated as Refined Factor (RF) and Time Ratio (TR). All grids are fully simulated with fields and particles, whose shape function is tailored on the local grid size for momentum conservation purposes [8]. Information exchange operations ensure proper coupling between the different grid levels. They consist of field and particle boundary condition exchange from the coarse to the refined grid (red and yellow dotted arrows in Figure 1) and electric field projection from the refined to the coarse grid (magenta arrow) [1]. The projection operation affects the area in the coarse grid simulated by both the coarse and the refined grid, the so called "overlap area" shaded in pink in Figure 1. It produces a "mixed grid" electric field obtained by merging the CG and RG electric field solution as recapped in Section 3.2. In this paper, we will study in particular the effects of this projection operation in terms of grid coupling.

The Implicit Moment Method (IMM) is used to advance fields and particles at all resolution levels. Equation 1 is the stability condition for the IMM: as long as the dt and dx are changed together, quite high jumps in resolution between the grids may be achieved. In [3], $RF = 12$ and $TR = 1 \rightarrow 6$ are used in simulations of magnetic reconnection where a realistic ion to electron mass ratio $m_r = 1836$ is used. The aim is to resolve ion scale processes on the CG and electron scale processes on the RG. The symbols Δ and δ stand for the coarse and refined grid resolution, d_i and d_e are the ion and electron skin depths, related to the ion to electron mass ratio m_r as $d_i/d_e = \sqrt{m_r}$, and ω_{pi} is the ion plasma frequency. On the CG, the resolution used is $\Delta x/d_i = 0.078$, $\omega_{pi}\Delta t = 0.05 \rightarrow 0.3$, on the RG is $\delta x/d_e = 0.28$, $\omega_{pi}\delta t = 0.05$. A reduced ion to electron mass ratio compresses ion and electron scales and allows to perform shorter and smaller simulations [28]. This is obviously an advantage, which may however be paid for in terms of inaccurate results: since the growth rate of different instabilities scales differently with the mass ratio, the risk exists of altering the hierarchy of the growth rates of competing instabilities [29].

For this reason, realistic mass ratios are our target.

Figure 2 illustrates a measure of performance for the MLMD method. The execution times of MLMD simulations (solid line) and single level simulations (dotted line) are compared as a function of the RF. At a fixed RF, the size of the domain is the same for the two simulations. The MLMD simulations are initialised with the RF indicated and $TR = 1$. A fraction $1/RF^2$ of the entire domain is simulated with higher resolution $\delta x = \Delta x/RF$, $\delta y = \Delta y/RF$. In the corresponding single level simulation, the RG resolution is used in the entire domain. The aim is to compare how much time is gained with a MLMD simulation if the higher resolution $\Delta x/RF$ is needed only in a fraction $1/RF^2$ of the domain. The benchmark is the single level alternative, i.e. using high resolution everywhere in the domain. Since, at increasing RF, the RG resolution increases, so does the execution time of the single level simulations. The execution time of the MLMD simulations increases only slightly, since the number of particles in the system is independent on the RF (the computational cost is proportional to the number of particles in the simulation). The increase is due to the higher impact of grid exchange information on the total execution time with increasing RFs. The savings in computational time granted by the MLMD method are remarkable at the RF values normally used in our simulations. At $RF = 14$, the MLMD execution time is about 70 time shorter that its single level counterpart. Additional information on performance evaluation can be found in [3].

The MLMD method is used in [2, 3] to simulate magnetic reconnection. Increased resolution is used in a small area surrounding the Electron Diffusion Region [26]. It is demonstrated that electron scale features, such as electron jet speed [27], can be recovered in the higher resolution area at a moderate cost even for realistic mass ratios.

Recently, the MLMD method has been used to simulate turbulence produced by the Lower Hybrid Drift Instability, LHDI [30], again at a realistic mass ratio [31]. The LHDI is notorious for being unstable over a large wavenumber and frequency range and can therefore be regarded as a "turbulence generator" since it breaks large scale fields into smaller and smaller scales. The MLMD method has been used to build "mixed grid spectra", obtained joining the power spectra calculated on the grids simulated with higher and lower resolution. In "mixed grid spectra", the CG provides the low wavenumber solution. Large domains, low wavenumbers are needed in LHDI simulations to allow the development of the ion-ion kink instability, responsible for the well known current sheet kinking at late times [32]. The RG provides the high wavenumber part, when the electromagnetic and electrostatic LHDI branches develop. Notice that at realistic mass ratio the wavenumber of the ion-ion kink instability and that of the electromagnetic LHDI, usually compressed at reduced mass ratio, are well separated. Both LHDI and the ion-ion kink instability have wavenumber in the perpendicular direction with respect to the ambient magnetic field. Let's indicate with $k_\perp$ the perpendicular wavenumber and with ρ_i and ρ_e the ion and electron gyroradius. The ion-ion kink instability develops at $k_\perp L_H \sim 0.5$, the electromagnetic LHDI at $k_\perp \sqrt{\rho_i \rho_e} \sim 1$. With our choice of initial parameters, they correspond to $k_\perp d_i \sim 1$ and $k_\perp d_i \sim 20$. The coarse and refined grid spectra are shown to join seamlessly in the wavenumber range reliably simulated at both resolution levels. Using a RG in conjunction with a CG extends of a factor RF the wavenumber range covered, at a minimum computational cost. In fact, the cost increases only of about two times with respect to the execution time of the CG alone. A 2D3V (two spatial components, three velocity components) single level simulation reaching the same maximum wavenumber would cost RF^2 times more.

3. Grid coupling mechanism in the MLMD method

In this Section, the evolution law for the mixed grid electric field is derived. We define the mixed grid electric field as the electric field in the overlap area (pink in Figure 1) *after* projection.

First, in Section 3.1 we will show how the equation solved at each level of the MLMD system, Equation 2, is derived from Maxwell's equations. This constitutes an excellent exercise for the calculations shown later. Then, the details of the projection operation in the MLMD method are recalled in Section 3.2. Finally, the evolution law for the mixed grid electric field is derived and discussed in Section 3.3.

3.1. Derivation of the single grid equation for electric field advancement

Each resolution level of the MLMD system solves the following equation to advance in time the electric field **E** [33]:

$$\frac{1}{c}\frac{\partial}{\partial t}\left(\frac{4\pi}{c}\mathbf{J}+\frac{1}{c}\frac{\partial \mathbf{E}}{\partial t}\right)=\nabla^2\mathbf{E}-\nabla(4\pi\rho), \tag{2}$$

where **J** is the total current, ρ the total density and c the speed of light. It is obtained by applying the gradient operator to Gauss' law,

$$\nabla\cdot\mathbf{E}=4\pi\rho, \tag{3}$$

and then using the vector identity

$$\nabla\times(\nabla\times\mathbf{E})=\nabla(\nabla\cdot\mathbf{E})-\nabla^2\mathbf{E} \tag{4}$$

to substitute the $\nabla(\nabla\cdot\mathbf{E})$ term. $\nabla\times\mathbf{E}$ is obtained from Maxwell-Faraday equation

$$\nabla\times\mathbf{E}=-\frac{1}{c}\frac{\partial \mathbf{B}}{\partial t}, \tag{5}$$

with **B** the magnetic field. The resulting $\nabla\times\mathbf{B}$ term in

$$\frac{1}{c}\frac{\partial}{\partial t}\nabla\times\mathbf{B}=\nabla^2\mathbf{E}-\nabla(4\pi\rho) \tag{6}$$

is then substituted away using Ampere's law

$$\nabla\times\mathbf{B}=\frac{4\pi}{c}\mathbf{J}+\frac{1}{c}\frac{\partial \mathbf{E}}{\partial t}. \tag{7}$$

Equation 2 describes the time evolution of **E** as a function of **J** and ρ. The magnetic field, which is conveniently eliminated from the evolution law, is then obtained from the electric field through Maxwell-Faraday. The implicit temporal discretization of Equation 2 and the IMM approximation allow to recover the familiar-looking Equation 26 in [1], solved in IMM in general [34] and in the MLMD method in particular. [33] remark that the time discretisation of Equation 2 leads to a system of linear equation which is better conditioned than alternative formulations of the evolution law for **E**. Also, it is noticed there that initial perturbations of Gauss' law are damped in Equation 2. We immediately think here of perturbations related to numerical noise.

3.2. Projection in the MLMD method

To analyse the coupling of the coarse and refined grid solution in the overlap area (pink in Figure 1), it is convenient to preliminarily recall the details of the projection operation (magenta line). The three components of the electric field are projected from the refined (RG) to the coarser (CG) grids as in

$$\mathbf{E}_{P,CG}=\frac{1}{2}\left(\mathbf{E}_{N,CG}+\mathbb{P}^{RG\rightarrow CG}\left(\mathbf{E}_{N,RG}\right)\right), \tag{8}$$

where $\mathbf{E}_P$ denotes the projected electric field and $\mathbf{E}_N$ the electric field calculated natively on the grid. The symbol $\mathbb{P}^{RG\rightarrow CG}$ is a short notation for the projection operator, implemented as in

$$\mathbb{P}^{RF\rightarrow CG}\left(\mathbf{E}_{N,RG}\right)=\frac{\sum_{RG}\mathbf{E}_{N,RG}W_{CG}(\mathbf{x}_{CG}-\mathbf{x}_{RG})}{\sum_{RG}W_{CG}(\mathbf{x}_{CG}-\mathbf{x}_{RG})}. \tag{9}$$

The same interpolation function $W(\mathbf{x}_{CG} - \mathbf{x})$ is used in the coarse grid for grid to particle and particle to grid interpolation. The "mixed grid electric field" mentioned earlier in the paper is indicated as $\mathbf{E}_{P,CG}$. One can see that, neglecting the operations necessary to interpolate the RG solution to the different gridding used in the CG, the projection operation is essentially an average between the coarse and the refined grid electric field values. A formulation of Equation 8 *in the continuum* (i.e., neglecting the spatial discretisation) therefore reads:

$$\mathbf{E}^\star = \frac{\mathbf{E} + \mathbf{E}_{RG}}{2}, \tag{10}$$

where $\mathbf{E}^\star$ (which corresponds to $\mathbf{E}_{P,CG}$ in Equation 9) is the mixed grid solution in the overlap area, $\mathbf{E}$ is now used to label the "native" coarse grid solution (the CG notation is dropped for easier reading) and $\mathbf{E}_{RG}$ is the refined grid solution.

3.3. *Evolution law for the mixed grid electric field*

In the MLMD method, Equation 2 is solved to obtain the native electric field $\mathbf{E}_{N,CG}$. Therefore, substituting Equation 10 into Equation 2 one gets

$$\frac{1}{c^2}\frac{\partial^2}{\partial t^2}\left(2\mathbf{E}^\star - \mathbf{E}_{RG}\right) = \nabla^2\left(2\mathbf{E}^\star - \mathbf{E}_{RG}\right) - \nabla(4\pi\rho) - \frac{4\pi}{c^2}\frac{\partial \mathbf{J}}{\partial t}, \tag{11}$$

where Gauss' law is assumed to hold in the CG. One should keep in mind that the symbol $\mathbf{E}$ in Section 3.1 labels the single grid electric field, while it is now used for the native coarse grid solution. Using the vector identity Equation 4 for $\mathbf{E}^\star$ in Equation 12, one obtains

$$\frac{1}{c^2}\frac{\partial^2}{\partial t^2}\left(2\mathbf{E}^\star - \mathbf{E}_{RG}\right) = 2\left(\nabla\cdot(\nabla\cdot\mathbf{E}^\star) - \nabla\times\nabla\times\mathbf{E}^\star\right) - \nabla^2\mathbf{E}_{RG} - \nabla(4\pi\rho) - \frac{4\pi}{c^2}\frac{\partial \mathbf{J}}{\partial t}. \tag{12}$$

Assuming that Gauss's law and Maxwell-Faraday equation hold for the mixed grid solution, i.e.

$$\nabla\cdot\mathbf{E}^\star = 4\pi\rho \tag{13}$$

and

$$\nabla\times\mathbf{E}^\star = -\frac{1}{c}\frac{\partial \mathbf{B}}{\partial t}, \tag{14}$$

Equation 12 becomes

$$\frac{1}{c^2}\frac{\partial^2}{\partial t^2}\left(2\mathbf{E}^\star - \mathbf{E}_{RG}\right) = \nabla(4\pi\rho) + 2\nabla\times\left(\frac{1}{c}\frac{\partial \mathbf{B}}{\partial t}\right) - \nabla^2\mathbf{E}_{RG} - \frac{4\pi}{c^2}\frac{\partial \mathbf{J}}{\partial t}. \tag{15}$$

It is now useful to remember that Equation 2 is solved on the refined grid as well, hence $\nabla^2\mathbf{E}_{RG}$ can be written as:

$$\nabla^2\mathbf{E}_{RG} = \frac{1}{c^2}\frac{\partial^2\mathbf{E}_{RG}}{\partial t^2} + \nabla(4\pi\rho_{RG}) + \frac{4\pi}{c^2}\frac{\partial \mathbf{J}_{RG}}{\partial t}, \tag{16}$$

where ρ_{RG} and $\mathbf{J}_{RG}$ are the density and the current in the Refined Grid. Substituting Equation 16 into Equation 15 yields

$$\frac{1}{c^2}\frac{\partial^2}{\partial t^2}\left(2\mathbf{E}^\star - \mathbf{E}_{RG}\right) = \nabla(4\pi\rho) + 2\nabla\times\left(\frac{1}{c}\frac{\partial \mathbf{B}}{\partial t}\right) - \left(\frac{1}{c^2}\frac{\partial^2\mathbf{E}_{RG}}{\partial t^2} + \nabla(4\pi\rho_{RG}) + \frac{4\pi}{c^2}\frac{\partial \mathbf{J}_{\mathbf{RG}}}{\partial t}\right) - \frac{4\pi}{c^2}\frac{\partial \mathbf{J}}{\partial t}, \tag{17}$$

and then

$$\frac{1}{c}\frac{\partial^2 \mathbf{E}^\star}{\partial t^2} = \nabla \times \left(\frac{\partial \mathbf{B}}{\partial t}\right) + \frac{4\pi c}{2}\nabla(\rho - \rho_{RG}) - \frac{1}{2}\frac{4\pi}{c}\frac{\partial}{\partial t}(\mathbf{J}_{RG} + \mathbf{J}) \tag{18}$$

after simple math. Integrating in time one obtains

$$\frac{1}{c}\frac{\partial \mathbf{E}^\star}{\partial t} = \nabla \times \mathbf{B} + \frac{4\pi c}{2}\int dt \nabla(\rho - \rho_{RG}) - \frac{1}{2}\frac{4\pi}{c}\left(\mathbf{J}_{RG} + \mathbf{J}\right), \tag{19}$$

which gives a measure of coupling between coarse and refined grid solution in the overlap area, again neglecting spatial discretisation issues. As a preliminary sanity check, one should notice that Equation 19 reduces to Ampere's law if $\mathbf{E}^\star \to \mathbf{E}$ (the mixed grid solution tends to the native coarse grid solution) and $\mathbf{J}_{RG} \to \mathbf{J}$, $\rho_{RG} \to \rho$ (the zeroth and first order refined grid moments tend to their coarse grid counterparts). Equation 19 provides an evolution law for the mixed grid electric field solution $\mathbf{E}^\star$ in terms of the native CG magnetic field solution $\mathbf{B}$ and, more interestingly, as function of the coarse and refined ρ and ρ_{RG} and $\mathbf{J}$ and $\mathbf{J}_{RG}$. Grid coupling in the overlap area depends on the coarse and refined grid densities and currents. The RG density and currents participate in determining the mixed grid electric field solution.

Notice that, apart from pathological cases, one expects the term $\int dt\nabla(\rho - \rho_{RG})$ not to diverge in time. However, differences in the coarse and refined grid densities (and between the mixed grid and refined grid electric field solution) are expected and have been observed and explained in the overlap area.

As an example, it is useful to revisit here the discussion on the mismatch between the CG and RG electric field and density traces observed in the 1D3V simulation of a Two Stream Instability, Section 6.3 of [1]. Figure 10 in [1] is reproduced here in Figure 3. The $\mathbf{E}^\star$ and $\mathbf{E}_{RG}$ fields are shown in panel a and b as a function of space (vertical axis) and time (horizontal axis). The symbols d_e and ω_{pe} indicate the electron skin depth and the electron plasma frequency. The RG overlaps the CG at $3.966 < x/d_e < 6.95$. The contours of the CG density are superimposed to the electric field in both panels. One can notice a mismatch in $\mathbf{E}^\star$ and $\mathbf{E}_{RG}$ especially in the area marked with the red rectangle. In [1] the mismatch is related to different electron hole merging patterns in the phase spaces of the CG and RG. At later times the CG, RG differences subside in longitudinal electric fields, densities, phase spaces.

The possibility of accommodating slightly different evolutions in the different grids, possibly in response to processes at different scales simulated differently by the two grids, is considered one of the strengths of the MLMD method and has been amply documented in [2, 3].

Of particular relevance for this discussion is the case of grid coupling documented in Figure 9 of [2], reproduced here as Figure 4. Two reconnection points are simulated in the CG. One of the two reconnection points, Figure 4 panel (a), is simulated with a RG also, while in Figure 4 panel (b) the "control reconnection point", simulated only with the CG, is shown. The presence of the RG makes the mixed grid solution (panel a) stable against the finite grid instability which develops instead in the control reconnection point (panel b) due to the absence of field smoothing. We now interpret this stabilising effect as a result of the influence of the RG density and currents on the mixed grid solution through the projection operation.

4. Conclusions

The MLMD method, a semi-implicit adaptive method for fully kinetic plasma simulations, relies on two pillars: the capability of the CG to drive RG evolution through CG-to-RG boundary conditions and good RG-CG coupling in the overlap area. The latter property is achieved through electric field projection from the refined to the coarse grid. In this paper, we investigate the mechanism of grid coupling though projection. We neglect spatial discretisation at the different grid levels and work in the continuum, where the CG electric field after projection is

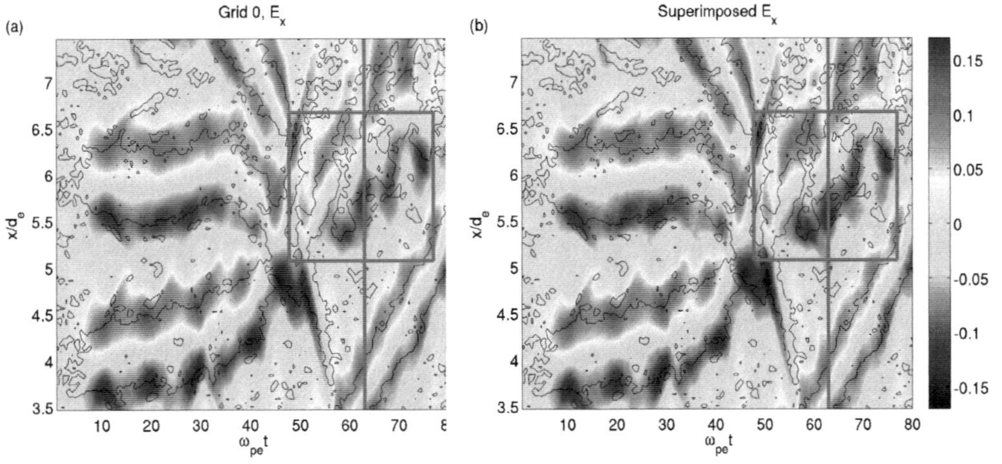

Figure 3. The $E_x^\star$ (panel a) and $E_{x,RG}$ (panel b) fields as a function of space (vertical axes) and time (horizontal axis) in a 1D3V simulation of Two Stream Instability. Coarse grid density contours are superimposed to both panels. Reproduced from [1].

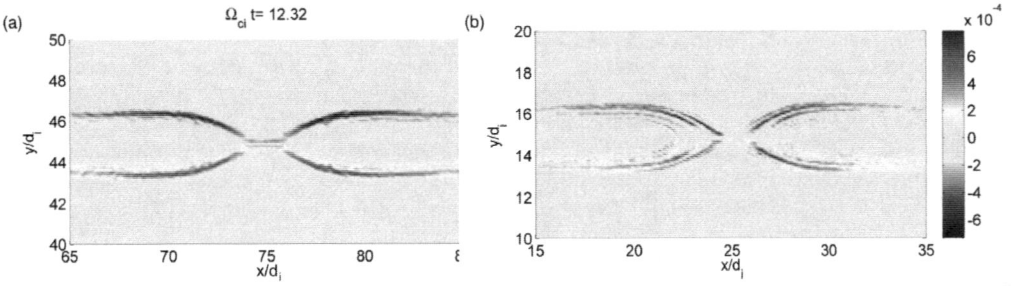

Figure 4. E_y component of the mixed grid solution (panel a) vs the coarse grid solution (panel b) in a simulation of magnetic reconnection where the absence of field smoothing results in the development of the finite grid instability at the X point simulated only with the CG. The presence of the Refined Grid stabilises the mixed grid solution. Reproduced from [2].

simply a mixed grid solution $\mathbf{E}^\star$ obtained by averaging the native CG solution $\mathbf{E}$ and the RF solution $\mathbf{E}_{RG}$. We write the evolution law for $\mathbf{E}^\star$ in terms of CG and RG quantities, Equation 19. Equation 19 shows that CG-RG coupling is achieved through densities and currents: both coarse grid and refined grid currents and densities contribute to the time evolution of $\mathbf{E}^\star$. These results give a theoretical framework to the cases of excellent CG-RG coupling documented in [1, 2, 3].

Acknowledgements

We acknowledge financial support by the Leverhulme Trust, Research project grant 2014-112. M.E.I. is funded by an FWO (Fonds Wetenschappelijk Onderzoek) post-doctoral fellowship, grant number 12O5215N.

References

[1] Innocenti M, Lapenta G, Markidis S, Beck A and Vapirev A 2013 *Journal of Computational Physics* **238** 115 – 140 ISSN 0021-9991 URL `http://www.sciencedirect.com/science/article/pii/S0021999112007590`
[2] Beck A, Innocenti M, Lapenta G and Markidis S 2014 *Journal of Computational Physics* **271**

430 – 443 ISSN 0021-9991 frontiers in Computational Physics Modeling the Earth System URL `http://www.sciencedirect.com/science/article/pii/S0021999113008152`

[3] Innocenti M, Beck A, Ponweiser T, Markidis S and Lapenta G 2015 *Computer Physics Communications* **189** 47 – 59 ISSN 0010-4655 URL `http://www.sciencedirect.com/science/article/pii/S001046551400410X`

[4] Moore G E 1965 *Electronics* **19** 114–117

[5] Kogge P 2008 Exascale computing study: Technology challenges in achieving exascale systems `http://www.cse.nd.edu/Reports/2008/TR-2008-13.pdf`

[6] Mallon D A, Eicker N, Innocenti M E, Lapenta G, Lippert T and Suarez E 2012 On the scalability of the clusters-booster concept: a critical assessment of the deep architecture *Proceedings of the Future HPC Systems: the Challenges of Power-Constrained Performance* (ACM) p 3

[7] Hockney R and Eastwood J 1988 *Computer simulation using particles* (Taylor & Francis)

[8] Birdsall C K and Langdon A B 2004 *Plasma physics via computer simulation* (Taylor & Francis)

[9] Chen G, Chacón L and Barnes D C 2011 *Journal of Computational Physics* **230** 7018–7036

[10] Markidis S and Lapenta G 2011 *Journal of Computational Physics* **230** 7037 – 7052 ISSN 0021-9991 URL `http://www.sciencedirect.com/science/article/pii/S0021999111003445`

[11] Mason R 1981 *Journal of Computational Physics* **41** 233

[12] Vu H and Brackbill J 1992 *Computer physics communications* **69** 253–276

[13] Lapenta G, Brackbill J and Ricci P 2006 *Physics of plasmas* **13** 055904

[14] Cohen B, Langdon A, Hewett D and Procassini R 1989 *Journal of Computational Physics* **81** 151–168

[15] Friedman A 1990 *Journal of Computational Physics* **90** 292 – 312 ISSN 0021-9991 URL `http://www.sciencedirect.com/science/article/pii/002199919090168Z`

[16] Hewett D W and Langdon A B 1987 *Journal of Computational Physics* **72** 121 – 155 ISSN 0021-9991 URL `http://www.sciencedirect.com/science/article/pii/0021999187900751`

[17] Tanaka M 1988 *Journal of Computational Physics* **79** 209 – 226 ISSN 0021-9991 URL `http://www.sciencedirect.com/science/article/pii/0021999188900125`

[18] Brackbill J U 1993 *Journal of Computational Physics* **108** 38

[19] Delzanno G L, Chacón L, Finn J M, Chung Y and Lapenta G 2008 *Journal of Computational Physics* **227** 9841–9864

[20] Chacón L, Delzanno G L and Finn J M 2011 *Journal of Computational Physics* **230** 87–103

[21] Lapenta G 2011 *Journal of Computational Physics* **230** 4679 – 4695

[22] Vay J L, Colella P, Friedman A, Grote D, McCorquodale P and Serafini D 2004 *Computer physics communications* **164** 297–305

[23] Fujimoto K and Sydora R D 2008 *Computer Physics Communications* **178** 915–923

[24] Innocenti M, Beck A, Markidis S and Lapenta G 2016 *Journal of Computational Physics* **312** 14 – 18 ISSN 0021-9991 URL `http://www.sciencedirect.com/science/article/pii/S0021999116000905`

[25] Brackbill J, Forslund D, Quest K and Winske D 1984 *Physics of Fluids (1958-1988)* **27** 2682–2693

[26] Daughton W, Scudder J and Karimabadi H 2006 *Physics of Plasmas* **13** 072101

[27] Drake J F, Shay M A and Swisdak M 2008 *Physics of Plasmas (1994-present)* **15** 042306 URL `http://scitation.aip.org/content/aip/journal/pop/15/4/10.1063/1.2901194`

[28] Pritchett P 2000 *Plasma Science, IEEE Transactions on* **28** 1976–1990 ISSN 0093-3813

[29] Bret A and Dieckmann M E 2010 *Physics of Plasmas (1994-present)* **17** 032109 URL `http://scitation.aip.org/content/aip/journal/pop/17/3/10.1063/1.3357336`

[30] Daughton W 2003 *Physics of Plasmas (1994-present)* **10** 3103–3119

[31] Innocenti M E, Norgren C, Newman D, Goldman M, Markidis S and Lapenta G submitted *Physics of Plasmas*

[32] Karimabadi H, Daughton W, Pritchett P and Krauss-Varban D 2003 *Journal of Geophysical Research: Space Physics (1978–2012)* **108**

[33] Ricci P, Lapenta G and Brackbill J 2002 *Journal of computational physics* **183** 117–141

[34] Markidis S, Lapenta G *et al.* 2010 *Mathematics and Computers in Simulation* **80** 1509–1519

Electrostatic PIC with adaptive Cartesian mesh

Vladimir Kolobov[1,2] **and Robert Arslanbekov**[1]

[1] CFD Research Corporation, Huntsville, AL, USA
[2] The University of Alabama in Huntsville, Huntsville, AL 35899, USA

E-mail: vladimir.kolobov@cfdrc.com

Abstract. We describe an initial implementation of an electrostatic Particle-in-Cell (ES-PIC) module with adaptive Cartesian mesh in our Unified Flow Solver framework. Challenges of PIC method with cell-based adaptive mesh refinement (AMR) are related to a decrease of the particle-per-cell number in the refined cells with a corresponding increase of the numerical noise. The developed ES-PIC solver is validated for capacitively coupled plasma, its AMR capabilities are demonstrated for simulations of streamer development during high-pressure gas breakdown. It is shown that cell-based AMR provides a convenient particle management algorithm for exponential multiplications of electrons and ions in the ionization events.

1. Introduction

Particle-in-Cell (PIC) is a mature technology widely used for plasma simulations. There are three versions of PIC: an electrostatic (EC) PIC uses Poisson equation for calculating electric field, an electromagnetic (EM) PIC employs full-wave Maxwell solver for calculations of electric and magnetic fields, and a quasi-static (QS) PIC uses a quasi-static version of the Maxwell equations (such as Darwin model) for calculation of the EM fields. Surprisingly, PIC solvers with adaptive mesh refinement (AMR) are relatively rare. It is worth mentioning the WARP code for accelerator applications [1], and an explicit EM-PIC code of Fujimoto [2] for astrophysics applications, both using block-structured AMR.

In this paper, we describe an initial implementation of an ES-PIC module in our Unified Flow Solver (UFS) framework [3]. UFS is designed for hybrid simulations of partially-ionized collisional plasmas using adaptive Cartesian mesh and automatic selection of kinetic and fluid solvers for transport processes. This is called Adaptive Mesh *and* Algorithm Refinement (AMAR) methodology. A cell-by-cell selection of most suitable solvers for different plasma species is based on continuum-breakdown criteria that can be specified in advance by the user.

The PIC module in UFS supplements already developed fluid models, mesh-based kinetic solvers (Boltzmann, Vlasov, Fokker-Plank) and the particle-based Direct Simulation Monte Carlo (DSMC) and Photon Monte Carlo (PMC) solvers previously developed for simulations of mixed rarefied-continuum flows, radiation transport and plasma dynamics.

Compared to neutral gases, plasma simulations pose extra challenges due to the disparity of time and spatial scales associated with small electron mass [4]. Criteria for selecting kinetic and fluid models for electrons, ions and neutrals are different, and strongly depend on plasma conditions. In collisionless magnetized plasmas, ions are often treated kinetically and electrons are assumed to be a fluid. In gas discharge physics, electrons are commonly treated kinetically, and fluid models applied

[1] To whom any correspondence should be addressed.

for ions and neutrals [5]. First steps towards adaptive kinetic-fluid plasma simulations have been made. Two-way coupling of a global Hall magnetohydrodynamics with a local implicit PIC model (MHD with Embedded PIC regions (MHD-EPIC)) has been recently demonstrated for space plasmas.[6]

In the present paper, we first provide some details of the UFS methodology and discuss parallel strategy for the AMAR codes. Then, we describe an initial implementation of the ES-PIC module in UFS focusing on challenges associated with AMR. Finally, validation of the ES-PIC solver for capacitively coupled plasma (CCP) and demonstration of its AMR capabilities for simulations of streamer breakdown are described. We show that cell-based AMR technique provides a convenient adaptive particle management algorithm for exponential multiplication of charged particles in PIC codes.

2. Unified Flow Solver

Basic architecture of UFS is shown in Figure 1. The AMAR core is implemented on top of Gerris Flow Solver (GFS) - an open source computing environment for solving partial differential equations with AMR [7]. GFS provides automatic mesh generation for complex geometries, portable parallel support using the MPI library, dynamic load balancing, and parallel offline visualization. GFS physics includes time-dependent incompressible variable-density Euler, Stokes or Navier-Stokes equations with volume of fluid method for interfacial flows. A coarse-grained parallelization algorithm is based on the "Forest of Trees" methodology [8].

UFS enables the higher degree of adaptation by using different physical models in different parts of computational domain. In particular, the computational domain could be decomposed into kinetic and fluid cells using physics-based continuum breakdown criteria. This methodology was first implemented for mixed rarefied-continuum flows and later extended for hybrid simulations of radiation transport coupling a Photon Monte Carlo (PMC) solver with a diffusion model of radiation transport selected based on the local photon mean free path.

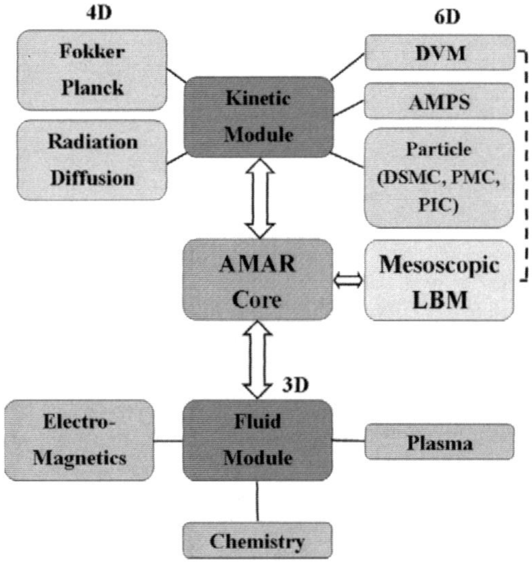

Figure 1: UFS Architecture.

The Kinetic Module in UFS can solve Boltzmann, Vlasov, and Fokker-Planck kinetic equations using Discrete Velocity Method (DVM). The recently developed Adaptive Mesh in Phase Space (AMPS) methodology [9] can adapt mesh in both physical and velocity spaces. In addition, the Kinetic Module has the particle-based DSMC [10] and PMC solvers [11]. The mesoscopic Lattice Boltzmann Method (LBM) uses a minimal set of discrete velocities as a subset of the DVM kinetic solvers [12].

The Fluid Module in UFS contains Euler and Navier-Stokes solvers for reacting gas mixtures using different numerical schemes. Flow solvers with adaptive Cartesian mesh offer a viable alternative to the conventional CFD methods using body-conforming mesh. The Immersed Boundary Method helps improving calculations of skin friction and heat fluxes for supersonic viscous flows [13].

For charged species (electrons and ions), multi-temperature drift-diffusion models are used, which are coupled to the Poisson solver for the electrostatic field [4].

2.1. Parallel Algorithms for Accelerated Heterogeneous Computing

Parallel capability is a must of today's codes. Modern systems often use GPU-accelerators for massively parallel tasks. Porting adaptive kinetic-fluid solvers to heterogeneous accelerator-based systems poses several challenges associated with irregular data structures of AMR, vastly different costs of computing in fluid and kinetic cells, and dynamic balancing of CPU and GPU loads during grid adaptations and model refinement.

UFS uses two methods of domain decomposition for parallel computing: a fine-grained method using Space Filling Curves (SFC) and a coarse-grained using Forest of Trees (FOT). The SFC algorithm is currently based on Morton order. After ordering of cells, a weight is assigned to each cell based on the node time required to perform computations in this cell. Kinetic cells have a much larger weight compared to the fluid cells. Furthermore, the array of cells modified with corresponding weights, is subdivided into sub-arrays equal to the number of nodes, in such a way that the weights of each sub-arrays are approximately the same. This method allows very precise and efficient domain decomposition and dynamic load balancing between different nodes.

The FOT algorithm consists of dividing the computational domain into boxes of fixed size with a Cartesian tree built in each box. Each tree becomes the smallest parallel subdomain. The trees are connected through their common boundaries. Graph partitioning algorithms are used for domain decomposition and Dynamic Load Balancing (DLB). This represents a coarse-grain parallelization algorithm since the CPU load balancing can be done only in terms of the building (root) boxes. Each node operates on a given number of boxes, and the computational grid is stored only on the node it belongs to. During grid adaptation these boxes are exchanged between the nodes for DLB. Each box has its own ID number and a set of boxes on each host node share the same Processor or node ID (PID) number. The box-based DLB has the benefit of efficient memory allocation because only the trees that belong to a node need to be kept in the node's memory.

It is important to emphasize key differences between computational load of cells in particle-based and grid-based kinetic solvers. In the particle-based solvers, the number of particles per cell should be kept approximately constant and the cell size should be about or smaller than the local mean free path or Debye length (for explicit PIC codes). Thus, the mesh adaptation for particle-based solvers does not increase the computational load if the particle density remains approximately the same throughout the computational domain. The computational load of mesh-based kinetic solvers is directly proportional to the number of cells in physical space. The weight of each kinetic cell (in physical space) is proportional to the number of cells in velocity space. Fluid cells have small weight compared to the kinetic cells. Adaptation of spatial mesh changes the cost of computations directly proportional to the number of cells in physical space with corresponding weights for the kinetic and fluid cells. These differences are of principal importance for the development of parallel algorithms for adaptive kinetic-fluid solvers based on particle- and mesh-based schemes.

So far, we have used the simplest approach where GPUs were applied for the kinetic cells whereas the fluid cells were computed on CPUs. This methodology has been recently demonstrated for simulations of mixed rarefied-continuum flow over a hypersonic scramjet on a CPU-GPU cluster using DVM Boltzmann and Navier-Stokes solvers [12].

3. An electrostatic Particle-in-Cell Module in UFS

We have implemented an electrostatic PIC module in UFS with a Monte Carlo Collision (MCC) algorithm describing collisions of electrons and ions with neutrals. Our multi-grid Poisson solver calculates the electric potential at cell centers. We first implemented the particle-field coupling via the nearest grid point (NGP) method to assign the particle space charges at cell centers. For calculation of the electric field at particle location, $\vec{E}_n(\vec{x}_n)$, we used bi-linear interpolation based on components of the electric field vector on cell nodes/corners. The AMR capabilities of the PIC module using the first-order accurate NGP technique are demonstrated below.

Both explicit and implicit PIC schemes have been implemented. Explicit PIC requires grid spacing to be smaller than ~3 Debye lengths ($\Delta x < 3\lambda_D$) and the time step to be smaller than the inverse electron plasma frequency. Implicit PIC can handle larger cell sizes and time steps. Larger cell sizes also allow using smaller numbers of particles while keeping acceptable particle per cell (PPC) parameter. We have chosen the direct implicit method of DIPIC [14]. In order to implement the DIPIC scheme in UFS, we have extended the Poisson solver with the implicit susceptibility. The implicit scheme allowed simulating higher density plasma with smaller number of particles (see below).

Having implemented and tested the NGP scheme, we used the more accurate cloud-in-cell (CIC) approximation to reduce noise (improve statistics). Both the momentum conserving scheme (CIC-EM) and energy-conserving scheme (CIC-EC) [15] have been implemented for uniform mesh. Challenges of CIC implementation on adaptive Cartesian mesh are discussed below.

3.1. Adaptive Cartesian Mesh

In terms of particle handling on the AMR grid, our PIC code follows the DSMC code previously described in Ref. [10]. A single grid is used for particle movement, collisions and visualization. The quad/octree data structure allows straightforward and efficient data management during dynamic grid refinement/coarsening and makes possible seamless parallelization of the code. However, particle movement/tracking becomes more involved compared to a uniform Cartesian grid. Particles are assigned to each (leaf) cell using cell-based pointers and the global particle-list is used only for easy bookkeeping purposes. The particle data storage in UFS-PIC is thus cell-based, which means that each cell has a list of particles assigned to it. When a particle crosses a face of a cell, it is removed from the list of its old cell and added to the list of a neighbour cell. Particles are moved inside each computational cell according to Newton's laws until they hit a cell face or a solid panel face. Particle tracing is efficient since all full cells are square or cubic. In cut-cells at non-conforming solid boundaries, the solid panel face is checked for possible reflection, and appropriate boundary conditions are applied when this face is crossed. Since a cell neighbour can be at a higher level, a new particle location is assigned depending on which part of the face has been hit by the particle.

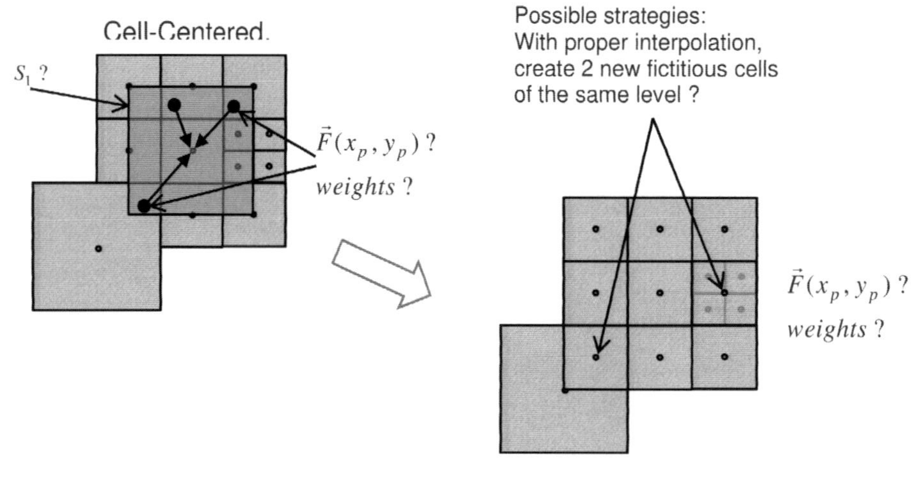

Figure 2. Illustration of the CIC method for a uniform grid (left) and adaptive grid (right). S_1 is the shape function used in the scatter step and $\vec{F}(x_p, y_p)$ function is used in the gather step.

No special treatment (such as assigning different weights) is carried out to account for different volumes of cells on the adapted grid. Moreover, no special treatment of small cut cells is performed.

The latter can lead to significant scattering of the surface data. To tackle this issue, a special routine is implemented which traverses all cut cells at a given coarse level (usually corresponding to initial uniform grid). Inside each such cell, all leaf cut cells are visited and the surface data are weight averaged with solid panel surface area. This procedure significantly reduces statistical scatter in the calculated surface fluxes, and provides similar advantages to a decoupling of surface elements from the grid structure. During refinement of a coarse cell, particles are stored in new cells according to their position in this coarse cell. During cell coarsening, particles in fine cells are assigned to a new coarse cell. Statistics counters are reset to zero when the AMR grid changes.

The implementation of CIC for cell-based adaptive Cartesian mesh is a subject of current research To derive the interpolation functions for the CIC approach with momentum and energy conservation constrains, as well as variable particle weights when particles travels between cells of different refinement levels, we have considered some possible strategies to application of the existing technologies to the adapted grids, which consists in creating fictitious (ghost) cells (see Figure 2) to mimic a uniform grid behavior with some proper field and particle interpolations. Such cells can be created in neighbor (both direct and diagonal) cells as finer (+1) and coarser (−1) level cells.

The charge conservation method requires using the same shape function during the scatter and gather steps. For a uniform grid, the shape function covers the region shown in Figure 2, left. For AMR, we use a similar shape function, which involves direct and diagonal neighbor cells of different levels of refinement (see Figure 2, right). Implementation of a better interpolation technique developed in [16] in our PIC-AMR code is expected to provide better accuracy. This technique allows a linear function defined on cell centers of the AMR grid to be ideally restored for any neighbor cell configuration (location and cell level).

3.2. MPI parallelization of the PIC module using FOT algorithm

The MPI parallelization of the PIC module follows closely that of the DSMC module. The FOT methodology is used with extra data exchange between space charges and acceleration vectors. Thus, good scaling and processor load have been observed using PIC with increasing number of processors for close-to-uniform spatial distributions of charged particles. For highly non-uniform spatial distributions of the particles and variable PPC number, implementing good DLB requires new graph partitioning algorithm taking into account different weight of cells (which is proportional to PPC number). It is expected that the FOT algorithm is preferable to SFC for the PIC code, because of multigrid method used by the Poisson solver.

In future work, we plan to implement DLB capabilities for particle-based kinetic solvers, which are required for adaptive grids and varying PPC number. The DLB module will use the box-exchange method and the graph partitioning algorithm currently available for the fluid solvers and is now solely based on the number of cells per partition. We will extend this algorithm for cells with different weights by implementing additional criteria for the graph partitioning algorithm.

3.3. Validation of the ES-PIC solver for CCP discharges

We have validated the UFS-PIC module for a Capacitively Coupled Plasma (CCP) benchmark described in a topical review [17]. This review discusses several issues of the PIC codes (sufficient representation of energetic particles, statistical noise, and numerical heating). The benchmark is a CCP discharge in Ar gas pressures 50 and 100 mTorr between two planar electrodes separated by a 3 cm gap driven by an RF voltage at a frequency of 15.36 MHz and voltage amplitude of 100 V.

The fidelity of PIC-MCC simulations strongly depends on the number of super-particles. The number of super-particles in our simulations for converged results was found to be ~150K for 50 mTorr and ~25–50K for 100 mTorr, which makes the spatially average PPC parameter of ~2,000 and ~300–600, respectively.

Figure 3 shows results of UFS-PIC simulations in terms of RF-cycle averaged spatial distributions of electron and ion densities, electron temperature and electrostatic potential, as well the electron energy distribution function (EEDF). Also shown are data from Ref. [17] for direct comparison. One can see that the agreement for all plasma parameters (electron and ion densities, electron temperatures and electrostatic potential) is very good for both pressures of 50 and 100 mTorr, in spite of the differences in the included electrode processes in these simulations. In our simulations, no electron emission from electrodes was assumed. The simulations in Ref. [17] have taken into account secondary electron emission from electrodes due to ion bombardment with a coefficient of 0.2. The latter can lead to the presence of energetic electrons in the space-charge sheath and to slightly different profiles of plasma densities and especially electron temperature in the sheaths. One can also observe that the electron temperature profiles at 50 and 100 mTorr are drastically different, which can be explained by the CCP operational mode transition taking place between 50 and 100 mTorr. The predicted RF-cycle averaged EEDFs are in excellent agreement with those obtained in Kim et al.[17].

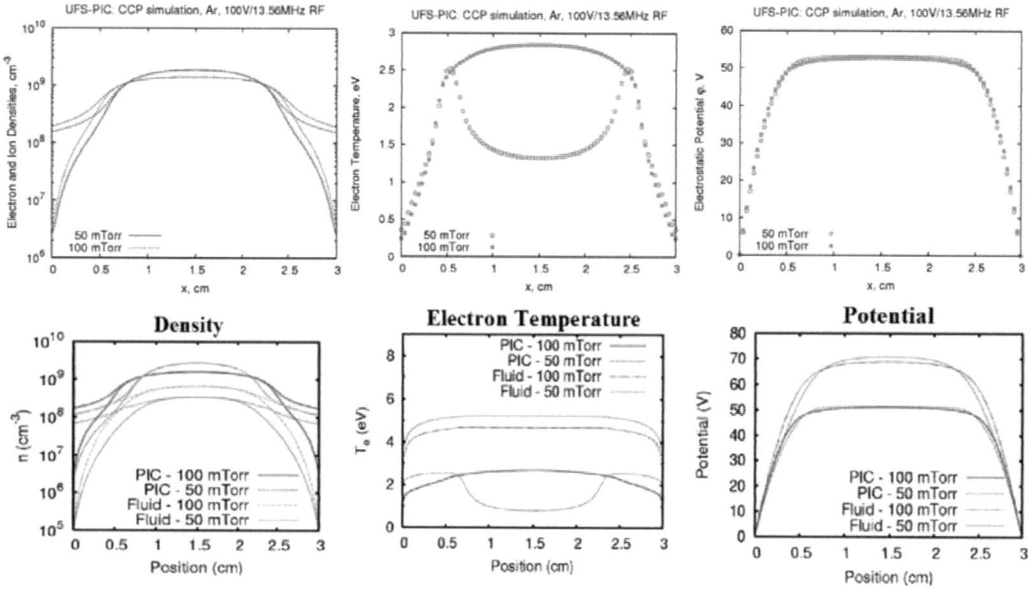

Figure 3. Comparison UFS-PIC results (top) with Kim et al.[17] results (bottom) for a benchmark CCP problem: RF-cycle averaged electron and ion densities (left), temperatures (center) and electric potential (right).

The accuracy of a PIC-MCC simulation strongly depends on the number of super-particles [17]. The best agreement with experimental data in Kim et al. [17], was obtained when using a sufficiently large number of super-particles of ~150K. This number of super-particles corresponds very well to our findings in the conducted convergence studies. Recall that we have also used ~150K super-particles for our 50 mTorr CCP simulations presented above. In Figure 4, we compare the UFS-PIC results obtained at different (spatially-average) PPC number of 2,000 (total number of particles ~150K), 750 and 350. One can see that only when (spatially average) PPC number reaches ~2,000 do the results converge and agree with the data in Kim et al. [17].

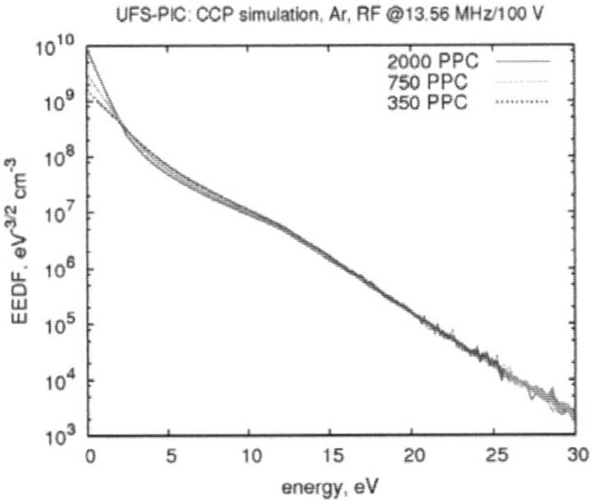

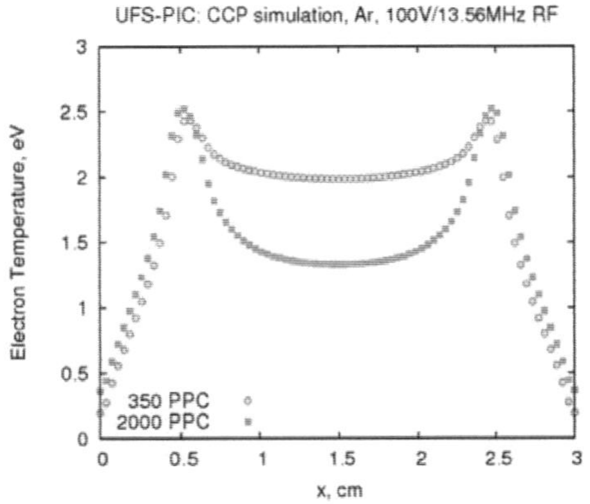

Figure 4. UFS-PIC simulation of CCP with different PPC parameters: RF-cycle averaged EEDF at plasma centre (left) and profiles of electron temperature (right).

The fact that PPC parameters of the order of a thousand are required provides a clear proof for the necessity of a hybrid approach for multi-dimensional simulations of realistic systems. Simulations of CCP discharges by a PIC method with such a large PPC make any realistic 3D and even 2D simulation impractical. It is expected that the low-energy electrons, controlling slow plasma processes (e.g., ambipolar diffusion, plasma densities, etc), can be described by alternative models (such as fluid models or Fokker-Planck kinetic solvers) while the high-energy electrons responsible for ionization can be described by the PIC methods. For a Maxwellian distribution with a temperature T_e, the ratio of high- to low energy electrons is proportional to $\exp(-\varepsilon_i / T_e)$, where ε_i is the ionization threshold.

4. Simulations of High-Pressure Gas Breakdown

We illustrate below 3D simulations of gas breakdown in Ar at 100 Torr with inter-electrode distance of 40 mm inspired by the experimental studies of high-pressure gas breakdown between a needle-like cathode and a planar anode [18] shown in Figure 5. It is seen how an ionization wave propagating from a pin cathode reaches the anode after ~10 ns, and then returns to the cathode after ~25 ns. We were able to simulate the initial stage of this process with our PIC and fluid models.

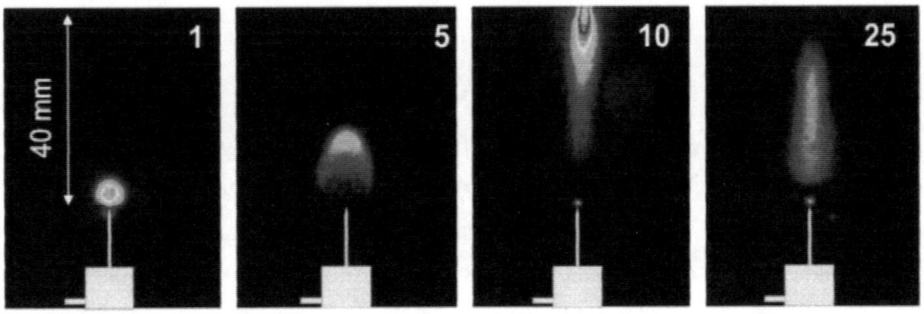

Figure 5. ICCD images of nanosecond discharge developing in air initiated by 400-ns HV pulse with short (11 ns) rise time. Numbers correspond to camera gate delay (in nanoseconds) after the pulse appearance on the HV electrode. Air, 680 Torr (adapted from Ref. [18]).

4.1. PIC simulations with AMR

Figure 6 shows results of 3D PIC simulations for applied voltages of 3.25 and 5 kV with the numbers of super-particles about 8M and 15M, respectively. At the high applied voltage, the breakdown occurs faster (~35 ns vs ~55 ns) and higher electron densities are attained, as expected. At the lower voltage, there is almost no branching, whereas quite significant branching occurs at the higher applied voltage. Similar branching was reported in simulations with uniform Cartesian mesh [19]. The AMR capabilities of the PIC are crucial for efficient simulations of the breakdown dynamics. However, even with AMR, we were not able to simulate breakdown at pressures above 100 Torr, and could not simulate the return strike observed in the experiments due to the very high computational cost of the PIC method under these conditions.

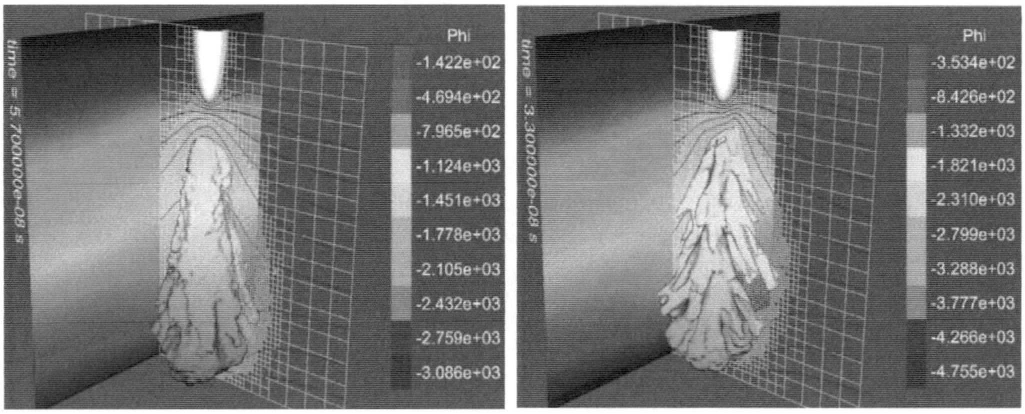

Figure 6. UFS-PIC simulation of breakdown for applied voltage of 3.25 kV (left) and 5 kV (right). Shown are isosurfaces of electron density (at 10^{16} m^{-3} and 5×10^{16} m^{-3} levels), adapted grids, and isolines of electrostatic potential at 57 (left) and 33 (right) ns.

4.2. Fluid Simulations

In addition to the PIC simulations described above, we have simulated the same problem using the fluid plasma model. The fluid model results shown in Figure 7 are quite different compared to the PIC results in Figure 6.

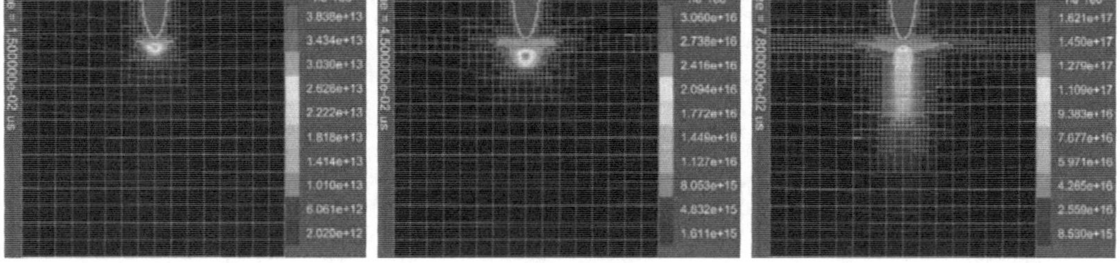

Figure 7. Fluid simulation of plasma breakdown in 100 Torr Ar gas with inter-electrode gap of 40 mm. Shown are profiles are electron density, adapted grids, and isolines of electrostatic potential at 3 time instances of 15, 45 and 78 ns.

In particular, the plasma channel predicted by the fluid model is narrower, and the time when the ionization wave reaches the anode surface (~90 ns) is longer than that predicted by the PIC. At 100

Torr, kinetic effects (such as non-local ionization by fast electrons, electron energy transport, etc.) are expected to play an important role in the breakdown process. Kinetic effects should also play a role at higher pressures in the high-field regions near the streamer front where the high-energy electrons are produced. Thus, hybrid PIC-fluid models are highly desirable for simulations streamer-spark-arc discharges. Using PIC only for the fast electrons in small areas near the steamer front and fluid model elsewhere can drastically reduce the computational cost without sacrificing the accuracy of simulations [19].

4.3 Discussion

One of the main problems in developing PIC-AMR codes is related to a decrease of the PPC number in the refined regions with a corresponding increase of the numerical noise.[20] The remedy for this problem may be the particle splitting-coalescence processes.[21] These processes should conserve such quantities as the charge and current densities, the total charge and mass of particles, the total momentum and energy of particles, and the distribution function of particles. For example, in the splitting algorithm in a 2D system, a particle is replaced by four particles. The particle merging algorithm has been refined and improved in Ref [22] using a k-d tree method for searching the nearest neighbors. Having efficient quad/octree generation and management (such as fast traversal/search) algorithms in the UFS framework, it would be relatively straightforward to implement such algorithms in our code. However, our particle handling on the AMR grid (see Section 3.1) could make this unnecessary.

Recently, a particle merging algorithm for problems where the number of particles grows exponentially in time has been proposed [23]. Such conditions commonly occur during gas breakdown. In our simulations of high-pressure streamer breakdown, the number of charged particles quickly increased from a few thousands to 10-20M during the breakdown process. Using dynamic AMR, the PPC parameter was kept at the same level (~500-1000). As often happens in plasma systems, refined cells are located in regions with small Debye length where the plasma density is large (AMR helps resolving the Debye length for PIC). Therefore, in most cases, the refined cells have sufficient number of particles per cell, and the particle splitting is not needed. In these cases, the AMR technique provides a convenient adaptive particle management algorithm.

In future work, we plan to use a hybrid solver where PIC model is used only in selected areas of computational domain whereas fluid model is used in other parts of the domain. Criteria for selecting kinetic patches could be based on the ratio of characteristic plasma scale (based on local gradients of plasma density) and the energy relaxation length for electrons (taking into electron collisions with atoms and the Coulomb interactions among charges particles).

5. Conclusions

We have described an initial implementation of an electrostatic Particle-in-Cell module with adaptive Cartesian mesh in our Unified Flow Solver framework. ES-PIC-AMR has been validated for a capacitively coupled plasma benchmark. Its AMR capabilities have been demonstrated for simulations of streamer development during high-pressure gas breakdown. The cell-based AMR technique provided a convenient adaptive particle management algorithm for problems with exponential multiplication of particles.

With the addition of the PIC module, we have now a full set of fluid and kinetic solvers (both grid-based and particle-based) that can be applied for hybrid plasma simulations. Future work will be devoted to identifying physics-based criteria for selecting appropriate kinetic and fluid solvers for different plasma components in different parts of the computational domain for applications requiring adaptive kinetic-fluid solvers.

Acknowledgments

This work was partially supported by the DARPA SBIR Project W31P4Q-15-C-0047 and by the US Department of Energy Office of Fusion Energy Science Contract DE-SC0001939.

References

1 Vay J-L, Grote D P, Cohen R H and Friedman A 2012, Novel methods in the Particle-In-Cell accelerator code-framework Warp, Computational Science & Discovery **5** 014019
2 Fujimoto K and Machida S 2006, Electromagnetic full particle code with adaptive mesh refinement technique: application to the current sheet evolution, J. Comput. Phys. **214** 550
3 Kolobov V I, Arslanbekov R R, Aristov V V, Frolova A A and Zabelok S A 2007, Unified solver for rarefied and continuum flows with adaptive mesh and algorithm refinement, J. Comput. Phys. **223** 589
4 Kolobov V I and Arslanbekov R R 2012, Towards adaptive kinetic-fluid simulations of weakly ionized plasmas, J. Comput. Phys. **231** 839
5 Kolobov V I 2013, Advances in electron kinetics and theory of gas discharges, Phys. Plasmas **20** 101610
6 Daldorff L K S *et al* 2014, Two-way coupling of a global Hall magnetohydrodynamics model with a local implicit particle-in-cell model, J. Comput. Phys. **268** 236
7 Gerris Flow Solver: http://gfs.sourceforge.net/
8 Agbaglah G *et al* 2011, Parallel simulation of multiphase flows using octree adaptivity and the volume-of-fluid method, Comptes Rendus Mécanique **339** 194
9 Arslanbekov R R, Kolobov V I and Frolova A A 2013, Kinetic solvers with adaptive mesh in phase space, Phys. Rev. E **88** 063301
10 Arslanbekov R R, Kolobov V I, Burt J and Josyula E 2012, Direct Simulation Monte Carlo with octree Cartesian mesh, AIAA 2012-2990
11 Mehta R, Kolobov V I, Arslanbekov R R et al 2011, Hybrid radiation transport model for weakly ionized plasma, AIAA-2011-3767
12 Zabelok S A, Arslanbekov R R and Kolobov V I 2015, Adaptive kinetic-fluid solvers for heterogeneous computing architectures, J. Comput. Phys. **303** 455
13 Arslanbekov R R, Kolobov V I and Frolova A A 2011, Analysis of compressible viscous flow solvers with adaptive Cartesian mesh, AIAA 2011-3381
14 Wang H, Jiang W and Wang Y 2010, Implicit and electrostatic particle-in-cell/Monte Carlo model in two-dimensional and axisymmetric geometry: I. Analysis of numerical techniques, Plasma Sources Sci. Technol. **19** 045023
15 Wang H *et al* 2014, On the energy conservation electrostatic particle-in-cell/Monte Carlo simulation: Benchmark and application to the radio frequency discharges, Chinese Phys. B **23** 035204
16 Borovikov D, Sokolov I V and Tóth G 2015, An efficient second-order accurate and continuous interpolation for block-adaptive grids, J. Comput. Phys. **297** 599
17 Kim H C, Iza F, Yang S S *et al* 2005, Particle and fluid simulations of low temperature plasma discharges: benchmarks and kinetic effects, J. Phys. D: Appl. Phys. **38** R283
18 Nudnova M M and Starikovskii A Yu 2008, Development of streamer flash initiated by HV pulse with nanosecond rise time, IEEE Trans. Plasma Science **36** 896
19 Li C *et al* 2012, A comparison of 3D particle, fluid and hybrid simulations of negative streamers, Plasma Sources Sci. Technol. **21** 055019
20 Fujimoto F 2011, A new electromagnetic PIC model with adaptive mesh refinement for high-performance parallel computation, J. Comput. Phys. **230** 8508
21 Lapenta G 2002, Particle rezoning for multidimensional kinetic particle-in-cell simulations, J. Comput. Phys. **181** 317
22 Teunissen J and Ebert U 2014, Controlling the weights of simulation particles: adaptive particle management using k-d trees, J. Comput. Phys. **259** 318
23 Luu P T, Tückmantel T and Pukhov A, Voronoi particle merging algorithm for PIC codes, http://arxiv.org/abs/1504.00636

A New Scheme To Solve The Two-Fluid Cosmic-Ray Magnetohydrodynamic Equations

Yuki Kudoh[1] **and Tomoyuki Hanawa**[2]

[1]Graduate School of Science, Chiba University, Yayoi-cho 1-33, Inage-ku, Chiba, 263-8522. Japan
[2]Center for Frontier Science, Chiba University, Yayoi-cho 1-33, Inage-ku, Chiba, 263-8522. Japan

E-mail: kudoyk.aplab@chiba-u.jp

Abstract. We examine the two-fluid cosmic-ray magnetohydrodynamic (CR MHD) equations to take account of the dynamical effects of cosmic rays (CRs) in the simplest form. For simplicity we assume that the pressure of the CRs is proportional to the energy density, $P_{\rm cr} = (\gamma_{\rm cr} - 1)E_{\rm cr}$, where $\gamma_{\rm cr}$ denotes the adiabatic index of CRs. We find the fully conservative form of the CR MHD equations and derive the Rankine-Hugoniot relation for a shock. One component of the CR MHD equations describes conservation of total number of CRs where the number density is defined as $\rho_{\rm cr} = P_{\rm cr}^{1/\gamma_{\rm cr}}$. We also find the Riemann solution, which provides us approximate Riemann fluxes fluxes for numerical solutions. We have also identified origin of spurious oscillation originating from the pressure balance mode across which the CR pressure has a jump while the total pressure is continuous. We propose the two step method to solve the CR MHD equations. We use 1D shock tube problems to compare our method with others.

1. Introduction

Cosmic rays, referred to CRs in the following, are important constituents in many astronomical objects such as the galactic disks, solar winds and supernova remnants. The energy density of CRs are comparable to those of magnetic field, turbulence and thermal gas. Thus CRs are expected to influence the dynamics of the interstellar gas. As pointed out by Parker [1,2], CRs inflate magnetic flux tubes to rise from the galactic plane. They also contribute to expansion of supernova remnants [3] and interaction of the solar system with the interstellar medium flow in the heliosphere [4].

The dynamical effects of CRs have been incorporated into the hydrodynamic (HD) equations under the fluid approximation in which CRs are assumed to have only pressure and energy density. The pressure of CRs is often assumed to be proportional to the energy density, $P_{\rm cr} = (\gamma_{\rm cr} - 1)E_{\rm cr}$, where $E_{\rm cr}$ and $\gamma_{\rm cr}$ denote the energy density and adiabatic index, respectively. The time evolution of CRs is described by the advection-diffusion equation. They are extended to the CR magnetohydrodynamic (MHD) equations by including the magnetic force and the induction equation. These CR MHD equations are a powerful tool in analyzing the dynamics of interstellar medium, though anisotropy and energy spectrum of CRs cannot be taken into account.

Although it has passed more than thirty years since the first formulation by Drury and Völk [5], some basic properties of the CR HD equations remain still unknown. When the diffusion term is omitted, the CR HD equations are hyperbolic and hence should have the Riemann solution, which are still unknown. Even the Rankine-Hugoniot relation at a shock front has not been obtained yet.

The deficit in our knowledge on the CR MHD equations are serious in solving them numerically since modern numerical technics are based on the Riemann solution. First the fully conservative form of the CR HD equations have not been known in the literature, although we solve the ordinary MHD equations in the conservative form. The work by the CR pressure has been treated as a source term in the literature, though it contains a spatial derivative. Second, the pressure balance mode, which appears as a new wave in the CR HD equations, is known to show a suspicious oscillation in the numerical solution [6]. We need the fully conservative form and the Riemann solution to find a smart way to solve them.

In this paper we derive the fully conservative form of the CR evolution. The newly derived equation describes the conservation of the *number* of CRs. We define the number density of CRs to be the $1/\gamma_{\rm cr}$-th power of $P_{\rm cr}$. Then we derive the Rankine-Hugniot relation at the shock front from the conservative form. We find that the ratio of the number density of CRs to the gas density is continuous at the shock front. We also find that numerical diffusion of CRs generates the oscillation from the pressure balance mode. Based on these new findings we propose a new scheme to solve the CR HD equations.

This paper is organized as follows. We analyze the CR MHD equations and derive the conservation form in §2. The extension to MHD is tedious but straight forward. The Rankine-Hugoniot relation and approximate Riemann solvers for CR HD equations are also derived in §2. The numerical results of 1D shock tube problems are shown in §3.

2. The two-fluid CR-MHD system

2.1. Basic equations

When the pressure of CRs, $P_{\rm cr}$, is taken into account, the MHD equations [7] are expressed as

$$\frac{\partial \rho}{\partial t} + \boldsymbol{\nabla} \cdot (\rho \boldsymbol{v}) = 0, \tag{1}$$

$$\frac{\partial}{\partial t}(\rho \boldsymbol{v}) + \boldsymbol{\nabla} \cdot \left[\rho \mathbf{vv} + \left(P_{\rm g} + \frac{B^2}{2}\right)\mathbf{I} - \mathbf{BB}\right] = -\boldsymbol{\nabla} P_{\rm cr}, \tag{2}$$

$$\frac{\partial}{\partial t}\left(\frac{P_{\rm g}}{\gamma_{\rm g}-1} + \frac{1}{2}\rho v^2 + \frac{B^2}{2}\right) + \boldsymbol{\nabla} \cdot \left[\left(\frac{\gamma_{\rm g}}{\gamma_{\rm g}-1}P_{\rm g} + \frac{1}{2}\rho v^2\right)\boldsymbol{v} - (\boldsymbol{v} \times \boldsymbol{B}) \times \boldsymbol{B}\right] = -\boldsymbol{v} \cdot \boldsymbol{\nabla} P_{\rm cr}, \tag{3}$$

$$\frac{\partial \boldsymbol{B}}{\partial t} = \boldsymbol{\nabla} \times (\boldsymbol{v} \times \boldsymbol{B}). \tag{4}$$

Here, ρ denotes the gas density, $\boldsymbol{v}$ does the velocity, $\boldsymbol{B}$ does the magnetic field, and $P_{\rm g}$ does the gas pressure. The gas is assumed to an ideal gas with the specific heat ratio, $\gamma_{\rm g} = 5/3$. The pressure force of CRs appear in equations (2) and (3), i.e., in the momentum conservation and energy conservation. The equation of continuity and induction equations are unchanged.

CRs are assumed to have no mass and to have the constant specific ratio, $\gamma_{\rm cr} = 4/3$. Then the energy conservation of CRs is expressed as

$$\frac{\partial}{\partial t}\left(\frac{P_{\rm cr}}{\gamma_{\rm cr}-1}\right) + \boldsymbol{\nabla} \cdot \left(\frac{\gamma_{\rm cr}}{\gamma_{\rm cr}-1}P_{\rm cr}\boldsymbol{v}\right) = \boldsymbol{v} \cdot \boldsymbol{\nabla} P_{\rm cr} + \boldsymbol{\nabla} \cdot \boldsymbol{F}_{\rm diff}, \tag{5}$$

$$\boldsymbol{F}_{\rm diff} = \kappa_\perp \boldsymbol{\nabla} P_{\rm cr} + \frac{(\kappa_\parallel - \kappa_\perp)}{|\boldsymbol{B}|^2}(\boldsymbol{B} \cdot \boldsymbol{\nabla} P_{\rm cr})\boldsymbol{B}, \tag{6}$$

by taking account of anisotropic diffusion. Here the symbols, $\kappa_\parallel$ and $\kappa_\perp$, denote the diffusion coefficients in the direction parallel and perpendicular to the magnetic field, respectively. For simplicity, we omit the diffusion in the following analysis, since it is evaluated separately in a numerical analysis, i.e., by use of an operator splitting method.

While equations (1), (2), and (4) are written in the conservation form, equations (3) and (5) are not. From the sum of equations (3) and (5), we obtain the total energy conservation,

$$\frac{\partial}{\partial t}\left(\rho H - P_{\rm T} + \frac{\boldsymbol{B}^2}{2}\right) + \boldsymbol{\nabla}\cdot\left[\rho H \boldsymbol{v} - (\boldsymbol{v}\times\boldsymbol{B})\times\boldsymbol{B}\right] = 0, \tag{7}$$

in the conservation form, where

$$H = \frac{\boldsymbol{v}^2}{2} + \frac{\gamma_{\rm g}}{\gamma_{\rm g}-1}\frac{P_{\rm g}}{\rho} + \frac{\gamma_{\rm cr}}{\gamma_{\rm cr}-1}\frac{P_{\rm cr}}{\rho}, \tag{8}$$

$$P_{\rm T} = P_{\rm g} + P_{\rm cr}. \tag{9}$$

Equation (5) is equivalent to

$$\frac{\partial}{\partial t}\rho_{\rm cr} + \boldsymbol{\nabla}\cdot(\rho_{\rm cr}\boldsymbol{v}) = 0, \tag{10}$$

where

$$\rho_{\rm cr} = P_{\rm cr}^{1/\gamma_{\rm cr}}. \tag{11}$$

Equation (10) denotes the conservation of CR particle *numbers*. Equation (11) implies that energy of each CR particle increases in proportion to $\rho_{\rm cr}^{\gamma_{\rm cr}-1} = \rho_{\rm cr}^{1/3}$ when compressed. It is consistent with the assumption that CRs consist of ultra-relativistic particles.

By replacing equations (3) and (5) with (7) and (10), we can express all the components of the CR MHD equations in the conservation form.

2.2. Rankine-Hugoniot Relation

The Rankine-Hugoniot relation, the jump condition at a discontinuity, is derived easily from the CR MHD equations in the conservation form; the fluxes should be constant across the discontinuity when seen in the comoving frame. In the following we consider a standing shock of which wave plane is normal to the x-axis. Then the Rankine-Hugoniot relation is expressed as

$$\rho v_x = \text{const.}, \tag{12}$$

$$\rho v_x^2 + P_{\rm T} + \frac{B_y^2 + B_z^2 - B_x^2}{2} = \text{const.}, \tag{13}$$

$$\rho v_x v_y - B_x B_y = \text{const.}, \tag{14}$$

$$\rho v_x v_z - B_x B_z = \text{const.}, \tag{15}$$

$$\rho H v_x + \left(B_y^2 + B_z^2\right) v_x - \left(v_y B_y + v_z B_z\right) B_x = \text{const.}, \tag{16}$$

$$v_x B_y - v_y B_x = \text{const.}, \tag{17}$$

$$v_x B_z - v_z B_x = \text{const.}, \tag{18}$$

$$\rho_{\rm cr} v_x = \text{const.}. \tag{19}$$

Combining equations (12) and (19) we obtain

$$\frac{\rho_{\rm cr}}{\rho} = \text{const.}, \tag{20}$$

across the shock. Here, the subscripts, x, y, and z specify the x-, y-, and z-components, respectively. We can derive also

$$\frac{D}{Dt}\left(\frac{\rho_{\rm cr}}{\rho}\right) = 0, \tag{21}$$

and hence

$$\frac{D}{Dt}\left(\frac{P_{\rm cr}}{\rho^{\gamma_{\rm cr}}}\right) = 0, \tag{22}$$

from equations (1) and (10). The ratio of the CR number density to the gas remains constant in a gas element.

The effects of CRs on the shock are summarized as follows. First CRs contribute to the normal component of the momentum flux and the energy flux. Thus the fast and slow modes are modified, while the Alfvén waves are not. The propagation speeds of the fast and slow modes are modified according to the change in the sound speed defined as

$$c_s \; = \; \left(\frac{dP_{\rm T}}{d\rho}\right)^{1/2} = \left(\frac{\gamma_{\rm g} P_{\rm g} + \gamma_{\rm cr} P_{\rm cr}}{\rho}\right)^{1/2} . \tag{23}$$

Second, the contact discontinuity is degenerate; both ρ and $\rho_{\rm cr}$ can be discontinuous while $P_{\rm T}$, $\boldsymbol{v}$ and $\boldsymbol{B}$ should be continuous. We define the entropy wave as the discontinuity in ρ and the pressure balance mode as that in $\rho_{\rm cr}/\rho$. The entropy is defined as

$$s = \ln P_{\rm g} - \gamma_{\rm g} \ln \rho, \tag{24}$$

in this paper.

2.3. Approximate Riemann Solvers

We can apply the Godunov method, the current standard in the grid based simulations, to the CR MHD equations since they have the conservation form as shown in the previous section. The HLL Riemann solver [8] , the simplest one among the Godunov type, can be easily derived from the conservation form. However, the HLL Riemann solver provides a diffusive solution as shown later. Thus we have developed a Roe-type solver [9] to reduce diffusion around the contact discontinuity. Although a solution of the latter is less diffusive, it suffers from an artificial oscillation originating from the pressure balance mode. The artificial oscillation can be suppressed when the advection of the pressure mode is evaluated separately. We describe the Roe-type Riemann solver and its modification to avoid the artificial oscillation.

In this subsection we show the Roe-type solver not for the CR MHD equations but for the CR HD in order to avoid complexity due to the existence of magnetic field. Inclusion of CRs changes nothing but the equation of state. Extension to the CR MHD equations is straight forward.

The 1D CR HD equations are expressed in the vector form,

$$\frac{\partial \boldsymbol{U}}{\partial t} + \frac{\partial \boldsymbol{F}}{\partial x} = 0, \quad \boldsymbol{U} = \begin{bmatrix} \rho \\ \rho v_x \\ \rho H - P_{\rm T} \\ \rho_{\rm cr} \end{bmatrix}, \quad \boldsymbol{F} = \begin{bmatrix} \rho v_x \\ \rho v_x^2 + P_{\rm T} \\ \rho H v_x \\ \rho_{\rm cr} v_x \end{bmatrix}, \tag{25}$$

where $\boldsymbol{U}$ and $\boldsymbol{F}$ denote the state and flux vectors, respectively. The Jacobi-matrix, $\boldsymbol{A} \equiv \partial \boldsymbol{F}/\partial \boldsymbol{U}$, has four eigenvalues,

$$\lambda_1 = v_x + c_s, \quad \lambda_2 = v_x, \quad \lambda_3 = v_x, \quad \lambda_4 = v_x - c_s, \tag{26}$$

and corresponding right eigenvecotrs,

$$\boldsymbol{r}_1 = \begin{bmatrix} 1 \\ v_x + c_s \\ H + c_s v_x \\ \dfrac{\rho_{\rm cr}}{\rho} \end{bmatrix}, \quad \boldsymbol{r}_2 = \begin{bmatrix} 1 \\ v_x \\ \dfrac{v_x^2}{2} + \dfrac{\rho_{\rm cr}}{\rho}\alpha \\ \dfrac{\rho_{\rm cr}}{\rho} \end{bmatrix}, \quad \boldsymbol{r}_3 = \begin{bmatrix} 0 \\ 0 \\ \alpha \\ 1 \end{bmatrix}, \quad \boldsymbol{r}_4 = \begin{bmatrix} 1 \\ v_x - c_s \\ H - c_s v_x \\ \dfrac{\rho_{\rm cr}}{\rho} \end{bmatrix}, \tag{27}$$

where the subscript, k, specifies the waves; $k = 1$ and 4 for the sound waves, $k = 2$ for the entropy wave, and $k = 3$ for the pressure balance mode [6]. The entropy wave and pressure balance mode denote the advection of s and $\rho_{\rm cr}/\rho$, respectively. We use these eigenvectors to decompose the spatial change in the state vector, $\boldsymbol{U}$.

The Roe-type numerical flux is expressed as

$$\boldsymbol{F}^{\rm Roe}_{j+1/2} = \frac{1}{2}\left(\boldsymbol{F}_{j+1} + \boldsymbol{F}_j - \sum_k w_k |\lambda_k| \boldsymbol{r}_k\right), \tag{28}$$

where

$$w_{1,4} = \frac{1}{2\bar{c}_s^2}(\Delta P_{\rm T} \pm \bar{\rho}\bar{c}_s \Delta v_x), \quad w_2 = \Delta\rho - \frac{\Delta P_{\rm T}}{\bar{c}_s^2}, \quad w_3 = \Delta\rho_{\rm cr} - \frac{\bar{\rho}_{\rm cr}}{\bar{\rho}}\Delta\rho. \tag{29}$$

Here the symbols with subscript j denote the values at the center of the j-th cell. The symbol with Δ denote the differences between the adjacent cells, while those with bar denote the Roe average,

$$\bar{\rho} = \sqrt{\rho_{j+1}\rho_j}, \quad \bar{v}_x = \frac{\sqrt{\rho_{j+1}}v_{x,j+1} + \sqrt{\rho_j}v_{x,j}}{\sqrt{\rho_{j+1}} + \sqrt{\rho_j}}, \quad \bar{H} = \frac{\sqrt{\rho_{j+1}}H_{j+1} + \sqrt{\rho_j}H_j}{\sqrt{\rho_{j+1}} + \sqrt{\rho_j}},$$

$$\bar{\rho}_{\rm cr} = \frac{\sqrt{\rho_{j+1}}\rho_{{\rm cr},j} + \sqrt{\rho_j}\rho_{{\rm cr},j+1}}{\sqrt{\rho_{j+1}} + \sqrt{\rho_j}}, \quad \bar{P}_{\rm T} = \frac{\sqrt{\rho_{j+1}}P_{{\rm T},j} + \sqrt{\rho_j}P_{{\rm T},j+1}}{\sqrt{\rho_{j+1}} + \sqrt{\rho_j}}.$$

The sound speed is defined as

$$\bar{c}_s^2 = (\gamma_{\rm g} - 1)\left(\bar{H} - \frac{\bar{v}_x^2}{2} - \frac{\bar{\rho}_{\rm cr}}{\bar{\rho}}\alpha\right), \quad \alpha = \frac{\gamma_{\rm g} - \gamma_{\rm cr}}{(\gamma_{\rm g} - 1)(\gamma_{\rm cr} - 1)}\frac{\Delta P_{\rm cr}}{\Delta \rho_{\rm cr}}. \tag{30}$$

The numerical flux given by equation (28) satisfies the property U,

$$\boldsymbol{U}_{j+1} - \boldsymbol{U}_j = \sum_{k=1}^{4} w_k \boldsymbol{r}_k, \quad \boldsymbol{F}_{j+1} - \boldsymbol{F}_j = \sum_{k=1}^{4} \lambda_k w_k \boldsymbol{r}_k. \tag{31}$$

We obtain the solution of the first order accuracy,

$$\boldsymbol{U}_j^{n+1} = \boldsymbol{U}_j^n - \frac{\Delta t}{\Delta x}\left(\boldsymbol{F}^{\rm Roe}_{j+1/2} - \boldsymbol{F}^{\rm Roe}_{j-1/2}\right), \tag{32}$$

where $\boldsymbol{U}_j^{n+1}$ and $\boldsymbol{U}_j^n$ denote the state vector at the $n+1$-th and n-th time steps, respectively. From $\boldsymbol{U}_j^n$ we obtain the primitive variables, ρ, v_x, $P_{\rm g}$, and $P_{\rm cr}$ at the n-th time step.

As mentioned earlier and shown later, a small oscillation of numerical origin arises from the pressure balance mode in a solution given by equation (32). In oder to remove the oscillation we propose a two step approach in which the advection of CRs is taken into account in the second step. In the first step we use the modified numerical flux,

$$\boldsymbol{F}^*_{j+1/2} = \frac{1}{2}\left(\boldsymbol{F}_{j+1} + \boldsymbol{F}_j - \sum_k w_k |\lambda_k| \boldsymbol{r}'_k\right), \tag{33}$$

$$\boldsymbol{r}'_2 = \begin{bmatrix} 1 \\ \bar{v} \\ \dfrac{\bar{v}^2}{2} + \dfrac{\bar{\rho}_{cr}}{\bar{\rho}}\alpha \\ 0 \end{bmatrix}, \quad \boldsymbol{r}'_3 = \begin{bmatrix} 0 \\ 0 \\ \alpha \\ 0 \end{bmatrix}, \quad \boldsymbol{r}'_k = \boldsymbol{r}_k \text{ (otherwise)}. \tag{34}$$

Using this modified numerical flux, we obtain the intermediate state,

$$\boldsymbol{U}^*_j = \boldsymbol{U}^n_j - \frac{\Delta t}{\Delta x}\left(\boldsymbol{F}^*_{j+1/2} - \boldsymbol{F}^*_{j-1/2}\right) = \begin{bmatrix} \rho^{n+1}_j \\ \rho^{n+1}_j v^{n+1}_{x,j} \\ \dfrac{\rho^{n+1}_j}{2}\left(v^{n+1}_{x,j}\right)^2 + \dfrac{P^{n+1}_{\mathrm{g},j}}{\gamma_{\mathrm{g}} - 1} + \dfrac{P^{n+1}_{\mathrm{cr},j}}{\gamma_{\mathrm{cr}} - 1} \\ \rho^*_{\mathrm{cr},j} \end{bmatrix}. \tag{35}$$

In the first step we obtain $\rho^*_{\mathrm{cr},j}$ and $v^{n+1}_{x,j}$ from $\boldsymbol{U}^{n+1}_j$.

In the second step we evaluate $P^{n+1}_{\mathrm{cr},j}$ and then $P^{n+1}_{\mathrm{g},j}$ and $\rho^{n+1}_{\mathrm{cr},j}$ by the following procedures. First we obtain

$$\rho^\dagger_{\mathrm{cr},j} = \rho^*_{\mathrm{cr},j} - \frac{\Delta t}{2\Delta x}\left(\bar{v}_{x,j+1/2} q_{j+1/2} + \bar{v}_{x,j-1/2} q_{j-1/2}\right), \tag{36}$$

$$q_{j+1/2} = \rho_{\mathrm{cr},j+1} - \rho_{\mathrm{cr},j} - \frac{\bar{\rho}_{\mathrm{cr},j+1/2}}{\bar{\rho}_{j+1/2}} \frac{\left(P_{\mathrm{T},j+1} - P_{\mathrm{T},j}\right)}{\bar{c}^2_{s,j+1/2}}. \tag{37}$$

to evaluate the change in the CR density by advection. Note that equation (37) denotes the spatial change in ρ_{cr} to be advected. We evaluate the advection of CRs by using

$$P^{n+1}_{\mathrm{cr},j} = \left(\rho^\dagger_{\mathrm{cr},j}\right)^{\gamma_{\mathrm{cr}}} - \frac{\Delta t}{\Delta x}\left[\min(\bar{v}_{x,j+1/2}, 0) q'_{x,j+1/2} + \max(\bar{v}_{x,j-1/2}, 0) q'_{j-1/2}\right], \tag{38}$$

$$q'_{j+1/2} = q_{j+1/2} \frac{P_{\mathrm{cr},j+1} - P_{\mathrm{cr},j}}{\rho_{\mathrm{cr},j+1} - \rho_{\mathrm{cr},j}}. \tag{39}$$

In other words we respect the conservation of CR energy rather than that of CR numbers about the advection. We evaluate $\rho^{n+1}_{\mathrm{cr},j}$ from $P^{n+1}_{\mathrm{cr},j}$ and $P^{n+1}_{\mathrm{g},j}$ from the third component of $\boldsymbol{U}^*_j$. This new scheme is referred to the two step method in the following.

3. Numerical Tests

We use the 1D shock tube problem to examine the validity of the numerical methods shown in the previous section. We use the explicit scheme of the first order accuracy for time integration to show the dependence on the Riemann solver directly. All the test problems are solved on a uniform grid and with a constant time step. Subsections 3.1 and 3.2 are dedicated to CR HD and CR MHD problems, respectively.

3.1. CR HD Shock Tube Problems

First we examine the case when the left side has a higher pressure and a higher density. Table 1 summarizes the gas pressure, the CR pressure and the density in the initial state. This problem is similar to the shock tube problem in pure hydrodynamics but the the CR pressure is taken into account. The Riemann solution is illustrated in Figure 1. The shock wave propagates with the phase speed, $\lambda_1 = 2.36684$ and the head of the rarefaction wave with that of $\lambda_4 = -2.16025$. The contact discontinuity and pressure balance mode move at $\lambda_2 = \lambda_3 = 1.56214$.

Table 1. The initial state in the CR HD shock tube problem.

	ρ	v_x	P_g	P_{cr}
left ($x < 0$)	1.0	0.0	2.0	1.0
right ($x \geq 0$)	0.2	0.0	0.02	0.1

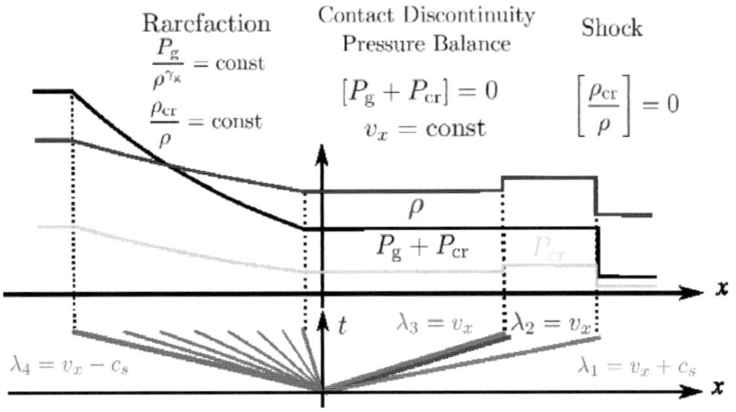

Figure 1. Illustration of the Riemann solution for the CR HD shock tube problem.

Figure 2 shows the numerical solution obtained with the two step method on the grid of $\Delta x = 1/128$. The time step is constant at $\Delta t = 2.0 \times 10^{-3}$ so that the CFL number is 0.61 at the shock front. The thick curves and dots denote the numerical solution while the thin lines do the analytic Riemann solution. The upper panels denote the density (ρ), velocity (v_x), and entropy (s) at $t = 0.1$, i.e. 50 time steps from the initial, as a function of x from left to right. The lower left panel denote the gas pressure (P_g) and the total pressure (P_T). The lower middle and lower right panels denote the CR pressure (P_{cr}) and the ratio of the CR number density to gas density (ρ_{cr}/ρ), respectively. The numerical solution reproduces the analytic Riemann solution well with the first order accuracy in space and in time. The velocity at the contact discontinuity is $v_x = 1.565$ in the numerical solution and $v_x = 1.562$ in the analytic solution. The total pressure at the contact discontinuity is $P_T = 0.8609$ in the numerical solution and $P_T = 0.8595$ in the analytic solution. The relative errors are about 0.2 %. Both v_x and P_T are fairly flat between the the shock front and the tail of the rarefaction. Note that the CR number density to the gas density, ρ_{cr}/ρ is constant except for the jump at the pressure balance mode. Also the gas entropy is constant except for the jump at the shock front and pressure balance mode.

We have solved the same problem with various methods. The results are shown in Figure 3. All the solutions are obtained on the same grid with the same time step for fair comparison. The

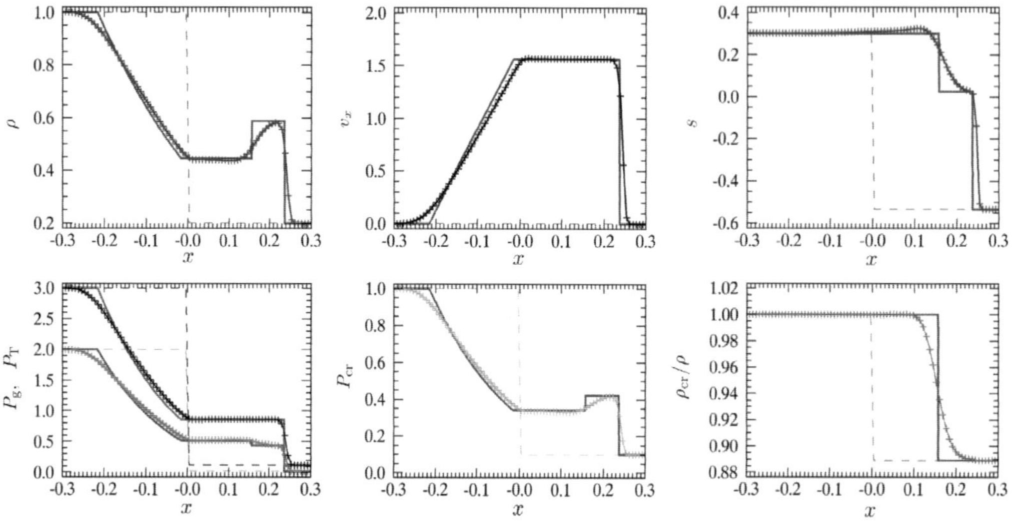

Figure 2. The numerical solution of the CR HD shock tube problem compared with the analytic Riemann solution (thin black solid curves). The numerical solution is obtained with the two step method. The upper panels show ρ, v_x and s, while the lower do P_g, P_T, P_{cr} and ρ_{cr}/ρ. The dashed lines denote the initial values.

upper panels show P_g (left) and s (right), while the lower do P_{cr} (left) and ρ_{cr}/ρ (right). The red and green curves denote the solutions obtained with the CR HD equations in the conserved form while the blue and purple do those obtained with the CR HD equations with some source terms, i.e., in the original form. The source term is $\boldsymbol{v} \cdot \boldsymbol{\nabla} P_{cr}$ in the solution denoted by the blue curves while it is $P_{cr} \boldsymbol{\nabla} \cdot \boldsymbol{v}$ in that denoted by the purple ones. The former scheme is similar to Kuwabara et al. [10], while the latter is similar to Hanasz et al. [11], Rasera and Chandran [12], and Yang et al [13]. The red curves denote the solution obtained with the two step method, while the other solutions are obtained with the HLL flux. All the panels show only the interval of $-0.05 \leq x \leq 0.30$ and the rarefaction waves are omitted in the panels.

The deviation from the Riemann solution is most serious in the solution denoted by the blue curves. The propagation of the shock wave is slowest mainly due to numerical diffusion of CR particles from the pressure balance mode. The CR HD equations denotes the adiabatic compression of CRs while they do not take account of diffusive shock acceleration (DSA). DSA term should be taken into account separately. The numerical diffusion of CRs is also serious in the solution denoted by the purple curves.

The HLL solution of the CR HD equations in the conserved form (green) provides a better estimate on the shock propagation. However it shows a spurious hump in P_g around the contact discontinuity. The hump is smaller in the solution obtained with the two step method (red).

Roe type flux and the two step method give us almost the same result for the test problem shown in Figure 2. This is because the pressure balance mode has only a small amplitude in the test problem. The second step used in the two step method is important to solve the pressure balance mode. In order to demonstrate the importance we have solved the advection of the pressure balance mode. Both the left and right side has the same total pressure in this advection test while the CR pressure is higher in the left. The velocity is taken to be uniform at $v_x = 1.0$ as summarized in Table 2.

Figure 4 shows the numerical solutions at the first three time steps in the advection test. The

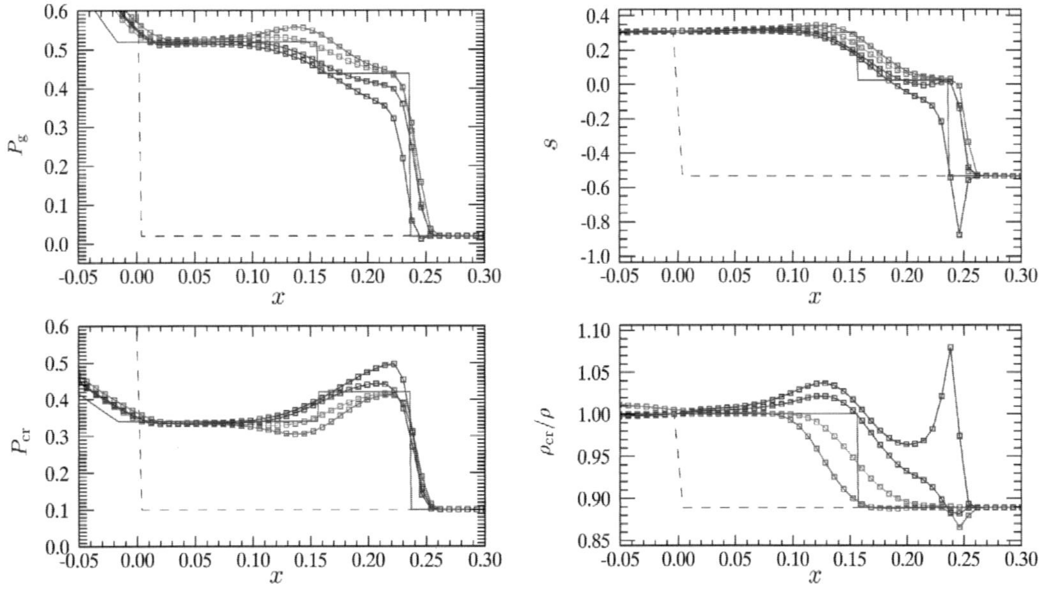

Figure 3. Comparison of the numerical solutions of the CR HD shock tube problem. The green and red curves denote the solutions obtained with the CR HD equations in the conservation form while the blue and purple curves do those obtained with the equations with source terms. The spatial grid and time step are the same as those used in the solution shown in Fig. 2

Table 2. The initial state in the test problem of the advection of pressure balance mode.

	ρ	v_x	P_g	P_{cr}
left ($x < 0$)	1.0	1.0	0.1	1.0
right ($x \geq 0$)	1.0	1.0	1.0	0.1

upper, middle and lower panels denote the solutions at the initial, the first, and the second time steps, respectively. The left panels denote P_g and P_{cr}. The middle and right panels denote P_T and v_x, respectively. The green, purple and red lines denote the solutions obtained with HLL, Roe and the two step method, respectively. HLL and Roe schemes produce spurious increase in P_T at the first step and spurious change in v_x at the second step.

The spurious increase in P_T is due to numerical diffusion in P_{cr}. The inevitable numerical diffusion modifies equation (10) into

$$\frac{\partial}{\partial t}\rho_{cr} + \nabla \cdot (\rho_{cr}\boldsymbol{v}) = \nabla \cdot (\eta \nabla \rho_{cr}), \tag{40}$$

where η denotes the effective diffusion coefficient. We can obtain

$$\frac{\partial}{\partial t}\left(\frac{P_{cr}}{\gamma_{cr}-1}\right) + \nabla \cdot \left[\frac{\gamma_{cr}}{\gamma_{cr}-1}P_{cr}\boldsymbol{v} - \eta\nabla\left(\frac{P_{cr}}{\gamma_{cr}-1}\right)\right] = \boldsymbol{v}\cdot\nabla P_{cr} - \frac{\eta}{\gamma_{cr}P_{cr}}\left|\nabla P_{cr}\right|^2, \tag{41}$$

from equation (40). The second term in the right hand side denotes decrease in P_{cr} and hence that in the CR energy density in equation (41). Since the total energy is conserved, the decrease

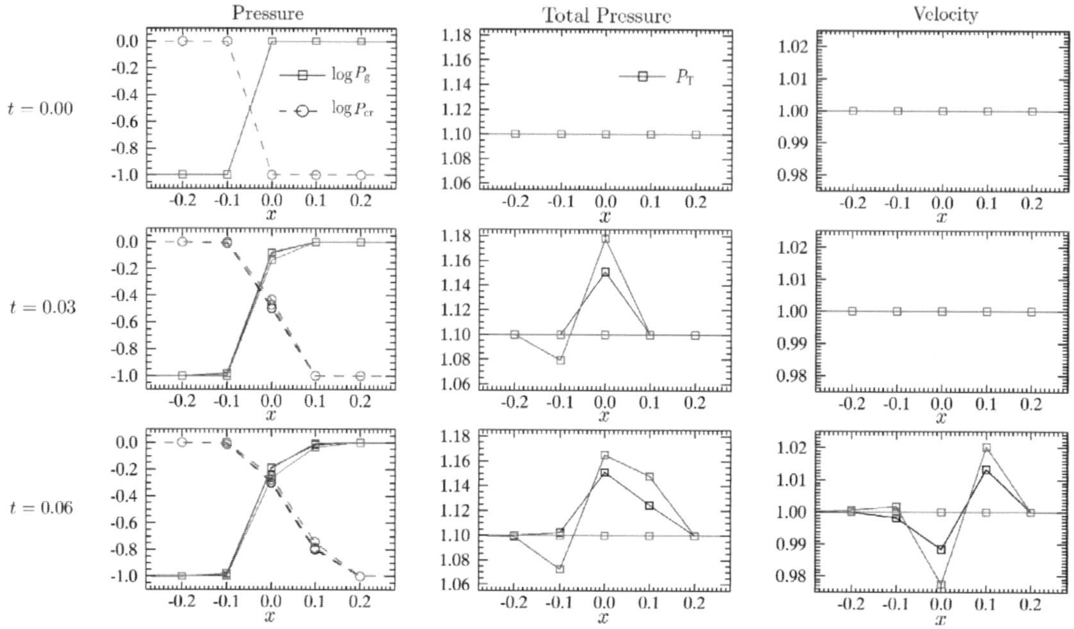

Figure 4. The first three time steps are shown from top to left in the advection test. The left panels denote P_g (solid) and P_{cr} (dashed). The middle and right panels denote P_T and v_x, respectively. The colors indicate the numerical schemes used. The green, black and red lines denote HLL, Roe and the two step method, respectively.

in the CR energy is supplemented by increase in the gas energy. Hence the gas pressure increases to increase the total pressure. Figure 5 illustrates this mechanism. Equation (10) gives a good solution for a shock but not for the pressure balance mode. Thus we need the two step method.

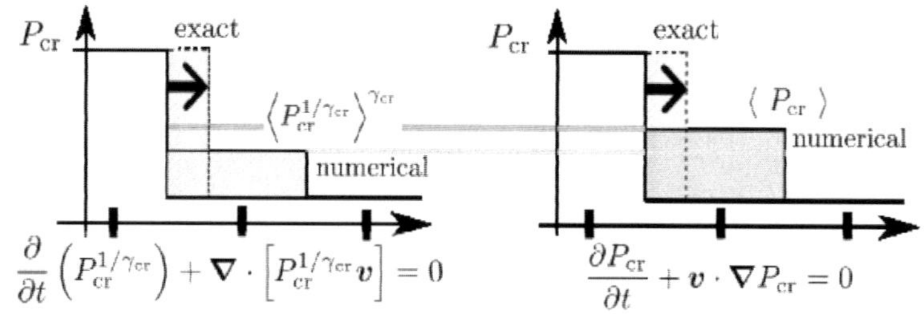

Figure 5. Illustration of the mechanism for the decrease in P_{cr} due to numerical diffusion.

3.2. MHD Shock Tube Test

We can extend the two step method for CR MHD equations. In this subsection we demonstrate that our two step method works well in a MHD shock tube problem. The initial state in our MHD shock tube problem is summarized in Table 3. It is similar to the problem of Brio & Wu [14], while the CR pressure is taken into account in ours.

Figure 6 shows the solution obtained with the two step method. The solution is obtained on the uniform grid of $\Delta x = 1/512$ with the constant time step of $\Delta t = 4 \times 10^{-4}$. The solid

Table 3. Initial left- and right-state in the shock tube problem of modified Brio & Wu [14].

	ρ	v_x	v_y	v_z	B_x	B_y	B_z	P_g	P_{cr}
Left ($x < 0$)	1.0	0.0	0.0	0.0	1.0	1.0	0.0	1.0	0.4
Right ($x \geq 0$)	0.125	0.0	0.0	0.0	1.0	-1.0	0.0	0.1	0.04

curves denote the solution at $t = 0.08$, while the dashed lines do the initial state ($t = 0$). We find the fast rarefaction propagating rightward in the range of $0.26 \leq x \leq 0.35$ and that propagating leftward in $-0.18 \leq x \leq -0.05$. The slow shock located at $x = 0.14$ is obtained without numerical oscillation. The gas entropy (s) remains constant except at the slow shock, contact discontinuity ($x = 0.07$) and the edge of the slow compound ($x = -0.04$). The ratio of the CR density to gas density (ρ_{cr}/ρ) changes only at the contact discontinuity. Also the transverse velocity (v_y) and magnetic field (B_y) are obtained successfully without any numerical oscillations.

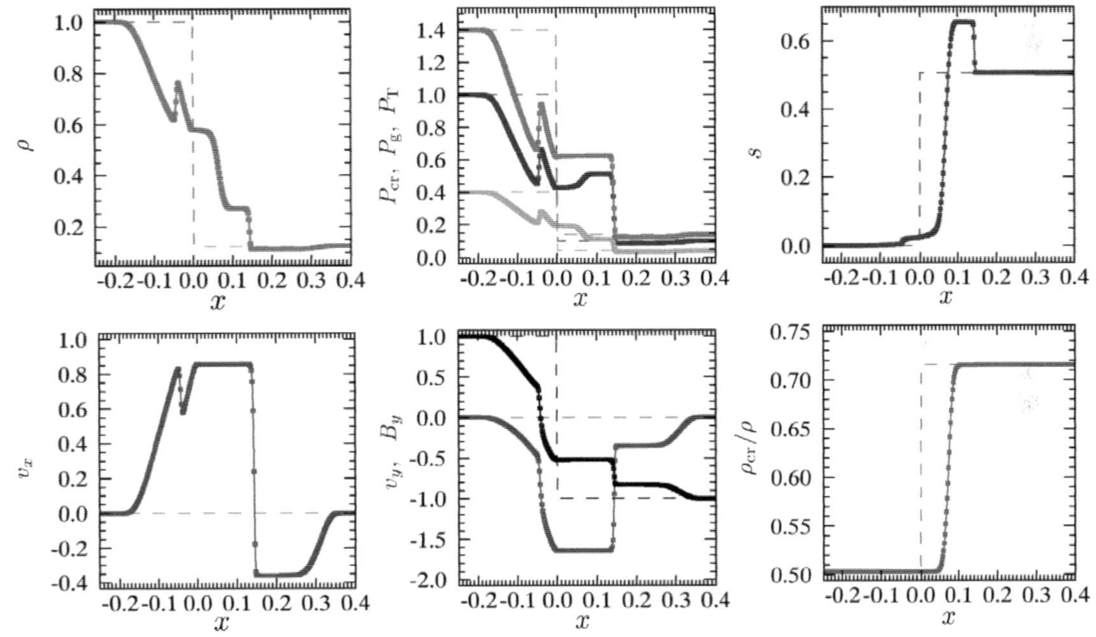

Figure 6. An example of CR MHD shock tube problem. The upper panels denote ρ, P_g, P_{cr}, P_T, and s. The lower panels denote v_x, v_y, B_y, and ρ_{cr}/ρ.

4. Conclusion

We have examined the CR MHD equations and found the Rankine-Hugoniot relation for a shock. The CR MHD equations have the fully conservative form which includes the conservation of CR numbers. It means that they describe *adiabatic* compression of CRs at the shock. We need to another term in the CR MHD equations if we take account of the diffusive shock acceleration. The Riemann solution and Roe-type Riemann solver are obtained for the CR MHD equations. We find that numerical diffusion of CRs generates spurious oscillation from the pressure balance mode. The oscillation can be removed by our two step method.

We thank Ryoji Matsumoto and Yosuke Matsumoto for fruitful discussions and continuous encouragement.

References

[1] Parker E N 1966 *ApJ* **145** 811
[2] Parker E N 1967 *ApJ* **149** 535
[3] Girichidis P, Naab T, Walch S, and Hanasz M 2014 arXiv:1406.4861
[4] Fahr H-J, Kausch T, and Scherer H 2000 *A&A* **357** 268
[5] Drury L O and Völk H J 1981 *ApJ* **248** 344
[6] Webb G M, Brio M, Zank G P, and Story T 1995 *ApJ* **422** 822
[7] Berezinskii V S, Bulanov S V, Dogiel V A, and Ptuskin V S 1990 *Astrophysics of cosmic rays* (Amsterdam: North-Holland)
[8] Harten A, Lax P D, and van Leer B 1983 *SIAM Rev.* **25** 35
[9] Roe P L 1981 *J. Comput. Phys.* **43** 357
[10] Kuwabara T, Nakamura K, and Ko C M 2004 *ApJ* **607** 82
[11] Hanasz M, Wóltański D, and Kowalik K 2009 *ApJL* **706** L155
[12] Rasera Y and Chandran B 2008 *ApJ* **685** 105
[13] Yang H-Y K, Ruszkowski M, Ricker P M, Zweibel E, and Lee D 2012 *ApJ* **761** 185
[14] Brio M, and Wu C C 1988 *J. Comput. Phys.* **75** 400

SpectralPlasmaSolver: a Spectral Code for Multiscale Simulations of Collisionless, Magnetized Plasmas

Juris Vencels[1], Gian Luca Delzanno[1], Gianmarco Manzini[1], Stefano Markidis[2], Ivy Bo Peng[2], Vadim Roytershteyn[3]

[1]Los Alamos National Laboratory, Los Alamos, NM, USA
[2]KTH Royal Institute of Technology, Stockholm, Sweden
[3]Space Science Institute, Boulder, CO, USA

E-mail: {vencels,delzanno,manzini}@lanl.gov,{markidis, bopeng}@kth.se, vroytershteyn@spacescience.org

Abstract. We present the design and implementation of a spectral code, called *SpectralPlasmaSolver* (SPS), for the solution of the multi-dimensional Vlasov-Maxwell equations. The method is based on a Hermite-Fourier decomposition of the particle distribution function. The code is written in Fortran and uses the PETSc library for solving the non-linear equations and preconditioning and the FFTW library for the convolutions. SPS is parallelized for shared-memory machines using OpenMP. As a verification example, we discuss simulations of the two-dimensional Orszag-Tang vortex problem and successfully compare them against a fully kinetic Particle-In-Cell simulation. An assessment of the performance of the code is presented, showing a significant improvement in the code running-time achieved by preconditioning, while strong scaling tests show a factor of 10 speed-up using 16 threads.

1. Introduction

Collisionless, magnetized plasmas are ubiquitous in a variety of laboratory, space and astrophysical environments. They are described by kinetic theory and are characterized by high-dimensionality, since the particle distribution function is defined in the six-dimensional phase space obtained by the three spatial and three velocity coordinates, and strong anisotropy, since transport properties can be very different along and across the magnetic field [1]. In addition, collisionless plasmas are characterized by extreme separation of spatial and temporal scales. This occurs already at the kinetic level due to the large mass difference between electrons and ions. However, the scale separation widens even further if one compares the kinetic scales with typical macroscopic scales of the system: several orders of magnitude in spatial and temporal scale separation are common for instance in magnetic fusion energy experiments or in the Earth's magnetosphere. Even with the most advanced supercomputers, fully-kinetic simulations of these systems are unfeasible today and in the foreseeable future.

On the other hand, fluid models treat the plasma as a macroscopic fluid and are therefore suitable for system-scale simulations. They are obtained from kinetic theory by taking moments of the distribution function. A kinetic closure is necessary to close the moment equations but it can be rigorously justified only in very restrictive limits [2], which are often not met in practice. Indeed, performing system-scale simulations of collisionless (or weakly collisional) plasmas that properly include the kinetic/microscopic physics is a major challenge of computational plasma physics. One of the recent approaches to this problem is to embed a kinetic solver in a small part of the computational domain of a fluid simulation [3, 4, 5].

The most common approaches for the numerical solution of the kinetic equations handle the discretization of phase space by either introducing macro-particles (as in the Particle-In-Cell (PIC) approach [6, 7]) or a computational grid (Eulerian-Vlasov [8, 9]) and therefore treat the full distribution function everywhere in the computational domain. Adaptive mesh refinement techniques can be used effectively to reduce the number of degrees of freedom for these kinetic solvers [10, 11, 12]. A third approach is spectral, namely the velocity part of the distribution function is expanded in basis functions (typically Hermite or Fourier) [13]. In certain cases, such as for a proper Hermite basis [14, 15] or the Legendre basis [16], it can be shown that the low-order moments of the expansion correspond to the fluid moments and the kinetic physics is recovered by simply adding more moments to the expansion. Thus, for a suitable spectral basis the fluid-kinetic coupling is built-in.

In recent years there has been a renewed interest on spectral methods for the kinetic equations [15, 17, 18, 19, 20, 21]. A comparison between the one-dimensional, electrostatic, Hermite spectral method and PIC on standard kinetic theory problems such as Landau damping, two-stream instabilities and ion acoustic waves was presented in Ref. [19], where it was concluded that, at least on these problems, the spectral method was orders of magnitude more accurate or faster than PIC. Vencels *et al.* [15] used the same method and performed kinetic simulations where the number of Hermite modes could be changed dynamically at run-time, thus providing the transition between fluid and kinetic regimes. An extension of the spectral method to the multi-dimensional, fully electromagnetic case is presented in [20], where exact conservation laws in the discrete are derived and some preconditioning techniques are presented to speed-up the convergence of the Newton-Krylov solver adopted for the nonlinear equations. These previous works also point out that the optimization of the Hermite basis (in order to reduce the number of modes and improve the computation time and robustness of the method), parallelization and efficient preconditioning techniques are critical needs for these methods.

In this paper we present the Hermite-based spectral solver, called SpectralPlasmaSolver (SPS), for the multi-dimensional Vlasov-Maxwell equations and discuss the design and implementation of the code. SPS is written in Fortran and parallelized with OpenMP for shared-memory systems. It uses open-source libraries such as *PETSc* [22] for nonlinear solvers and preconditioning and FFTW [27] for the convolutions.

The paper is organized as follows. Section 2 presents the governing equations and the Hermite-Fourier method. Section 3 discusses the design and implementation of the SPS code. Section 4 focuses on the application of the method to a two-dimensional Orszag-Tang vortex problem, often used as a paradigm to study plasma turbulence. Scaling tests on a shared-memory system are presented in Section 5, while conclusions are drawn in Section 6.

2. The Hermite-Fourier Spectral Method

We model a collisionless, magnetized plasma by solving the Vlasov-Maxwell equations. We consider a Cartesian system of coordinates with the spatial domain defined by $[0, L_x] \times [0, L_y] \times [0, L_z]$ (L_x, L_y, L_z are the domain lengths in each spatial direction) and assume periodic boundary conditions. We use the following normalization: time is normalized to the electron plasma frequency $\omega_{pe} = \sqrt{\frac{e^2 n_0}{\varepsilon_0 m_e}}$ (where e is the elementary charge, m_e is the electron mass, ε_0 is vacuum permittivity, and n_0 is a reference density), velocity to the speed of light c, the spatial coordinate to the electron inertial length $d_e = c/\omega_{pe}$, the magnetic field to a reference magnetic field B_0, and the electric field to cB_0. The distribution function f is normalized as fc^3/n_0.

Accordingly, the dimensionless Vlasov equation reads:

$$\frac{\partial f_s}{\partial t} + \mathbf{v}\cdot\nabla_x f_s + \frac{q_s}{e}\frac{\omega_{cs}}{\omega_{pe}}\left(\mathbf{E}+\mathbf{v}\times\mathbf{B}\right)\cdot\nabla_v f_s = 0, \tag{1}$$

where s labels the plasma species ($s = e,\ i$ for electrons and ions), the cyclotron frequency is $\omega_{cs} = q_s B_0/m_s$ and $\mathbf{E}$ ($\mathbf{B}$) is the electric (magnetic) field. Equation (1) is coupled to Maxwell's equations via the electromagnetic field:

$$\frac{\partial \mathbf{B}}{\partial t} = -\nabla\times\mathbf{E}, \qquad \frac{\partial \mathbf{E}}{\partial t} = \nabla\times\mathbf{B} - \frac{\omega_{pe}}{\omega_{ce}}\left(\int_{-\infty}^{+\infty}\mathbf{v}f_i d\mathbf{v} - \int_{-\infty}^{+\infty}\mathbf{v}f_e d\mathbf{v}\right), \tag{2}$$

$$\nabla\cdot\mathbf{E} = \frac{\omega_{pe}}{\omega_{ce}}\left(\int_{-\infty}^{+\infty} f_i d\mathbf{v} - \int_{-\infty}^{+\infty} f_e d\mathbf{v}\right), \qquad \nabla\cdot\mathbf{B} = 0. \tag{3}$$

We expand the particle distribution function in Hermite basis functions,

$$f_s = \sum_{n=0}^{N_n-1}\sum_{m=0}^{N_m-1}\sum_{p=0}^{N_p-1} C^s_{n,m,p}(\mathbf{x},t)\Psi_n(\xi^s_x)\Psi_m(\xi^s_y)\Psi_p(\xi^s_z) \tag{4}$$

(where N_n, N_m and N_p are the total number of Hermite modes in the v_x, v_y and v_z directions, respectively), to obtain a set of non-linear partial differential equations (PDEs) for the coefficients of the expansion. We consider the asymmetrically-weigthed (AW) basis function, $\Psi_n(x) = (\pi 2^n n!)^{-1/2} H_n(x) e^{-x^2}$, where H_n is the Hermite polynomial of the n-th order [23]. The argument of the basis functions is defined as $\xi^s_\beta = \left(v_\beta - u^s_\beta\right)/\alpha^s_\beta$, where u^s_β and α^s_β are (constant) shift and scaling parameters that have to be supplied by the user and $\beta = x,\ y,\ z$. We further decompose the spatial part of the particle distribution function into Fourier basis functions,

$$C^s_{n,m,p}(\mathbf{x},t) = \sum_{k_x=-N_x/2}^{N_x/2}\ \sum_{k_y=-N_y/2}^{N_y/2}\ \sum_{k_z=-N_z/2}^{N_z/2} C^{k_x,k_y,k_z,s}_{n,m,p}(t)\exp\left[2\pi i\left(\frac{k_x x}{L_x}+\frac{k_y y}{L_y}+\frac{k_z z}{L_z}\right)\right], \tag{5}$$

where $N_\beta + 1$, $\beta = x,\ y,\ z$, is the total number of Fourier modes in each spatial direction.

Using the orthogonality properties of the Hermite and Fourier basis functions, the Vlasov-Maxwell equations can be rewritten as a set of ordinary differential equations (ODEs) [20]:

$$\begin{aligned}
\frac{dC^{k_x,k_y,k_z,s}_{n,m,p}}{dt} = &-\frac{2\pi i k_x}{L_x}\alpha^s_x\left(\sqrt{\frac{n+1}{2}}C^{k_x,k_y,k_z,s}_{n+1,m,p} + \sqrt{\frac{n}{2}}C^{k_x,k_y,k_z,s}_{n-1,m,p} + \frac{u^s_x}{\alpha^s_x}C^{k_x,k_y,k_z,s}_{n,m,p}\right) - \\
&\frac{2\pi i k_y}{L_y}\alpha^s_y\left(\sqrt{\frac{m+1}{2}}C^{k_x,k_y,k_z,s}_{n,m+1,p} + \sqrt{\frac{m}{2}}C^{k_x,k_y,k_z,s}_{n,m-1,p} + \frac{u^s_y}{\alpha^s_y}C^{k_x,k_y,k_z,s}_{n,m,p}\right) - \\
&\frac{2\pi i k_z}{L_z}\alpha^s_z\left(\sqrt{\frac{p+1}{2}}C^{k_x,k_y,k_z,s}_{n,m,p+1} + \sqrt{\frac{p}{2}}C^{k_x,k_y,k_z,s}_{n,m,p-1} + \frac{u^s_z}{\alpha^s_z}C^{k_x,k_y,k_z,s}_{n,m,p}\right) + \\
&\frac{q_s}{e}\frac{\omega_{cs}}{\omega_{pe}}\left\{\frac{\sqrt{2n}}{\alpha^s_x}[E_x * C_{n-1,m,p}]_{k_x,k_y,k_z} + \frac{\sqrt{2m}}{\alpha^s_y}[E_y * C_{n,m-1,p}]_{k_x,k_y,k_z} + \frac{\sqrt{2p}}{\alpha^s_z}[E_z * C_{n,m,p-1}]_{k_x,k_y,k_z}\right\} + \\
&\frac{q_s}{e}\frac{\omega_{cs}}{\omega_{pe}}\Bigg[B_x * \Bigg\{\sqrt{mp}\left(\frac{\alpha^s_z}{\alpha^s_y} - \frac{\alpha^s_y}{\alpha^s_z}\right)C^s_{n,m-1,p-1} + \sqrt{m(p+1)}\frac{\alpha^s_z}{\alpha^s_y}C^s_{n,m-1,p+1} - \\
&\sqrt{(m+1)p}\frac{\alpha^s_y}{\alpha^s_z}C^s_{n,m+1,p-1} + \sqrt{2m}\frac{u^s_z}{\alpha^s_y}C^s_{n,m-1,p} - \sqrt{2p}\frac{u^s_y}{\alpha^s_z}C^s_{n,m,p-1}\Bigg\}\Bigg]_{k_x,k_y,k_z} +
\end{aligned}$$

$$\frac{q_s}{e}\frac{\omega_{cs}}{\omega_{pe}}\left[B_y*\left\{\sqrt{np}\left(\frac{\alpha_x^s}{\alpha_z^s}-\frac{\alpha_z^s}{\alpha_x^s}\right)C^s_{n-1,m,p-1}+\sqrt{(n+1)p}\frac{\alpha_x^s}{\alpha_z^s}C^s_{n+1,m,p-1}-\right.\right.$$
$$\left.\left.\sqrt{n(p+1)}\frac{\alpha_z^s}{\alpha_x^s}C^s_{n-1,m,p+1}+\sqrt{2p}\frac{u_x^s}{\alpha_z^s}C^s_{n,m,p-1}-\sqrt{2n}\frac{u_z^s}{\alpha_x^s}C^s_{n-1,m,p}\right\}\right]_{k_x,k_y,k_z}+$$
$$\frac{q_s}{e}\frac{\omega_{cs}}{\omega_{pe}}\left[B_z*\left\{\sqrt{nm}\left(\frac{\alpha_y^s}{\alpha_x^s}-\frac{\alpha_x^s}{\alpha_y^s}\right)C^s_{n-1,m-1,p}+\sqrt{n(m+1)}\frac{\alpha_y^s}{\alpha_x^s}C^s_{n-1,m+1,p}-\right.\right.$$
$$\left.\left.\sqrt{(n+1)m}\frac{\alpha_x^s}{\alpha_y^s}C^s_{n+1,m-1,p}+\sqrt{2n}\frac{u_y^s}{\alpha_x^s}C^s_{n-1,m,p}-\sqrt{2m}\frac{u_x^s}{\alpha_y^s}C^s_{n,m-1,p}\right\}\right]_{k_x,k_y,k_z}, \qquad (6)$$

for the Vlasov equation, and

$$\frac{dB_\delta^{k_x,k_y,k_z}}{dt}=-2\pi i\varepsilon_{\beta\gamma\delta}\frac{k_\beta}{L_\beta}E_\gamma^{k_x,k_y,k_z},$$
$$\frac{dE_\delta^{k_x,k_y,k_z}}{dt}=2\pi i\varepsilon_{\beta\gamma\delta}\frac{k_\beta}{L_\beta}B_\gamma^{k_x,k_y,k_z}-\frac{\omega_{pe}}{\omega_{ce}}\sum_{s=e,\,i}\frac{q_s}{e}\alpha_x^s\alpha_y^s\alpha_z^s\left(\frac{\alpha_\delta^s}{\sqrt{2}}C_{a,b,c}^{k_x,k_y,k_z,s}+u_\delta^sC_{0,0,0}^{k_x,k_y,k_z,s}\right) \qquad (7)$$

for Maxwell's equations (written in index notation with summation over repeated indices implied). Here $\varepsilon_{\beta\gamma\delta}$ is the Levi-Civita tensor and $(a,b,c)=(1,0,0),\,(0,1,0),\,(0,0,1)$ for $\delta=x,\,y,\,z$, respectively. In Eq. (6), the convolution operator is defined as

$$[H*C_{n,m,p}]_{k_x,k_y,k_z}=\sum_{k_x'=-N_x/2}^{N_x/2}\sum_{k_y'=-N_y/2}^{N_y/2}\sum_{k_z'=-N_z/2}^{N_z/2}H^{k_x-k_x',k_y-k_y',k_z-k_z'}C_{n,m,p}^{k_x',k_y',k_z',s}, \qquad (8)$$

where H is an arbitrary function and the system is truncated by imposing $C^s_{n,m,p}=0$ for $n\geq N_n$, $m\geq N_m$, $p\geq N_p$. It is worth nothing that the Fourier spatial discretization preserves the divergence constrains (3) of Maxwell's equations: if these constraints are satisfied at $t=0$, they are satisfied at all times. In matrix form, Eqs. (6) and (7) become

$$\frac{d\mathbf{C}}{dt}=\mathbb{L}_1\mathbf{C}+\mathcal{N}(\mathbf{C},\mathbf{F}),$$
$$\frac{d\mathbf{F}}{dt}=\mathbb{L}_2\mathbf{C}+\mathbb{L}_3\mathbf{F}, \qquad (9)$$

with $\mathbf{C}$ representing the coefficients of the Hermite expansion and $\mathbf{F}$ the electromagnetic field. The matrix $\mathbb{L}_1$ corresponds to the linear advection operator on the right hand side of Eq. (6), while $\mathcal{N}$ is the non-linear operator associated with the convolutions. Finally, the matrices $\mathbb{L}_2$ and $\mathbb{L}_3$ are defined by the right hand side of Eqs. (7).

A common aspect of collisionless plasmas is the filamentation of phase space [24], i.e. the development of increasingly smaller phase-space structures. This implies that any numerical method would eventually run out of resolution. In the context of the Hermite basis, lack of resolution can lead to the well-known recurrence effect [25], and this is typically addressed by adding a collisional term to the right hand side of Eq. (6). We choose

$$\mathcal{C}[C_{n,m,p}^{k_x,k_y,k_z,s}]=-\nu\left[\frac{n(n-1)(n-2)}{(N_n-1)(N_n-2)(N_n-3)}+\frac{m(m-1)(m-2)}{(N_m-1)(N_m-2)(N_m-3)}+\right.$$
$$\left.\frac{p(p-1)(p-2)}{(N_p-1)(N_p-2)(N_p-3)}\right]C_{n,m,p}^{k_x,k_y,k_z,s}, \qquad (10)$$

essentially to damp the high-order modes of the Hermite series. Note that the operator in Eq. (10) does not act on the first three Hermite modes so that the conservation of total mass, momentum and energy remains valid in the discrete, as shown in Ref. [20].

Equations (6) and (7) are discretized in time by a fully-implicit, second-order accurate Crank-Nicolson scheme. In residual form, the resulting set of equations becomes

$$\begin{cases} \mathbf{R}_1\left(\mathbf{C}^{\theta+1},\mathbf{F}^{\theta+1}\right) = \mathbf{C}^{\theta+1} - \mathbf{C}^{\theta} - \Delta t\left[\mathbb{L}_1\mathbf{C}^{\theta+1/2} + \mathcal{N}\left(\mathbf{C}^{\theta+1/2},\mathbf{F}^{\theta+1/2}\right)\right] = 0 \\ \mathbf{R}_2\left(\mathbf{C}^{\theta+1},\mathbf{F}^{\theta+1}\right) = \mathbf{F}^{\theta+1} - \mathbf{F}^{\theta} - \Delta t\mathbb{L}_2\mathbf{C}^{\theta+1/2} - \Delta t\mathbb{L}_3\mathbf{F}^{\theta+1/2} = 0 \end{cases} \iff \mathbb{R}(\mathbf{X}) = 0, \tag{11}$$

where we have introduced the time step Δt, used superscript θ to label time $\mathbf{C}(t^{\theta}) = \mathbf{C}(\theta\Delta t) = \mathbf{C}^{\theta}$, $\mathbf{C}^{\theta+1/2} = \left(\mathbf{C}^{\theta+1} + \mathbf{C}^{\theta}\right)/2$ (and similarly for $\mathbf{F}$), $\mathbf{X} = \begin{bmatrix}\mathbf{C}\\ \mathbf{F}\end{bmatrix}$ and $\mathbb{R} = \begin{bmatrix}\mathbf{R}_1\\ \mathbf{R}_2\end{bmatrix}$.

2.1. Jacobian-Free Newton-Krylov solver

The most important part of the algorithm is the solution of the nonlinear Eqs. (11). For this, we use a Jacobian-Free Newton-Krylov (JFNK) method [26]. At each time step the initial guess $\mathbf{X}^0$ is iteratively updated with the Newton method , $\mathbf{X}^{n+1} = \mathbf{X}^n + \Delta\mathbf{X}^n$, where n is the nonlinear iteration number and ΔX^n is the solution update, which is obtained by solving the linear system

$$\mathbb{J}(\mathbf{X}^n)\Delta\mathbf{X}^n = -\mathbb{R}(\mathbf{X}^n) \tag{12}$$

with the Jacobian matrix $\mathbb{J}$ defined as $\mathbb{J}_{ij} = \dfrac{\partial\mathbb{R}_i(\mathbf{X})}{\partial X_j}$, where $\mathbb{R}_i$ labels the i-th row of $\mathbb{R}$. The linear system (12) is solved with the Generalized Minimal RESidual (GMRES) iterative method. It approximates the exact solution by minimizing the Euclidean norm of the residual $||\mathbb{J}(\mathbf{X}^n)\Delta\mathbf{X}^n + \mathbb{R}(\mathbf{X}^n)||$. The method requires the computation of matrix-vector products $\mathbb{J}\Delta\mathbf{X}^n$, which can be approximated by a directional derivative with a small step increment without actually storing $\mathbb{J}$ and computing the product.

2.2. Preconditioning

It is well known that the convergence of the linear part of the JFNK solver strongly depends on the problem's eigenvalue spectrum, and that GMRES typically does not scale well as the problem's size increases. Hence, applying a suitable preconditioner can significantly accelerate the convergence rate and reduce the simulation time.

By default PETSc uses left preconditioning, i.e. the preconditioner matrix $\mathbb{P}_L^{-1}$ is applied to the left of system (12):

$$\mathbb{P}_L^{-1}\mathbb{J}(\mathbf{X}^n)\Delta\mathbf{X}^n = -\mathbb{P}_L^{-1}\mathbb{R}(\mathbf{X}^n). \tag{13}$$

An efficient preconditioner can effectively reduce the number of linear iterations without being too costly to compute. In SPS, we follow [20] and construct the preconditioner by accounting for the convolutions in Eq. (11) via linearization relative to some reference solution

$$\mathbb{P}_L^{-1} = \begin{bmatrix} \mathbb{I} - \frac{\Delta t}{2}\mathbb{L}_1 - \frac{\Delta t}{2}\mathcal{N}\left(\cdot,\mathbf{F}^{ref}\right) & -\frac{\Delta t}{2}\mathcal{N}\left(\mathbf{C}^{ref},\cdot\right) \\ -\frac{\Delta t}{2}\mathbb{L}_2 & \mathbb{I} - \frac{\Delta t}{2}\mathbb{L}_3 \end{bmatrix}^{-1}, \tag{14}$$

where $\mathbb{I}$ is the identity matrix. In principle, the preconditioner can be applied at every time step, where the reference solution corresponds to the solution at the previous time step. In practice, we typically compute it only once at the beginning of the simulation (i.e. the reference solution is given by the initial condition of the system) and apply it at all time steps. While this can lead to a deterioration of the performance of the preconditioner as the simulation progresses, in the example discussed below this strategy was quite efficient.

3. Implementation of the SPS code

The SPS code has been implemented in Fortran to solve problems in a two-dimensional spatial geometry with three velocity components. Fortran was chosen for its native support for complex number operations that are widely used in the code. The SPS code consists of approximately 4,000 lines of code and is based on the PETSc [22] and FFTW [27] libraries, and on OpenMP.

We use the PETSc library to solve the discretized governing equations with the JFNK method. PETSc is an open-source library for the numerical solution of partial differential equations. It is written in C and has an interface to Fortran. Even though PETSc has been primarily designed to support programming on distributed systems, current PETSc versions use OpenMP and pthreads for matrix and vector operations.

The FFTW library [27] is used to perform the convolutions in Eq. (6) by means of two Fast Fourier Transforms. The convolution is carried out by first Fourier-transforming the input vectors from spectral space to real space, multiplying them in real space and transforming their result back to the Fourier space. The complexity of the FFT-based convolution is $\mathcal{O}(M\log M)$, where M is the input vector size, as opposed to the $\mathcal{O}(M^2)$ complexity of the direct calculation.

The OpenMP API is used to parallelize the SPS code on shared-memory systems. A profiling of the SPS performance revealed that the most compute-intensive part of the code is the evaluation of the residual function during the JFNK solver iterations. For this reason, we mainly use OpenMP to parallelize loops in this part of the calculation.

4. Verification of SPS: Orszag-Tang Vortex Test

The SPS code has been benchmarked on several problems, including Landau damping, two-stream instabilities, beam-plasma instabilities and ion acoustic waves. Here we briefly discuss a complex test case: simulations of the two-dimensional Orszag-Tang vortex. The Orszag-Tang vortex is an initial configuration often used in studies of 2D plasma turbulence (e.g. [28, 29]). Evolution of the system leads to the formation of current sheets, which are narrow regions of intense current density corresponding to large gradients of the magnetic field. If the system size is large enough, the system eventually transitions to turbulence. We focus on collisionless magnetized plasmas with moderate ratios of thermal to magnetic pressures $\beta_e \sim \beta_i \sim 1$, where the characteristic thickness of the current sheets is expected to be of the order of $d_e \sim \rho_e$. Here $\beta_s = 8\pi n_s T_s/B_0^2$, $\rho_e = v_{te}/\omega_{ce}$, $v_{ts} = \sqrt{T_s/m_s}/c$, $n_e \approx n_i$ are the electron and ion densities, and T_s is the effective temperature of species s.

We consider a domain of size $L_x \times L_y$, with the initial distribution function given by

$$f_s(\boldsymbol{x}, \boldsymbol{v}, t=0) = \frac{1}{(2\pi)^{3/2}\, v_{ts}^3} \exp\left(-\frac{[\boldsymbol{v} - \boldsymbol{U}_s(\boldsymbol{x})]^2}{2v_{ts}^2}\right) \tag{15}$$

and the initial magnetic field by $\boldsymbol{B} = \boldsymbol{e}_z + \delta\boldsymbol{B}(\boldsymbol{x})$, where $\boldsymbol{U}_s = U_0\left[-\sin(k_y y)\boldsymbol{e}_x + \sin(k_x x)\boldsymbol{e}_y\right] + U_{s,z}\boldsymbol{e}_z$ and $\delta\boldsymbol{B} = \frac{\delta B}{B_0}\left[-\sin(k_y y)\boldsymbol{e}_x + \sin(2k_x x)\boldsymbol{e}_y\right]$, with $\boldsymbol{e}_\beta$ the unit vector along β. Here $U_0 = V_A\frac{\delta B}{B_0}$, $V_A = B_0/\sqrt{4\pi n_0 m_i}/c$, $k_x = 2\pi/L_x$, $k_y = 2\pi/L_y$. To satisfy Ampere's law, the initial configuration includes the electron flow $U_{e,z} = -\frac{\delta B}{B_0}\frac{\omega_{ce}}{\omega_{pe}}\left[2k_x\cos(2k_x x) + k_y\cos(k_y y)\right]$. The ion flow is $U_{i,z} = 0$. The following physical parameters were used: domain size $L_x = L_y = 10 d_i$, $m_i/m_e = 25$, $T_e = T_i$, $\omega_{pe}/\omega_{ce} = 2$, $\beta_e = \beta_i = 0.25$, $\delta B/B_0 = 0.2$.

For the SPS simulations, the configuration (15) is expanded in Hermite-Fourier basis functions numerically. We choose $\alpha_\beta^s = \sqrt{2}v_{ts}$ and $u_\beta^s = 0$. Our representation with $u_\beta^s = \text{const}$ cannot capture Eq. (15) exactly with only one polynomial, therefore runs conducted with different values of $N = N_n = N_m = N_p$ correspond to slightly different approximations of the initial condition. Other parameters used for the simulations are: $N_x = N_y = 128$, $\omega_{pe}\Delta t = 1$ and $\nu = 1$, while N will be varied parametrically. The results of the SPS simulations are compared against a reference solution obtained using a fully kinetic particle-in-cell (PIC) simulation of the same initial conditions. The PIC simulation was conducted using the explicit relativistic simulation code VPIC [30]. The VPIC simulation with the parameters specified above had 880×880 cells with 4000 particles per cell per species at $t = 0$. The PIC time step was $\omega_{pe}\Delta t \approx 0.04$.

Figure 1 compares the current density j_z normalized to a characteristic value $en_0 v_{te}$ at $\omega_{pe}t = L_x/V_A$ in two SPS simulations obtained with $N = 4$ and $N = 6$, against the reference PIC simulation. At this time the system has not transitioned to fully developed turbulence yet. Despite a relatively small number of Hermite polynomials, the test cases successfully reproduce both the large-scale structure observed in the PIC simulation and the small, intense current sheets. Some of the secondary features are arguably reproduced better in the case with larger number of polynomials in the basis. The multi-scale nature of fluctuations excited in the system is apparent from an estimate of the omnidirectional power spectrum of magnetic fluctuations S_B presented in Fig. 2. The spectrum also clearly demonstrates potential advantages offered by noise-free spectral methods for turbulence simulations. While the spectrum at very high k is eventually dominated by discrete particle noise even in this highly resolved PIC simulation, SPS can provide clean estimates of power in large-k modes. It is interesting to note that both PIC and SPS simulations suggest that power spectra extend to relatively small scales $kd_e > 1$, with a possible transition to another power law at $k\lambda_D \sim 1$. A detailed investigation of energy dissipation and fluctuation spectra in these simulations will be presented elsewhere.

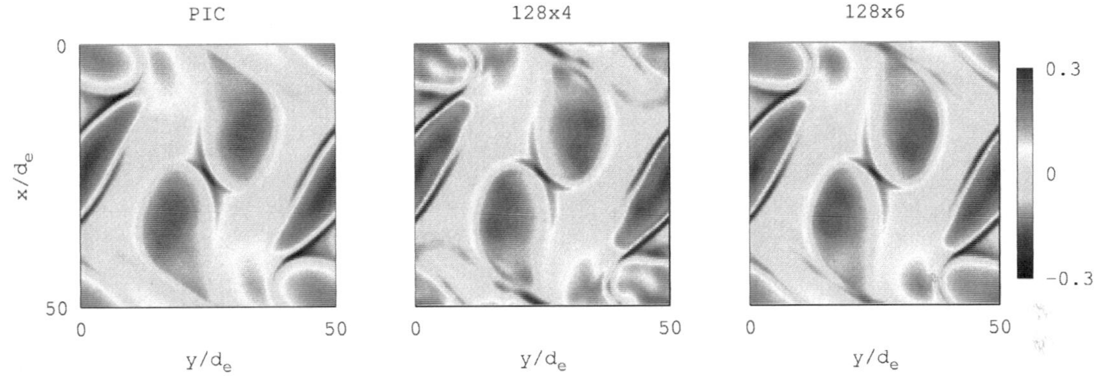

Figure 1. Normalized current density $\hat{j}_z = j_z/(en_0 v_{te})$ in the Orszag-Tang simulations at $\omega_{pe}t = L_x/V_A$. Left: Reference PIC simulation; Middle: SPS simulation with $N_x = N_y = 128$ and N=4; Right: SPS simulation with $N_x = N_y = 128$ and N=6.

The global energy balance in the Orszag-Tang vortex simulations is illustrated in Fig. 3. SPS reproduces well the changes in the different energy channels measured in the reference PIC simulation. We note that in contrast to fluid and MHD turbulence, dissipation of fluctuating energy in weakly collisional plasmas is due to collective interactions of particles with the electromagnetic field, even though the ultimate dissipation in the sense of entropy generation is provided by collisions. Understanding of the turbulent energy dissipation is thus quite challenging and has been a focus of intense research effort. The methodology presented here may make a significant contribution to the field by enabling accurate simulations of kinetic phenomena in turbulence, free of inherent limitations of other currently available methods.

5. Performance Results

In this section we discuss the performance of SPS. The simulations were performed on a node of the Wolf supercomputer at the Los Alamos National Laboratory. Each Wolf node has two 8-core Intel Xeon E5-2670 processors with 2.6 GHz CPU and 64 GB of memory.

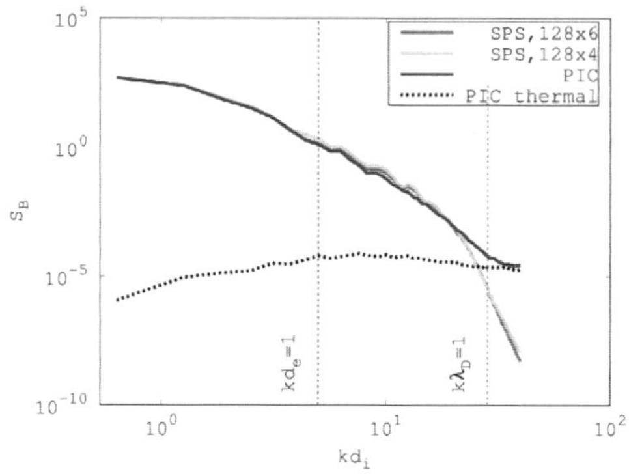

Figure 2. Omnidirectional power spectrum of magnetic fluctuations in the Orszag-Tang vortex simulations at the same time as in Fig. 1. The black dotted line shows the spectrum of thermal noise in the PIC simulation. The spectrum is normalized according to $\int B^2 dxdy = \int S_B(k)dk$.

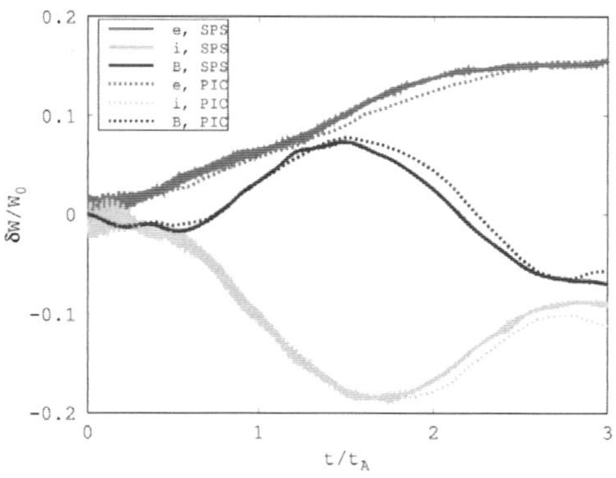

Figure 3. Changes in the kinetic energy of electrons (e), ions (i), and of the magnetic field energy (B) in the Orszag-Tang vortex simulations. Here $\delta W(t) = W(t) - W(t=0)$, $W_{\text{kin},s} = \int (m_s v^2/2) f_s \, d^3v d^2x$, $W_{\text{B}} = \int (B^2/2) \, d^2x$, and $\omega_{pe} t = L_x/V_A$. The energies are normalized to the energy of the initial perturbation $W_0 = \int \left[(m_i U_i^2/2 + m_e U_e^2/2 + \delta B^2/2\right] d^2x$. SPS simulation had $N_x = N_y = 128$ and $N = 6$.

The SPS code has been compiled with the Intel Fortran 15.0.5 compiler and linked to PETSc version 3.6.2 and to FFTW version 3.3.4. The relative and absolute tolerances of JFNK are set to 10^{-6}. PETSc Incomplete LU (ILU) factorization is used to invert matrix $\mathbb{P}_L$ for preconditioning. For preconditioned and unpreconditioned simulations, we set the PETSc options *-snes_mf_operator* and *-snes_mf*, respectively.

5.1. Performance of the preconditioner

Table 1 shows the running time and the average number of nonlinear and linear iterations per time step for the Orszag-Tang problem. The simulation parameters are described in Section 4, except that $N_x = N_y = 32$ and the simulation time corresponds to 100 time steps. Let us focus on the unpreconditioned case. For $N = 4$ the average number of linear iterations per time step is ~ 14 and the running time is 297 s. As we double N, the problem's size increases by a factor 7.7 but the running time is ~ 14 times longer since the average number of linear iterations doubles.

Going from $N = 8$ to $N = 12$ increases the problem's size by a factor 3.4 while the running time increases by a factor 5.3, again because of the higher number of linear iterations (now ~ 52) required for convergence. Let us now look at the preconditioned case. For $N = 4$, the average number of linear iterations is 5, about a third of those for the unpreconditioned case, translating into a 40% reduction of the running time. Up to $N = 16$ the number of linear iterations remains fairly constant and, consequently, the running time shows a fairly good linear scaling with the number of unknowns of the system. For $N = 12$ there is a factor of 4 speed-up relative to the unpreconditioned case and the gain will increase for larger problems. The preconditioner loses performance for $N = 20$, where the average number of linear iterations rises to 8.5 with a 20% increase in the running time relative to a linear scaling. This could arise from the problem not being exactly the same as N changes, due to the approximation of the initial condition and the nature of Eq. (10). Finally, the number of nonlinear iterations is insensitive to preconditioning.

Table 1. Running time and average number of nonlinear and linear iterations per time step for the Orszag-Tang problem with $N_x = N_y = 32$ and 16 threads.

		Preconditioned		Unpreconditioned	
N	**Unknowns**	**Time, s**	**Nonlin./ Lin. iter.**	**Time, s**	**Nonlin./ Lin. iter.**
4	145,926	178	2/5.1	297	2/14.4
8	1,121,670	1,475	2/5.4	4,138	2/30.8
12	3,770,118	5,165	2/5.6	22,120	2/51.6
16	8,927,622	13,130	2/5.9	–	–
20	17,430,534	31,200	2/8.5	–	–

5.2. Parallel performance on shared-memory systems

A strong scaling test, where we kept the problem size fixed and increased the number of threads, has been carried out for the Orszag-Tang problem. The parameters are the same as those used for the preconditioning studies in the previous subsection and we focus on the case with $N = 8$. Table 2 shows the parallel speed-up varying the number of threads in use: the strong scaling tests show a factor of 10 speed-up with 16 threads. In addition, Amdahl's law [31]:

$$\text{Speed-up} = \frac{1}{S + \dfrac{1-S}{p}} \tag{16}$$

is used to evaluate the performance of SPS. In Eq. (16), S is the fraction of time in the code that is not parallelized and p is the number of threads. For the Orszag-Tang test, $S = 0.033$ was calculated by running the code with 1 thread and using Intel Vtune 15.3.0 for profiling. Table 2 shows that there is good agreement between the theoretical and measured speed-up.

6. Conclusions

We have presented the design and implementation of the SPS code, which solves the Vlasov-Maxwell equations for a plasma in multiple dimensions. The code is based on a Hermite-Fourier spectral decomposition of the particle distribution function. The resulting set of non-linear ODEs is discretized in time with a second-order accurate Crank-Nicolson scheme and is solved numerically with a preconditioned JFNK solver.

SPS is written in Fortran and uses the PETSc library for solving the nonlinear equations and the FFTW library to handle the convolution operations efficiently. The code is parallelized for

Table 2. Strong scaling test: parallel speed-up varying the number of threads for the Orszag-Tang problem with $N_x = N_y = 32$, $N = 8$ and with preconditioning.

Threads	Speed-up	Amdahl's law
1	1	1
2	1.9	1.9
4	3.6	3.6
8	6.5	6.5
16	10.5	10.7

shared-memory machines using OpenMP. A comparison against a fully kinetic PIC simulation for a Orszag-Tang vortex problem shows that SPS can capture reasonably well the large-scale and small-scale current structure of the system, despite using a relative low number of Hermite modes. It also highlights the strength of the method and its ability to capture the high-k tail of the fluctuation spectrum which is normally obscured by noise in PIC simulations. Performance tests have also been presented on the same problem, showing a significant reduction of the number of linear iterations and of the running time obtained by preconditioning, and a factor of 10 speed-up for a strong scaling test with 16 threads.

Acknowledgments

This work was funded by the Laboratory Directed Research and Development (LDRD) program, under the auspices of the National Nuclear Security Administration of the U.S. Department of Energy by Los Alamos National Laboratory, operated by Los Alamos National Security LLC under contract DE-AC52-06NA25396. This research used resources provided by the Los Alamos National Laboratory Institutional Computing Program. VR acknowledges support from NASA grant NNX15AR16G. PIC simulations utilized resources provided by the NASA High-End Computing Program through the NASA Advanced Supercomputing Division at Ames Research Center.

References

[1] Goldston R J and Rutherford P H 1995 *Introduction to Plasma Physics* Plasma Physics Series (Institute of Physics Publications)
[2] Fitzpatrick R 2014 *Plasma Physics* (CRC Press)
[3] Kolobov V and Arslanbekov R 2012 *Journal of Computational Physics* **231** 839 – 869
[4] Markidis S, Henri P, Lapenta J, Rönnmark K, Hamrin M, Meliani Z and Laure E 2014 *Journal of Computational Physics* **271** 415–429 frontiers in Computational Physics Modeling the Earth System
[5] Daldorff L K S, Tóth G, Gombosi T I, Lapenta G, Amaya J, Markidis S and Brackbill J U 2014 *Journal of Computational Physics* **268** 236–254
[6] Birdsall C K and Langdon A B 2004 *Plasma Physics Via Computer Simulation* (Taylor & Francis)
[7] Hockney R and Eastwood J 1988 *Computer Simulation Using Particles* (Taylor & Francis)
[8] Cheng C Z and Knorr G 1976 *Journal of Computational Physics* **22** 330–351
[9] Filbet F, Sonnendrücker E and Bertrand P 2001 *Journal of Computational Physics* **172** 166–187
[10] Vay J L, Colella P, Kwan J W, McCorquodale P, Serafini D B, Friedman A, Grote D P, Westenskow G, Adam J C, Hron A and Haber I 2004 *Physics of Plasmas* **11**
[11] Fujimoto K and Machida S 2006 *Journal of Computational Physics* **214** 550 – 566
[12] Arslanbekov R R, Kolobov V I and Frolova A A 2013 *Phys. Rev. E* **88**(6) 063301
[13] Armstrong T P, Harding R C, Knorr G and Montgomery D 1970 *Methods Comput. Phys.* **9** 29
[14] Camporeale E, Delzanno G L, Lapenta G and Daughton W 2006 *Physics of Plasmas* **13** 092110
[15] Vencels J, Delzanno G L, Johnson A, Peng I B, Laure E and Markidis S 2015 *Procedia Computer Science* **51** 1148–1157
[16] Manzini G, Delzanno G L, Vencels J and Markidis S 2015 *J. Comput. Phys.* (Submitted)
[17] Parker J T and Dellar P J 2015 *Journal of Plasma Physics* **81**(02)
[18] Loureiro N F, Schekochihin A A and Zocco A 2013 *Physical Review Letters* **111**(2) 025002

[19] Camporeale E, Delzanno G, Bergen B and Moulton J 2016 *Computer Physics Communications* **198** 47–58
[20] Delzanno G L 2015 *Journal of Computational Physics* **301** 338–356
[21] Loureiro N F, Dorland W, Fazendeiro L, Kanekar A, Mallet A, Vilelas M S and Zocco A 2015 *arXiv:1505.02649*
[22] Balay S, Abhyankar S, Adams M F, Brown J, Brune P, Buschelman K, Dalcin L, Eijkhout V, Gropp W D, Kaushik D, Knepley M G, McInnes L C, Rupp K, Smith B F, Zampini S and Zhang H 2015 PETSc Web page URL `http://www.mcs.anl.gov/petsc`
[23] Holloway J P 1996 *Transport Theory and Statistical Physics* **25** 1–32
[24] Klimas A J 1987 *Journal of Computational Physics* **2022226** 202–226
[25] Joyce G, Knorr G and Meier H K 1971 *Journal of Computational Physics* **8** 53 – 63
[26] Knoll D A and Keyes D E 2004 *Journal of Computational Physics* **193** 357–397
[27] Frigo M and Johnson S G 2005 *Proceedings of the IEEE* **93** 216–231
[28] Orszag S A and Tang C M 1979 *Journal of Fluid Mechanics* **90** 129–143
[29] Biskamp D and Welter H 1989 *Physics of Fluids B* **1** 1964–1979
[30] Bowers K J, Albright B J, Yin L, Bergen B and Kwan T J T 2008 *Physics of Plasmas* **15** 055703
[31] Amdahl G M 1967 Validity of the single processor approach to achieving large scale computing capabilities *Proceedings of the April 18-20, 1967, spring joint computer conference* (ACM) pp 483–485

A fluid-particle hybrid framework for the PLUTO code: applications to non-thermal emission in jets.

B. Vaidya[1], **A. Mignone**[1], **G. Bodo**[2] **and S. Massaglia**[1]
[1]Dipartimento di Fisica, University di Torino, via Pietro Giuria 1, I-10125 Torino, Italy
[2]INAF, Osservatorio Astronomico di Torino, Strada Osservatorio 20, I-10025 Pino Torinese, Italy

E-mail: bvaidya@unito.it

Abstract. We present an implementation of a fully parallel hybrid framework for the evolution of Lagrangian particles coupled to a MHD fluid for the PLUTO code. For the applications of interest, particles represent ensembles of electrons whose spectral energy distribution is governed by a kinetic transport equation that takes into account different physical processes such as diffusive shock acceleration, synchrotron emission and adiabatic expansion. An application to model non-thermal emission from extragalactic jets shows the effectiveness and strength of the approach in describing not only the dynamics but also the radiation properties of jets and, in general, of high-energy astrophysical plasma environments.

1. Introduction

Many astrophysical and space physics applications require increasingly complex physics to be included for a more realistic modeling. Depending on the scale of interest, the numerical modeling can be tackled by means of different techniques such as Particle-in-Cell (PIC), multi-fluid or single-fluid MHD.

Recently, hybrid implementations combining both particles and grid-based fluid descriptions have been proposed in order to reduce the large scale gap between macro- and micro-physics. The MHD-PIC approach of [1], for example, describes the interaction between collisionless cosmic ray particles and a thermal plasma. Similarly, [2] proposed a hybrid approach for the BATS-R-US code that combines Hall-MHD and PIC methods in order to capture small-scale kinetic effects in magnetosphere simulations. Other hybrid formulations have been considered to model, e.g., dust (in the Athena code, [3]), collisionless components (in the Enzo code, [4]), Lagrangian test particles (in the Flash code, [5]).

Here we focus on a hybrid particle/fluid approach suitable for treating non-thermal radiation processes from astrophysical plasmas. Previous attempts to include optically thin radiation processes have been proposed by [6] who used passive scalars (fluid tracers) to model synchrotron emission from pulsar wind nebulae and by [7] in the context of jets from radio-galaxies. This approach, however, may become extremely consuming in terms of memory usage as more accurate spectral modeling is needed. A more convenient approach using particles has been proposed by [8] in the context of relativistic jet modeling. Here particles are evolved taking into account radiative losses and their spectral energy distribution allows non-thermal electrons to be injected at shocks. A similar approach was used by [9].

Here we present a fully time-dependent numerical implementation to treat Lagrangian particles describing a population of non-thermal electrons (or protons) dynamically coupled to a MHD fluid. The implementation is fully parallel and part of a more general particle module being developed for the PLUTO code [10] and that will be described in a forthcoming paper. Here we only focus on the static grid implementation. Our formulation is applied to investigate non-thermal emission in extragalactic jets at kilo parsec (kpc) scales where several radio and X-rays observations show the typical imprint of synchrotron emission from highly energetic electrons [11].

2. A Particle Module for the PLUTO Code

A Lagrangian macro-particle (MP) represents an ensemble of real particles (e.g., electrons) sufficiently close in physical space and associated with a sub-grid physical modeling. The position of each MP is determined by the simple transport equation,

$$\frac{d\boldsymbol{x}_p}{dt} = \boldsymbol{v}_p = \boldsymbol{v}\,, \tag{1}$$

where $\boldsymbol{v}$ is the gas velocity.

Each MP is characterized by a distribution function $N(\epsilon, t)$ representing the actual particle number density as a function of energy ϵ. For convenience, energies are distributed in equally-space (in log) bins ranging from $\epsilon_{\rm min}$ to $\epsilon_{\rm min}$ with total of $\mathcal{M}$ bins. The energy distribution changes in time according to several different physical processes, as described in the seminal paper by Kardashev [12]. Here we take into account, for simplicity, only adiabatic compression/expansion and synchrotron radiative losses so that our evolutionary equation becomes

$$\frac{\partial N(\epsilon,t)}{\partial t} = \frac{\partial}{\partial \epsilon}\Big[\left(\alpha(t)\epsilon + \beta(t)\epsilon^2\right) N\Big]\,, \tag{2}$$

where $\alpha(t) = \frac{1}{3}\nabla\cdot\boldsymbol{v}$ represents losses due to adiabatic effects and $\beta(t) \propto \boldsymbol{B}^2$ describes synchrotron losses. Magnetic field and velocity are interpolated from the fluid directly on the particle position. Equation (2) is solved by means of analytical methods by extending the approach of [12] to a general initial energy distribution (not only a power law). The solution is discretized on a *Lagrangian* energy grid and the update is given by the following step :

$$N^{n+1}(\epsilon^{n+1}) = N^n(\epsilon^n)e^a(1+c\epsilon^n)^2\,, \tag{3}$$

where N^n and N^{n+1} are, respectively, the particle densities at time levels t^n and t^{n+1} in the energy bins ϵ^n and ϵ^{n+1}. In a *Lagrangian* description the bin widths change in time according to

$$\epsilon^{n+1} = \frac{\epsilon^n e^{-a}}{1+c\epsilon^n}\,, \tag{4}$$

where, to second-order accuracy, a and c are given by

$$\begin{aligned} a &= \int_{t^n}^{t^{n+1}} \alpha(t)dt \approx \tfrac{1}{2}(\alpha(t^{n+1}) + \alpha(t^n))\Delta t. \\ c &= \int_{t^n}^{t^{n+1}} \beta(t)e^{-a}dt \approx \tfrac{1}{2}(\beta(t^{n+1})e^{-a} + \beta(t^n))\Delta t. \end{aligned} \tag{5}$$

If the initial distribution is defined over an interval $[\epsilon_{\rm min},\ \epsilon_{\rm max}]$, this approach does not require specifying explicit boundary conditions and automatically preserves the total number of particles, since

$$\int_{\epsilon^{n+1}_{\rm min}}^{\epsilon^{n+1}_{\rm max}} N^{n+1}(\epsilon)d\epsilon^{n+1} = \int_{\epsilon^{n}_{\rm min}}^{\epsilon^{n}_{\rm max}} N^{n}(\epsilon)d\epsilon^{n}. \tag{6}$$

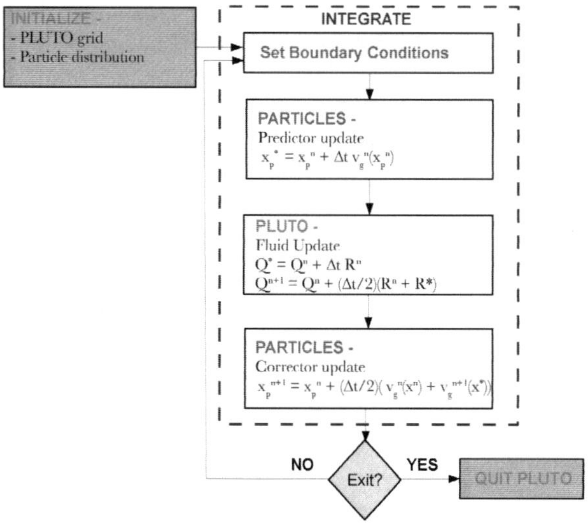

Figure 1. Flowchart for the hybrid Lagrangian particles/fluid update in the PLUTO code.

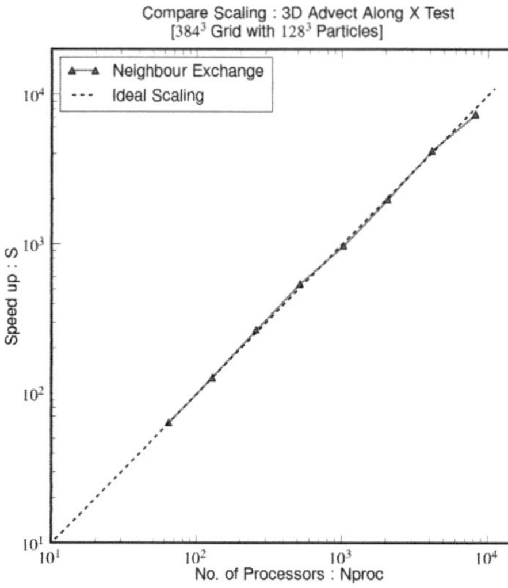

Figure 2. Parallel speedup (strong scaling) for 3D uniform advection problem. The *black dashed* line represents ideal scaling.

In presence of shock waves, the distribution function can change by taking into account diffusive shock acceleration (Fermi I process). However, since the details of the micro-physics are at the unresolved Larmor scale, we employ a sub-grid prescription, under an assumption of isotropy and that the acceleration time scale, $\tau_{\rm acc}$ is much less than dynamical time scale [9]:

$$N_+(\epsilon, t) = \frac{q}{r}\epsilon^{-q+2} \int_{\epsilon_{\rm min}}^{\epsilon} N_-(e, t) e^{q-3} de \tag{7}$$

with $q = 3r/(r-1)$ and the post shock spectrum, N_+, depends on the up-stream particle spectrum N_- and the compression ratio, $r = \rho_+/\rho_-$. Such an assumption is indeed valid for lower energy particles whose emitted radiation is in radio and optical bands. The spectrum update formulation depends on the strength of the shock which is governed by its compression ratio and its direction with respect to magnetic field.

3. Numerical Implementation

The system of fluid+macro-particles (MPs) is governed by the MHD equations (for the fluid) and Eq. (1) for the particle motion. MPs are initialized on the computational domain by prescribing their coordinates and the distribution function $N(\epsilon)$. The same time-marching scheme used for the fluid is also employed to update the particle position; for a $2^{\rm nd}$-order Runge-Kutta scheme, for example, a single time advance is comprised of the steps shown in Fig. 1. Notice that in both the predictor and corrector steps magnetic field and velocity must be interpolated from the fluid grid to the particle position.

The hybrid module has been successfully parallelized using the Message Parsing Interface (MPI) library. We follow a simple approach where each processor updates only the particles lying on its computational sub-domain. Particles must be transferred between neighbors when they cross a processor boundary: in such a way, each processor communicates only with its neighbors. The efficiency of this approach has been tested by timing a simple test consisting of a 3D uniform advection with 384^3 grid cells and 128^3 uniformly distributed MPs. Strong scaling is reported in Fig. 2 for 64 up to 8192 processor showing excellent efficiency (more than 93%).

4. Numerical Benchmarks

In this section, we test our implementation against simple validation benchmarks.

4.1. Stationary Vortex in 2D

The first test case is used to demonstrate the accuracy of our method in tracing the particle evolution along the velocity streamlines. For this purpose, we consider a 2D hydrodynamical stationary vortex configuration [13] for which an analytical solution is available. MPs are initialized in 8 concentric circles at different radii such that 64 particles cover each circumference uniformly. We then track the radial distance for each particle as it evolves and compute the error as a function of time. In Fig. 3 we show the error of a MP initially placed at $r = 2.5$ as a function of time t. They main source of error comes from poorly sampled fluid grid and hence we observe that the maximum error decreases with increasing grid size. At the resolution of 128^2 the errors in radial positions of particles fall below 10^{-7} (red line).

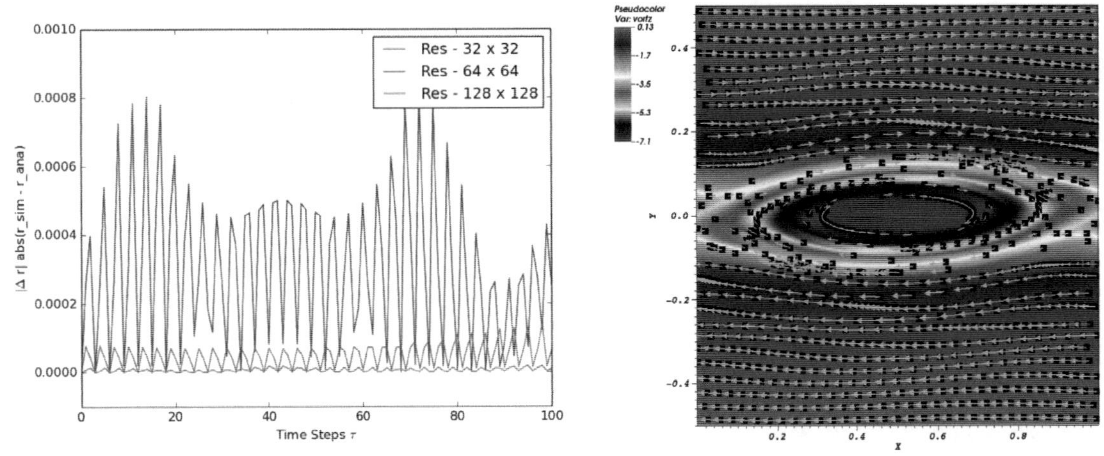

Figure 3. *Left*: error in the radial position for a single particle in the stationary vortex problem at different resolutions. *Right*: evolution of Lagrangian particles in the KH vortex problem. The colormap shows the vorticity while particles and their corresponding velocity are over-plotted.

4.2. Kelvin-Helmholtz Unstable Flow

In the next test we consider a dynamical configuration by simulating the evolution of a 2D shear layer unstable to Kelvin-Helmholtz (KH) instability using the HD module of the PLUTO code. The computational domain is the unit square covered by 512^2 zones with uniform density and pressure. A velocity shear $v_x = V_0 \tanh(y/l)$ with $V_0 = 0.5$ and $l = 0.0637$ is perturbed with a sinusoidal disturbance in v_y with amplitude of 0.01 and a total of 1024 MPs are initially placed uniformly. Periodic boundary conditions are applied in the x direction, while open boundaries are set along y direction. Fig. 3 shows the vorticity of the fluid at $t = 20$ where MPs are represented as black squares and the marked grey arrows (whose lengths are scaled by magnitude of particles velocity) point clearly in the direction of flow as expected.

4.3. Adiabatic Expansion in Sedov-Taylor Blast Wave.

A 2D Sedov-Taylor explosion is set on the unit square using 1024^2 cells and open boundary conditions. We adopt constant density ($\rho = 1$) throughout the domain while the internal energy

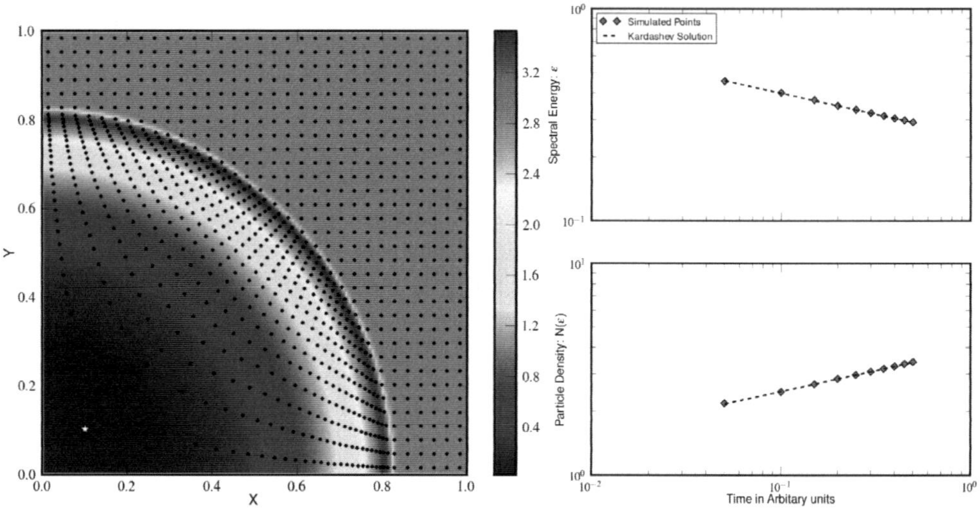

Figure 4. *Left Panel:* Density colormap overlaid with particle distribution (black dots) for the Sedov-Taylor problem at $t = 0.5$. The particle marked as a *white* star is used for further spectral analysis. *Right Panel:* spectral energy evolution for the marked particle in presence of adiabatic expansion only. The *red* squares represent simulation data and black dashed line is the analytical solution [12].

is set to $\rho e = 1.5 \times 10^{-5}$ everywhere except for a small region at the origin where $\rho e = 6.82 \times 10^{3}$. An ideal EoS with adiabatic index $\gamma = 5/3$ is used.

We place a total of 1024 MPs uniformly distributed within the domain and assign to each MP an initial power-law spectrum $N(\epsilon) = N_0 \epsilon^{-\gamma}$ with $\gamma = 3$ covering a range of 6 orders of magnitude in the energy distribution with $\mathcal{M} = 100$ equally-spaced logarithmic energy bins. We solve the hydrodynamical equations (no magnetic field) and, in order to quantify the effect of adiabatic effects only, we do not include shock-aceleration when solving Eq. (2) for the evolution of the distribution function. The solution at $t = 0.5$ is shown in the left panel of Fig. 4. Note that for a very strong shock, the post-shock velocity can be approximated as $v(r,t) \propto r v_s$ where $v_s \propto r_s/t$ and $r_s \propto t^{2/5}$ are the shock front velocity and position, respectively. From this argument, we see that the adiabatic expansion coefficient in Eq. (2) becomes $\alpha \propto 1/t$. Because of this, adiabatic effects modify all energy bins in the same way and the resulting spectral distribution involves a parallel shift of the spectrum, $N(\epsilon)$ without affecting the power-law index γ. This is shown in the right panel of Fig. 4 for a single MP. The analytical dependence of the electrons energy ϵ and of the distribution function $N(\epsilon)$ as a function of time (with $\beta = 0$, no magnetic fields) is presented in [12]. These solutions are over-plotted as *black dashed* lines in middle and bottom panels of Fig. 4. The *red* squares that depicts the value of one of the energy bins for the same particle behind the shock are consistent with the analytical estimates.

4.4. Magnetized Sedov-Taylor Blast Wave.

Next, we consider the effects of synchrotron losses on the evolution of the distribution function in a magnetized cylindrical blast wave simulation. The domain is initially filled with constant density and pressure, $\rho = 1$ and $p = 0.1$ while a constant magnetic field $(B_x, B_y) = (1, 1)$ is prescribed. An over-pressurized region is set inside the circle $r \leq r_0$ (with $r_0 = 0.1$) where $p = 10$. We solve the MHD equations on the 2D Cartesian square $x, y \in [-1, 1]$ using 512^2 zones and place 4096 MPs uniformly with the same spectral distribution used for the previous test.

For the sake of validation, we evolve the MP distribution function by solving Eq. (2) again by neglecting shock acceleration.

The left panel of Fig. 5 shows the particle distribution (black dots) along with the fluid density (colored map) at $t = 0.45$. Owing to the presence of magnetic field, the explosion is no longer symmetric but proceeds faster along magnetic field lines. The particle number density decreases at the center where a strong rarefaction takes place and is enhanced behind the oval-shaped slow shock along the main diagonal. In the same figure, two MPs of interest from the point of view of spectral evolution are marked as *green* and *red* diamonds. The spectral evolution for the *green* particle is shown in top right panel of Fig. 5 where we plot $\epsilon^2 N(\epsilon)$ as a function of the energy ϵ. The initial power-law distribution shows a systematic shift towards lower energies right after $t \approx 1$. This takes place as the shock wave traverses through this particle resulting in an expanding flow around it. The subsequent evolution of the distribution function is mainly determined by synchrotron emission losses which produces a steeper spectrum at higher energies.

The evolution of the highest energy bin for both marked MPs is shown in the left panel of Fig. 6 where the *black dashed* line is Kardashev's solution of Eq. (2) with $\alpha = 0$, implying losses only due to synchrotron emission. As expected, the highest spectral energy bin of the unshocked *red* MP is consistent with the analytical estimate as it is unaffected by any adiabatic effect. On the contrary, the distribution function of the *green* evolve under the combined effects of adiabatic expansion and synchrotron losses. For both of the marked MPs, the normalized particle density in the highest energy bin, $(N(\epsilon_{\rm max})/N_0)\epsilon^\gamma$ decays with time. The green particle decays at a higher rate as it suffers both synchrotron and adiabatic losses (see right panel of Fig. 6). This test clearly demonstrates the different effects of the two radiative loss terms that are considered in the present work.

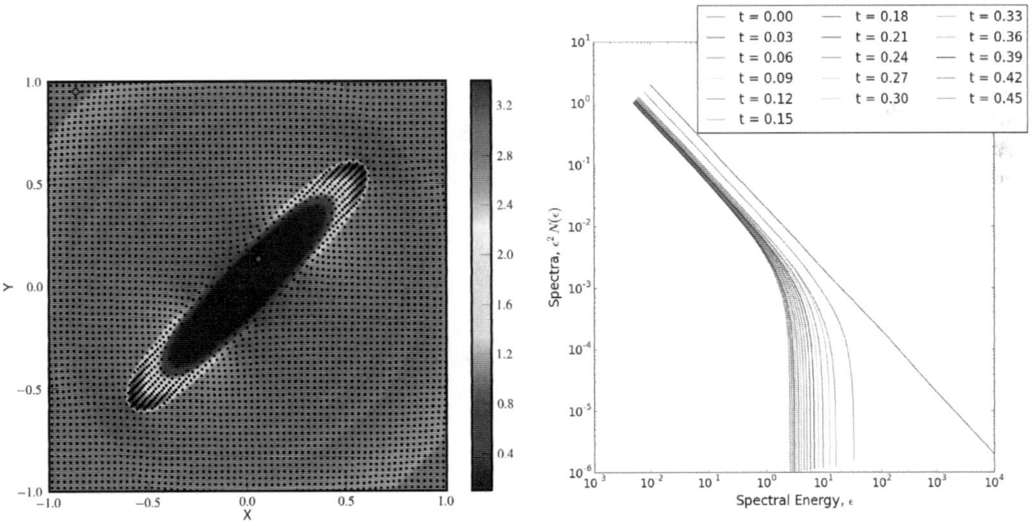

Figure 5. *Left Panel:* Density (colormap) and particle distribution (black dots) in the oblique MHD shock problem. Particles marked in *green* and *red* are used for further analysis. *Right Panel:* spectral evolution of the green-marked particle at different times.

4.5. Application to AGN Jets

A large body of observational evidence indicates that jets from active galactic nuclei (AGN) in FRI sources undergo significant deceleration at the kpc scale and slow down to mildly relativistic

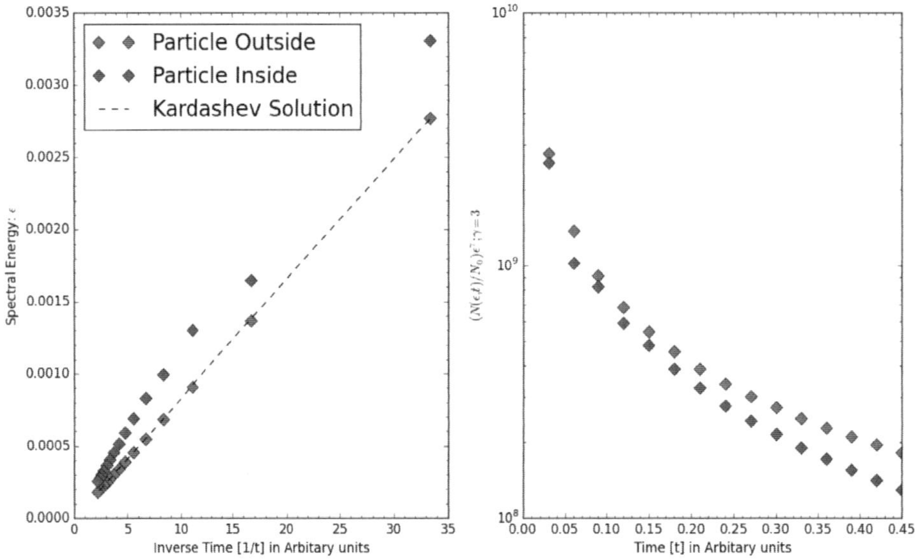

Figure 6. *Left panel:* time evolution of the largest energy bin for the marked particles (red and green squares) compared to the analytical expectation. *Right Panel:* time evolution of the normalized particle density for the largest energy bin, $(N(\epsilon)/N_0)\epsilon^\gamma$, with $\gamma = 3$ for the two marked particles.

speeds with a typical bulk Lorentz factor $\Gamma \sim 2$ [14]. We employ our particle module to investigate the physical processes responsible for the non-thermal emission of AGN jets observed at these scales.

To this purpose we now focus on the evolution of a magnetized jet in 2D cylindrical coordinates (r, z) and consider a computational domain with size of $r_{\rm max} = 3.2$ kpc and $z_{\rm max} = 12.8$ kpc and has a resolution of ~ 12.5 pc. The ambient medium is non-magnetized and stratified in density, $\rho = 1/(1 + R^2)$ where $R = \sqrt{r^2 + z^2}$ is the spherical radius. We prescribe constant pressure $p_0 = 1/\gamma$ so that the sound speed is unity at the origin. A pressure-matched supersonic beam with lower density ($\rho_j = 10^{-2}$) is injected through a nozzle of radius $r_j = 100$ pc at $z = z_{\rm min}$ with velocity $v_j = 4$. The jet carries a purely toroidal magnetic field $B_\phi = 100\mu G$, compatible with observational estimates. Macro-particles are initially placed inside the domain but also as injected at regular time intervals through the nozzle with an initial power-law distribution function of the form $N(\epsilon) \propto \epsilon^{-3}$. Energy bins cover a range of 6 orders of magnitude in energy from 10^{-4} to 100 ergs. so that the resulting emission spans from radio to X-ray wavelength. We follow their evolution through Eq. (1)-(2) and model shock acceleration through Eq. (7).

The left panel of Fig. 7 shows the jet tracer and the corresponding MPs distribution at $t \sim 0.5\,Myr$. At this time, a total number of $\gtrsim 1.3 \times 10^5$ MPs cover the whole jet. In the right panel of Fig. 7 we show the spectral evolution and the trajectory of a representative MP that is introduced through the jet nozzle at time $t \approx 26.3$. The initial power law spectrum (black line) undergoes strong synchrotron losses resulting in a steeper spectrum (red curve) at $t \approx 27$. After this time, the distribution function flattens (green curve with power-law index $\gamma \approx -2.3$) as the particle crosses an internal shock surface. Subsequently, the MP is pushed out the shocked region and encounters a turbulent magnetized region where its spectrum becomes steeper and the distribution function progressively moves to the radio wavelengths. Similar effects are seen in several other MPs, although particles that never enter the shock also continuously loose energy over time.

In order to investigate the collective effect of all particles on the emitted radiation we estimate the emissivity $J(\nu)$ by consistently employing the spectra of each MP at various instants during the simulation. The emissivity is related to the spectral distribution as follows:

$$J(\nu) = \int_0^\infty j(x)N(\epsilon)d\epsilon = 4\pi\frac{dI(\nu)}{dl}, \tag{8}$$

where $j(x)$ is the emissivity of a single electron, $x = \nu/\nu_c$ and $\nu_c = 3\pi\gamma_e^3\nu_0$ is the critical frequency function of the electron energy and orbital frequency ν_0. For an optically thin medium, the emissivity can be related to the observed intensity $I(\nu)$ as shown above with l being the length along line of sight. In Fig. 8 we show the distribution of $J(\nu)$ at two different frequencies, $\nu = 1.6$ GHz and $\nu = 15$ GHz typically used to observe AGN jets with radio interferometers such as VLA and VLBI. The left and middle panels show the distribution at 1.6 GHz without (left) and with (middle) shock acceleration, Eq. (7). Brighter emission in proximity of the bow shock and at the internal knots is observed when Fermi-I process is considered via sub-modelling. In addition, we see a large fraction of diffused faint emission from older population of MPs in the turbulent regions of the jet. This effect is more pronounced in the middle panel as particles tend to be re-accelerated at small scale shocks. The distribution of $J(\nu)$ for 15 GHz is evidently different from the one obtained at 1.6 GHz. The high energy radio band only tracks particles at the shock that are able to resist the cooling due to synchrotron emission.

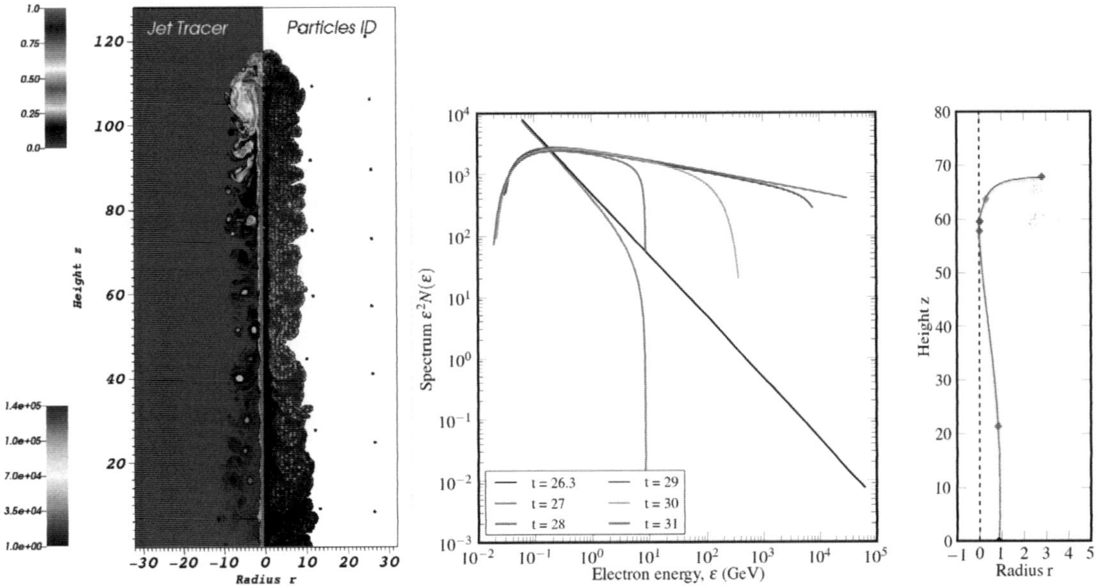

Figure 7. *Left Panel:* Jet tracer (*left half*) and particles ID (*right half*) for the reference run described in Sec. 4.5. Particles are labeled by increasing injection ID. Green particles have been injected more recently and are responsible for the bulk of emission. Purple particles have been introduced at earlier stages of the evolution and have already cooled down. *Middle panel:* spectral evolution of a representative particle. *Bottom panel:* particle trajectory. The same color key used in the middle panel is applied.

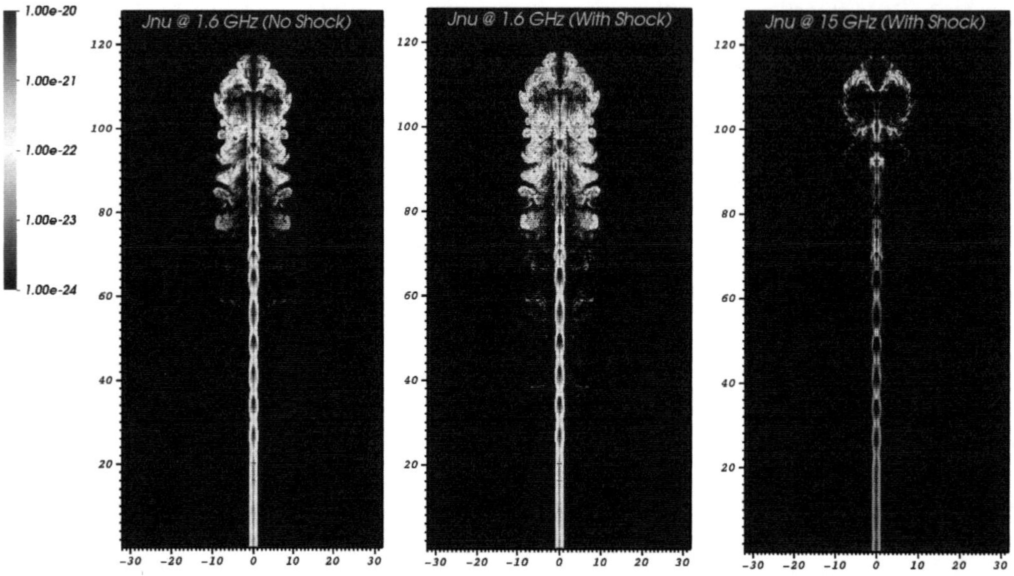

Figure 8. Emissivity, $J(\nu)$ [in ergs $s^{-1}Hz^{-1}cm^{-1}$], at 1.6 GHz (*left* and *middle* panels) and 15 GHz (*right* panel). The effects due to shock acceleration are switched off in the *left* panel.

5. Discussion & Summary

We have developed a fully parallel, hybrid module incorporating *Lagrangian* particles in the static grid version of the PLUTO code. A single macro-particle can be thought as a swarm of real particles (e.g. electrons) whose spectral energy distribution can be described using a convection-diffusion transport equation. For the present purpose, only adiabatic effect, synchrotron losses and shock-acceleration have been included. The numerical algorithm for fluid/particle evolution together with the parallelization strategy have been presented. Numerical benchmarks demonstrating the correctness of the implementation have been illustrated by a direct comparison with analytical solutions. The adaptive mesh implementations will be addressed in an upcoming work.

Furthermore, we have presented axisymmetric MHD simulations of jets aimed at studying the competition between the shock acceleration mechanism and various radiation losses in the context of kpc-scale extragalactic jets. We find that internal knots formed in the jet become efficient sites for particle acceleration capable of producing flat spectra with power-law indices that are consistent with the theory of diffusive shock acceleration. However, we also see that for typical magnetic fields inferred from observations, the losses eventually dominates producing a steeper spectrum at radio wavelengths. This suggests that additional mechanisms must operate to continuously re-accelerate particles in order to sustain the emission over longer time spans. In this sense, magnetic reconnection and turbulent scattering (Fermi II order) acting in the jet cocoon are among the most promising candidates. These possibilities will be explored in forthcoming papers along with 3D relativistic jet simulations.

References

[1] Bai X N, Caprioli D, Sironi L and Spitkovsky A 2015 *The Astrophysical Journal* **809** 55 (*Preprint* 1412.1087)
[2] Daldorff L K S, Tóth G, Gombosi T I, Lapenta G, Amaya J, Markidis S and Brackbill J U 2014 *Journal of Computational Physics* **268** 236–254
[3] Bai X N and Stone J M 2010 *Astrophysical Journal Suppl.* **190** 297–310 (*Preprint* 1005.4980)

[4] Bryan G L, Norman M L, O'Shea B W, Abel T, Wise J H, Turk M J, Reynolds D R, Collins D C, Wang P, Skillman S W, Smith B, Harkness R P, Bordner J, Kim J h, Kuhlen M, Xu H, Goldbaum N, Hummels C, Kritsuk A G, Tasker E, Skory S, Simpson C M, Hahn O, Oishi J S, So G C, Zhao F, Cen R, Li Y and Enzo Collaboration 2014 *Astrophysical Journal Suppl.* **211** 19 (*Preprint* 1307.2265)

[5] Dubey A, Daley C, ZuHone J, Ricker P M, Weide K and Graziani C 2012 *Astrophysical Journal Suppl.* **201** 27

[6] Del Zanna L, Volpi D, Amato E and Bucciantini N 2006 *A&A* **453** 621–633 (*Preprint* astro-ph/0603080)

[7] Tregillis I L, Jones T W and Ryu D 2001 *The Astrophysical Journal* **557** 475–491 (*Preprint* astro-ph/0104305)

[8] Mimica P, Aloy M A, Agudo I, Martí J M, Gómez J L and Miralles J A 2009 *The Astrophysical Journal* **696** 1142–1163 (*Preprint* 0811.1143)

[9] Micono M, Zurlo N, Massaglia S, Ferrari A and Melrose D B 1999 *A&A* **349** 323–333 (*Preprint* astro-ph/9906440)

[10] Mignone A, Bodo G, Massaglia S, Matsakos T, Tesileanu O, Zanni C and Ferrari A 2007 *Astrophysical Journal Suppl.* **170** 228–242 (*Preprint* astro-ph/0701854)

[11] Hardcastle M 2015 Kiloparsec-Scale AGN Jets *The Formation and Disruption of Black Hole Jets* (*Astrophysics and Space Science Library* vol 414) ed Contopoulos I, Gabuzda D and Kylafis N p 83

[12] Kardashev N S 1962 *Soviet Astronomy* **6** 317

[13] Chi-Wang S 2001 High order finite difference and finite volume weno schemes and discontinuous galerkin methods for cfd Tech. rep.

[14] Laing R A and Bridle A H 2013 *Monthly Notices in Royal Astronomy* **432** 1114–1132 (*Preprint* 1303.6664)

Journal of Physics: Conference Series **719** (2016) 012024 doi:10.1088/1742-6596/719/1/012024

Numerical Relativistic Magnetohydrodynamics with ADER Discontinuous Galerkin methods on adaptively refined meshes.

O Zanotti[1], M Dumbser[1] and F Fambri[1]

[1] Laboratory of Applied Mathematics, Via Mesiano 77, 38123 Trento, Italy

E-mail: olindo.zanotti@unitn.it

Abstract. We describe a new method for the solution of the ideal MHD equations in special relativity which adopts the following strategy: (i) the main scheme is based on Discontinuous Galerkin (DG) methods, allowing for an arbitrary accuracy of order N+1, where N is the degree of the basis polynomials; (ii) in order to cope with oscillations at discontinuities, an "a-posteriori" sub-cell limiter is activated, which scatters the DG polynomials of the previous time-step onto a set of 2N+1 sub-cells, over which the solution is recomputed by means of a robust finite volume scheme; (iii) a local spacetime Discontinuous-Galerkin predictor is applied both on the main grid of the DG scheme and on the sub-grid of the finite volume scheme; (iv) adaptive mesh refinement (AMR) with local time-stepping is used. We validate the new scheme and comment on its potential applications in high energy astrophysics.

1. Introduction

Discontinuous Galerkin (DG) methods, which have been applied to terrestrial physical problems for a long time, are still relatively unknown in astrophysics. The situation is now rapidly changing, and their usage is likely to increase in the next decade. Due to their robustness, high order of accuracy in smooth regions and very good scalability, these methods are attracting a lot of attention, particularly in the relativistic context [1, 2]. A particularly interesting field of research in high energy astrophysics is represented by the solution of the special relativistic magnetohydrodynamics (RMHD) equations, which govern the physics of peculiar systems such as extragalactic jets, gamma-ray bursts and magnetospheres of neutron stars. Unfortunately, Galerkin methods suffer from a strong limitation, which has prevented a widespread use due to the fact that they cannot escape the Gibbs phenomenon, thus producing oscillations at discontinuities. The common route to circumvent this problem is represented by artificial viscosity [3, 4], spectral filtering [5], (H)WENO limiting procedures [6, 7], and slope and moment limiting [8, 9]. However, an unavoidable consequence of these approaches is that they destroy the sub-cell resolution properties of the DG method. In [10] we have recently proposed a new solution to this longstanding problem, which is based on a sub-cell finite volume limiting approach, while preserving the high resolution capabilities of DG. If combined with Adaptive Mesh Refinement (AMR), this approach can guarantee un-precedented levels of accuracy [11] and their applications to the relativistic framework are very promising [12].

2. The numerical scheme

The equations of RMHD can be cast in conservative form as

$$\partial_t \mathbf{u} + \partial_i \mathbf{f}^i = 0\,, \tag{1}$$

where the definition of the conserved quantities $\mathbf{u}$ and of the corresponding fluxes $\mathbf{f}$ can be found in [13, 14]. The constraint on the divergence free character of the magnetic field can also be written formally as in Eq. (1), according to the *divergence-cleaning approach* proposed by [15], that we also follow here. Concerning the recovering of the primitive variables from the conserved ones, we adopt the third method reported in Sect. 3.2 of [16]. We adopt Cartesian coordinates over a domain Ω, which is formed by elements T_i as

$$\Omega = \bigcup_{i=1}^{N_E} T_i\,, \tag{2}$$

where the index i ranges from 1 to the total number of elements N_E. According to the DG philosophy, at any time t^n, the numerical solution of Eq. (1) is represented by polynomials of maximum degree $N \geq 0$, namely

$$\mathbf{u}_h(\mathbf{x}, t^n) = \sum_{l=0}^{N} \Phi_l(\mathbf{x})\hat{\mathbf{u}}_l^n\,, \tag{3}$$

where the coefficients $\hat{\mathbf{u}}_l^n$ are the so-called degrees of freedom. We can schematically describe our numerical scheme by distinguishing the following different steps:

- We apply a predictor step, in which Eq. (1) is solved within each element T_i through a locally implicit space-time discontinuous Galerkin scheme. In practice, the polynomials of Eq. (3) need to be evolved in time, locally for each cell, and subsequently used for the time integration of the numerical fluxes. This is the heart of the ADER approach, as formulated in the modern version by [17].
- We construct a one-step ADER discontinuous Galerkin (DG) scheme, which uses the information obtained by the predictor. This implies that we first multiply the governing equations (1) by a test function Φ_k and then we integrate over the space-time control volume $T_i \times [t^n; t^{n+1}]$. After integration by parts in space, we get

$$\int_{t^n}^{t^{n+1}}\int_{T_i} \Phi_k \frac{\partial \mathbf{u}_h}{\partial t} d\mathbf{x}dt + \int_{t^n}^{t^{n+1}}\int_{\partial T_i} \Phi_k\, \mathbf{f}\left(\mathbf{u}_h\right)\cdot \mathbf{n}\, dSdt - \int_{t^n}^{t^{n+1}}\int_{T_i} \nabla\Phi_k \cdot \mathbf{f}\left(\mathbf{u}_h\right) d\mathbf{x}dt = 0\,. \tag{4}$$

  If $\mathbf{u}_h$, expressed by (3), is inserted in the first term of Eq. (4), and if we use the result of the predictor step to compute the other terms of Eq. (4), we effectively obtain a one-step numerical scheme, which allows to obtain the new solution at time t^{n+1} with a high order of accuracy both in space and in time. We also note that the second term of Eq. (4) contains a surface integration, which involves the solution of a Riemann problem at the element boundary, thus guaranteeing that the final method is upwind. In our numerical tests we have used the simple Rusanov flux and the HLL solver.
- We introduce an a-posteriori sub-cell limiter. Since the scheme resulting from Eq. (4) is unlimited, it will necessarily produce oscillations at discontinuities. In order to solve this problem, we introduce a special kind of limiter, which acts a posteriori in the following way. First we identify a particular cell as troubled if it either violates physical conditions or it

violates a simple version of the discrete maximum principle [10]. In each troubled cell, a subgrid composed of $2N + 1$ sub-cells (along each spatial direction) is built and the DG polynomial at the previous time (when pathologies had not manifested yet) is projected over the subgrid in such a way to provide a piecewise constant data representation at the subgrid level. Having done that, it is possible to evolve the cell averages at the subgrid level via standard finite volume, or even a TVD scheme.

- We combine the new scheme with Adaptive Mesh Refinement (AMR) and local time-stepping. Our AMR approach can be referred to as a "cell-by-cell" refinement, according to which every cell T_i is individually refined with no creation of grid patches. The refinement factor along each direction is denoted by $\mathfrak{r}$ and the maximum level of refinement is $\ell_{\max}$. In particular, the two AMR operations of projection and averaging need to involve the sub-cell averages of the solution on the sub-grids. Further details can be found in [11, 18].

In the numerical tests that follow we refer to the new scheme as ADER-DG-$\mathbb{P}_N$ where N is the degree of the DG polynomial, and we effectively use a TVD sub-cell limiter.

3. Numerical tests

The convergence of the new ADER scheme, with a single step for the time update, has been verified by [12] up to the 5-th order. Here we show the performances of the new scheme by solving two representative and well known RMHD tests. In particular, we solve the first shock tube problem in the sample proposed by [13], using a coarse grid of 40×5 cells, which is adaptively refined in space and time according to $\mathfrak{r} = 3$ and $\ell_{\max} = 2$. Fig. 1 reports the results of the calculation, for which we have used the ADER-DG-$\mathbb{P}_3$ scheme and a Courant factor CFL $= 0.5$. In the two top panels of the figure, where 3D-views of the magnetic field B^y and of the rest mass density are shown, we have highlighted in red those cell which required the activation of the limiter, while the unlimited cells are colored in cyan. Our ADER-DG numerical scheme can capture all the significant waves of this Riemann problem, while suppressing all the spurious oscillations.

The second test that we consider is the cylindrical expansion of a blast wave in a plasma with an initially uniform magnetic field. The details about it can be found in [14]. The initial conditions assume that, within a radius $R = 1.0$, the rest-mass density and the pressure are $\rho = 0.01$ and $p = 1$. On the contrary, $\rho = 10^{-4}$ and $p = 5 \times 10^{-4}$ outside the cylinder. There is not velocity to start with, and the magnetic field is constant and oriented along the x-direction. In Fig. 2, we show the results at the final time $t = 4.0$ obtained for $B_x = 0.5$, which makes this test particularly severe. We have solved this problem over the computational domain $\Omega = [-6, 6] \times [-6, 6]$, with 40×40 elements on the coarsest refinement level, $\mathfrak{r} = 3$ and $\ell_{\max} = 2$. Moreover, the Rusanov Riemann solver has been chosen. The wave pattern of the configuration is formed by a weak external circular fast shock, visible in the magnetic pressure, and by a reverse shock which surrounds the region where the magnetic field is confined. The two bottom panels show the AMR grid and the limiter map, from which we deduce that only a relatively small number of cells required the activation of the sub-cell TVD scheme.

4. Conclusions

We have proposed a modification of the pure discontinuous Galerkin (DG) method that solves the longstanding problem of Gibbs oscillation with an a-posteriori limiter. The limiter acts on a sub-grid of each single troubled cell, and it recomputes the solution via a robust WENO or TVD finite volume scheme. This strategy, combined with AMR, allows for an unprecedented level of accuracy. The new method can solve successfully the equations of special relativistic magnetohydrodynamics and it is likely to contribute significantly in the numerical modeling of various high energy astrophysics phenomena. It has been already adopted in the relativistic

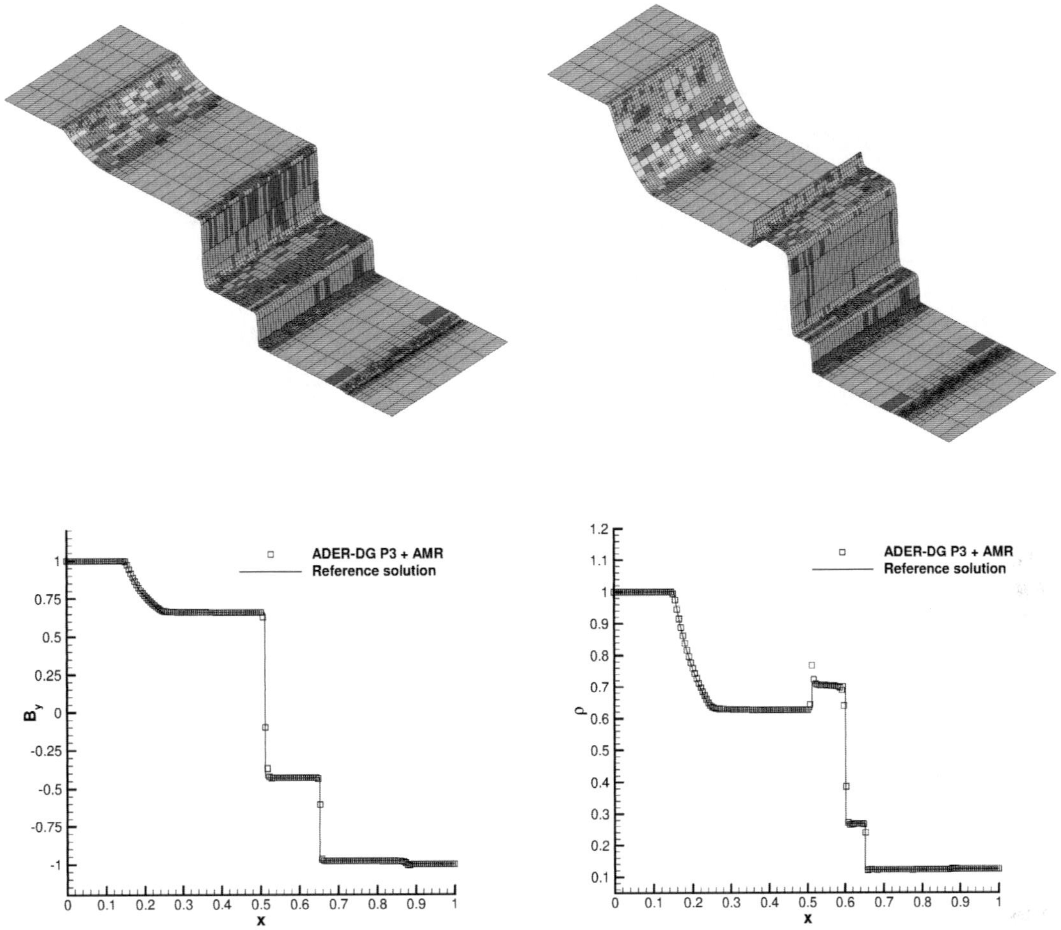

Figure 1. Solution of the first Riemann problem by [13] at $t_{\text{final}} = 0.55$, obtained with the ADER-DG-$\mathbb{P}_3$ scheme supplemented with the ADER-TVD sub-cell limiter. Top panels: 3D views of the magnetic field B^y (left) and of the rest-mass density (right). Bottom panels: one-dimensional profiles compared to the exact solution.

context by other groups [1] and it is likely to be applied also for the solution of the Einstein equations.

Acknowledgments

The research presented in this paper was financed by the European Research Council (ERC) under the European Union's Seventh Framework Programme (FP7/2007-2013) with the research project *STiMulUs*, ERC Grant agreement no. 278267.

References

[1] Bugner M., Dietrich T., Bernuzzi S., Weyhausen A., Bruegmann B., 2015, ArXiv e-prints
[2] Teukolsky S. A., 2015, ArXiv e-prints
[3] Persson P.-O. and Peraire J. *AIAA Paper 2006-112*, 2006.
[4] Cesenek J., Feistauer M., Horacek J., Kucera V., and Prokopova J. *Applied Mathematics and Computation*, 219:7139–7150, 2013.

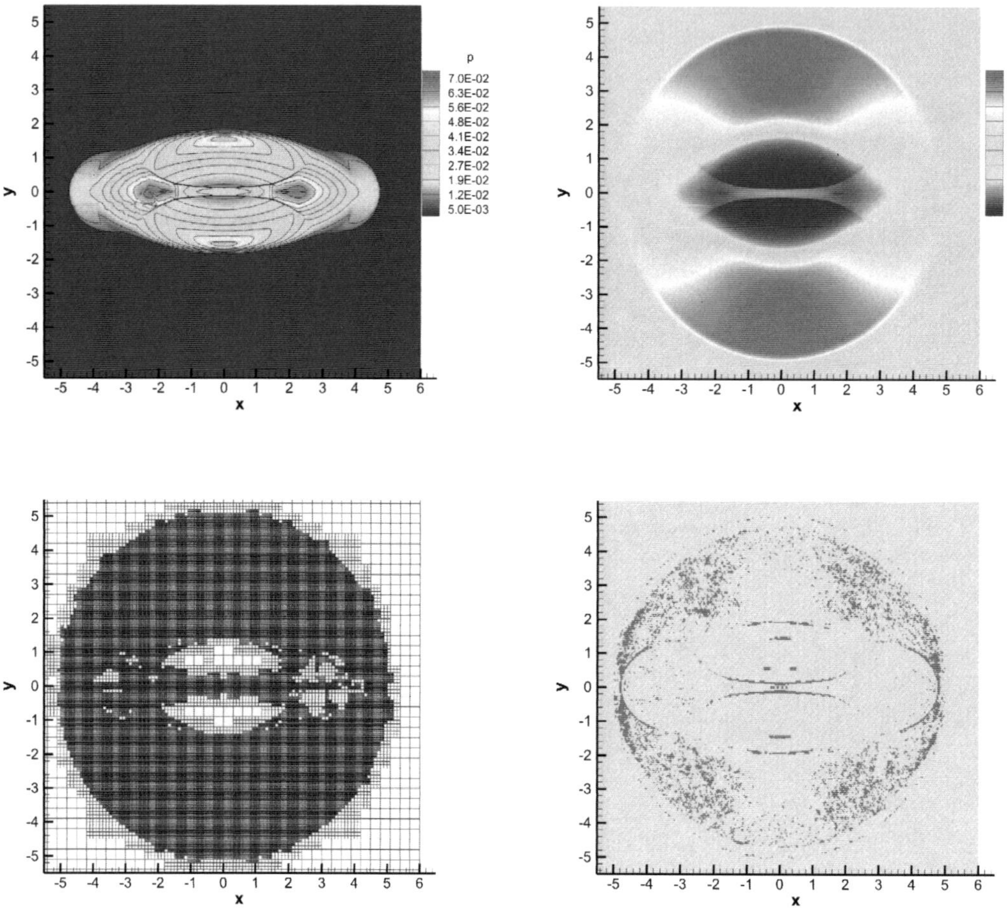

Figure 2. RMHD blast wave problem with $B_x = 0.5$ at time $t = 4.0$, obtained with the ADER-DG $\mathbb{P}_3$ scheme and the a-posteriori second order TVD sub-cell limiter. Top panels: gas pressure (left) and magnetic pressure (right). Bottom panels: AMR grid (left) and limiter map (right) with troubled cells marked in red and regular unlimited cells marked in cyan.

[5] Radice D. and Rezzolla L. *Phys. Rev. D*, 84(2):024010, July 2011.
[6] Qiu J. and Shu C.W.. *SIAM Journal on Scientific Computing*, 26:907–929, 2005.
[7] Shu C.W., Zhu J., Zhong X. and Qiu J. *J. Comp. Phys.*, 248:200–220, 2013.
[8] Cockburn B. and Shu C. W.. *Mathematics of Computation*, 52:411–435, 1989.
[9] D.Kuzmin. *Journal of Computational Physics*, 257, Part B(0):1140 – 1162, 2014.
[10] Dumbser M., Zanotti O., Loubère R., Diot S., 2014, Journal of Computational Physics, 278, 47
[11] Zanotti O., Fambri F., Dumbser M., Hidalgo A., 2015, Computers and Fluids, 118, 204
[12] Zanotti, O., Fambri F., Dumbser M. *Mon. Not. R. Astron. Soc.*, 452:3010–3029, September 2015.
[13] Balsara D., 2001a, Astrophysical Journal Suppl. Series, 132, 83
[14] Komissarov S. S., 1999, Mon. Not. R. Astron. Soc., 303, 343
[15] Dedner A., Kemm F., Kröner D., Munz C. D., Schnitzer T., and Wesenberg M.
[16] Del Zanna L., Zanotti O., Bucciantini N., Londrillo P., 2007, Astron. Astrophys., 473, 11 *Journal of Computational Physics*, 175:645–673, January 2002.
[17] Dumbser M., Enaux C., Toro E.F. *Journal of Computational Physics*, 227:3971–4001, 2008.
[18] Dumbser M., Zanotti Z., Hidalgo A., Balsara D.S. *Journal of Computational Physics*, 248:257–286, 2013.